T0275577

ANALYTIC PRO-p GROUPS

J.D. Dixon
Carleton University, Ottawa

M.P.F. du Sautoy
University of Cambridge

A. Mann
Hebrew University, Jerusalem

D. Segal
All Souls College Oxford

Analytic Pro-*p* groups

Second Edition

Revised and enlarged by

Marcus du Sautoy & Dan Segal

CAMBRIDGE
UNIVERSITY PRESS

PUBLISHED BY THE PRESS SYNDICATE OF THE UNIVERSITY OF CAMBRIDGE
The Pitt Building, Trumpington Street, Cambridge, United Kingdom

CAMBRIDGE UNIVERSITY PRESS
The Edinburgh Building, Cambridge CB2 2RU, UK
40 West 20th Street, New York NY 10011–4211, USA
477 Williamstown Road, Port Melbourne, VIC 3207, Australia
Ruiz de Alarcón 13, 28014 Madrid, Spain
Dock House, The Waterfront, Cape Town 8001, South Africa

http://www.cambridge.org

First published 1991
Second edition 1999
First paperback edition 2003

A catalogue record for this book is available from the British Library

ISBN 0 521 65011 9 hardback
ISBN 0 521 54218 9 paperback

Transferred to digital printing 2003

Contents

Contents

Preface

This second edition of *Analytic pro-p groups* has been prepared by two of the authors, Marcus du Sautoy and Dan Segal. We thank Avinoam Mann for contributing Appendix A, and both him and John Dixon for giving us a free hand.

Apart from minor changes in wording and presentation, the book differs from the first edition in the following respects.

- Most of the content of the original Chapter 6 has been redistributed among the other chapters and the 'Interludes'; as a result, the new Chapter n takes the place of the old Chapter $n + 1$ for $n = 6, 7, 8$ and 9.
- Chapters 10–13 are new, as are §§5.4, 6.5, 6.6, 7.4 and Interludes A, B, D and E (Interlude B is based on material from the old §§6.1 and 6.2).
- Some of the old exercises have been incorporated into the main text; many new exercises have been added.
- The following sections have undergone major revision: §§3.3, 4.4, 4.5, 7.1, 7.2, 8.1, 9.4 and 9.5.
- The old §6.6 has been left out, as has most of the old §6.2.
- The bibliography has been updated and expanded.

We wish to thank several readers, and especially one reviewer, of the first edition, who pointed out a number of errors – typographical and mathematical. We are grateful to the American Mathematical Society for permission to reproduce Interlude C, which first appeared in the *Bulletin of the American Mathematical Society*.

Introduction

And the end of all our exploring
Will be to arrive where we started
And know the place for the first time
T.S. Eliot: Little Gidding

The origin of this book was a seminar held at All Souls College, Oxford, in the Spring of 1989. The aim of the seminar was to work through Michel Lazard's paper *Groupes analytiques p-adiques* [L], at least far enough to understand the proof of 'Lubotzky's linearity criterion' (Lubotzky 1988). In fact, Lubotzky's proof combined Lazard's characterisation of p-adic analytic groups with some recent results of Lubotzky and Mann (1987b) on 'powerful' pro-p groups. We found that by reversing the historical order of development, and starting with powerful pro-p groups, we could reconstruct most of the group-theoretic consequences of Lazard's theory without having to introduce any 'analytic' machinery. This was a comforting insight for us (as group theorists), and gave us the confidence to go on and develop what we hope is a fairly straightforward account of the theory of p-adic analytic groups.

The first edition was divided (like Gaul) into three parts. Parts I and II were essentially linear in structure. The point of view in Part I was group-theoretic; in Part II, more machinery was introduced, such as normed algebras and formal power series. Between Parts I and II was an *Interlude* (Chapter 6): this consisted of a series of more or less independent digressions, describing applications of the results to various aspects of group theory.

This second edition is also in three parts. Parts I and II cover the same ground as before, with some additional material; however, the old Chapter 6 has been replaced by several shorter *Interludes*, and a new Part III has appeared. Readers seeking a simple introduction to pro-p

groups and to p-adic analytic groups will still find this in Parts I and II; the four new chapters that constitute Part III deal with a variety of topics that we felt deserved a place, but which are not necessary for an understanding of the basic theory. The book as a whole now gives a fairly comprehensive account of what is known about pro-p groups on the 'smallish' side; 'large' pro-p groups – free products, groups of tree automorphisms and the like – are discussed in the forthcoming book [DSS].

We now outline the contents in more detail. *Part I* is an account of pro-p groups of finite rank. **Chapter 1** is a leisurely introduction to profinite groups and pro-p groups, starting from first principles. **Chapter 2** is about finite p-groups. A finite p-group G is defined to be *powerful* if G/G^p is abelian (if p is odd; the case $p = 2$ is slightly different). The key results established in this chapter are due to Lubotzky and Mann (1987a): (i) *if G is powerful and can be generated by d elements, then every subgroup of G can be generated by d elements;* and (ii) *if G is a p-group and every subgroup of G can be generated by d elements, then G has a powerful normal subgroup of index at most* $p^{d(\log(d)+2)}$ (log to the base 2). **Chapter 3** returns to profinite groups. Here the *rank* of a profinite group is defined, in several equivalent ways. Defining a pro-p group G to be powerful if $G/\overline{G^p}$ is abelian (where $^{-}$ denotes closure, and the proviso regarding $p = 2$ still applies), we deduce from the above results that *a pro-p group has finite rank if and only if it has a powerful finitely generated subgroup of finite index* (Lubotzky and Mann 1987b). This is then used to give several alternative characterisations for pro-p groups of finite rank. **Chapter 4** continues with the deeper investigation of finitely generated powerful pro-p groups. These groups, being 'abelian modulo p', are in many ways rather like abelian groups. In particular, each such group contains a normal subgroup of finite index which is 'uniform'; we shall not define this here, but note that the uniform pro-p groups are among those studied by Lazard under the name 'groupes p saturables'. Following an exercise in [L], we show that *a uniform pro-p group G has in a natural way the structure of a finitely generated free $\mathbb{Z}_p$-module,* which we denote $(G, +)$ (here $\mathbb{Z}_p$ denotes the ring of p-adic integers). Defining an additional operation 'bracket' on this module, we also indicate how $(G, +)$ can be turned into a Lie algebra L_G over $\mathbb{Z}_p$: this is the first hint of a connection with Lie groups. It follows that the automorphism group of G has a faithful linear representation over $\mathbb{Z}_p$, and hence that G itself is 'linear modulo its centre'. Part I concludes with **Chapter 5**. Here we study the most familiar p-adic analytic group, namely $\mathrm{GL}_d(\mathbb{Z}_p)$, and show quite explicitly that a suitable congruence subgroup is a uniform pro-p group. Together with the results of Chapter 4, this is used to show that *the automorphism group of any pro-p group of finite rank is itself virtually* (i.e. up to finite index) *a pro-p group of finite rank.*

Interlude A is a summary of results, established throughout the book, that characterise the class of pro-p groups of finite rank, and that determine the *dimension* of such a group.

Although *Part II* is headed 'Analytic groups', these do not appear as such until Chapter 8. **Chapter 6** is utilitarian, giving definitions and elementary results about complete normed $\mathbb{Q}_p$-algebras which are needed later; also established here are relevant properties of the *Campbell–Hausdorff formula*. **Chapter 7** forms the backbone of Part II. In it, we show how to define a norm on the group algebra $A = \mathbb{Q}_p[G]$ of a uniform pro-p group G, in a way that respects both the p-adic topology on $\mathbb{Q}_p$ and the pro-p topology on G. (Readers familiar with Chapter 8 of the first edition will note that the construction of the norm has been streamlined, and that the troublesome case $p = 2$ has been tamed by modifying the norm in that case.) The completion $\widehat{A}$ of this algebra with respect to the norm serves two purposes. On the one hand, an argument using the binomial expansion of terms in $\widehat{A}$ is used to show that the group operations in G are given by *analytic functions* with respect to a natural co-ordinate system on G, previously introduced in Chapter 4. On the other hand, $\widehat{A}$ serves as the co-domain for the *logarithm* mapping $\log : G \to \widehat{A}$. We show that the set $\log(G)$ is a $\mathbb{Z}_p$-Lie subalgebra of the commutation Lie algebra on $\widehat{A}$, isomorphic via log to the Lie algebra L_G defined intrinsically in Chapter 4 (in fact, the proof simultaneously establishes that L_G satisfies the Jacobi identity, and that $\log(G)$ is closed with respect to the operation of commutation). An appeal to *Ado's Theorem*, in conjunction with the Campbell–Hausdorff formula, then shows that G has a faithful linear representation over $\mathbb{Z}_p$; it follows that *every pro-p group of finite rank has a faithful linear representation over* $\mathbb{Z}_p$.

The final section of Chapter 7 examines the structure of the completed group algebras $\mathbb{Z}_p[[G]]$ and $\mathbb{F}_p[[G]]$: the associated graded rings are shown to be *polynomial rings over* $\mathbb{F}_p$, which implies that $\mathbb{Z}_p[[G]]$ and $\mathbb{F}_p[[G]]$ are both Noetherian integral domains. These results are not needed for the theory of analytic groups developed in the rest of Part II, but have other important applications (for example in Chapter 12, and in the cohomology theory of p-adic analytic groups, see [DSS]). Under the name of the 'Iwasawa algebra', the completed algebra $\mathbb{Z}_p[[G]]$ plays an important role in the theory of cyclotomic fields (Washington 1982).

The linearity result proved in Chapter 7 is applied in **Interlude B** to establish 'Lubotzky's linearity criterion'. This amounts to the statement that *a finitely generated (abstract) group G has a faithful linear representation over a field of characteristic zero if and only if some pro-p completion of G has finite rank.*

p-adic analytic groups are defined in **Chapter 8**. Although we introduce p-adic manifolds, only a bare minimum of theory is developed (in contrast to Serre (1965), for example, there is no use of differentials). Using the results of Chapters 6 and 7, it is shown that *a pro-p group has a p-adic analytic structure if and only if it has finite rank,* and, more generally, that *every p-adic analytic group has an open subgroup which is a pro-p group of finite rank.* These major results are due to Lazard [L], except that he refers to finitely generated virtually powerful pro-p groups where we have 'pro-p groups of finite rank'.

In **Interlude C** we reprint an announcement by Marcus du Sautoy, where the preceding theory is applied to study the subgroup-growth behaviour of finitely generated groups.

Chapter 9 is concerned with some of the 'global' properties of p-adic analytic groups. The first main result here is that *every continuous homomorphism between p-adic analytic groups is an analytic homomorphism,* from which it follows that *the analytic structure of a p-adic analytic group is determined by its topological group structure.* Next, it is shown that *closed subgroups, quotients and extensions of p-adic analytic groups are again p-adic analytic;* these results now follow quite easily from the corresponding properties of pro-p groups of finite rank. Section 9.4 (which can be read independently of Chapter 8) establishes that *the correspondence $G \leftrightarrow L_G$ is an isomorphism between the category of uniform pro-p groups and the category of 'powerful' Lie algebras over $\mathbb{Z}_p$.* This is used in the final section of the chapter to establish the equivalence of the category of p-adic analytic groups (modulo 'local isomorphism') with the category of finite-dimensional Lie algebras over $\mathbb{Q}_p$.

The first chapter of *Part III*, **Chapter 10**, gives an account of the theory of *pro-p groups of finite coclass* (completing the brief discussion given in §6.4 of the first edition). This beautiful theory is central to the classification of finite p-groups. The chapter can be read directly after Chapter 4, as it depends on the theory of powerful pro-p groups but not on the analytic machinery developed in Part II.

The next two chapters discuss the *dimension subgroup series* in finitely generated pro-p groups. In **Chapter 11** this series is used to derive some more delicate characterisations of pro-p groups of finite rank, originally discovered by Lazard using a different method. **Chapter 12** investigates the *graded restricted Lie algebra* associated with the dimension series; after developing from first principles the necessary theory of restricted Lie algebras, the chapter proves some celebrated theorems of Jennings and of Lazard, about dimension subgroups and about the coefficients in the 'Golod–Shafarevich series' of a finitely generated pro-p group.

The next two *Interludes* give applications of Lazard's theorem on the Golod–Shafarevich series; **Interlude D** is devoted to the *Golod–Shafarevich inequality* in a large class of pro-p groups and abstract groups, while **Interlude E** presents Grigorchuk's theorem about *groups of sub-exponential growth*.

In the Introduction to the first edition, we wrote ' ... Analytic groups over other fields also deserve consideration; a Lie group in the usual sense (over $\mathbb{R}$ or $\mathbb{C}$) cannot be a pro-p group (except in the trivial, discrete, case), but some extremely interesting pro-p groups arise as analytic groups over fields of characteristic p: for example, suitable congruence subgroups in $\mathrm{SL}_n(\mathbb{F}_p[[t]])$. The theory of such groups poses some exciting challenges: they will have to be faced in a different book.' **Chapter 13** is a first step in facing those challenges. A *pro-p domain* is a commutative, Noetherian complete local integral domain R whose residue field is finite, of characteristic p. We propose a definition for 'groups analytic over R', and generalise one of the main results of Chapter 8 by showing that *every R-analytic group contains an open R-standard subgroup*; here, a group is called *R-standard* if it can be identified with $(\mathfrak{m}^k)^{(n)}$, where $\mathfrak{m}$ is the maximal ideal of R, and the group operation is given by a formal group law with coefficients in R. By their very nature, R-standard groups are pro-p groups. In general they do not have finite rank, but they resemble $\mathbb{Z}_p$-standard groups (which do have finite rank) in interesting ways. These are explored in the later part of the chapter, which establishes some results due to Lubotzky and Shalev (1994).

Appendix A gives a proof of the *Hall–Petrescu* commutator-collection formula, which is used in Chapters 10 and 11. **Appendix B** contains the proofs of some elementary facts about topological groups, used in Chapter 9.

The *Exercises* at the end of each chapter serve two important purposes, beyond the usual one of providing practice in the use of new concepts: some fill gaps in the proofs of the main text, and some lead the reader through the proofs of interesting and/or important results which didn't find a place in the main text, but deserve to be known. They are meant at least to be read, whether or not the reader actually wants to do them!

The brief *Notes* at the end of each chapter represent a very inadequate attempt to give credit where it is due. Because we have done things in our own quixotic way, it has not always been easy to attribute specific results.

To say that the language and much of the theory of profinite groups has passed into 'folklore' does an injustice to the creators; we apologise for this, giving the weak excuse that it's the best we can do. However, it does seem clear that Serre's 'Cohomologie galoisienne' (recently translated as Serre (1997)) was the first appearance in book form of much of the basic theory, particularly as far as pro-p groups are concerned.

The book as a whole should be seen as an exegesis, plus extended commentary, of the first four chapters of Lazard's magisterial work *Groupes analytiques p-adiques* [L]. (The fifth chapter of [L] deals with the cohomology of p-adic analytic groups; an adequate treatment of this important subject is beyond the scope of this book, and the competence of the present authors. For a recent account of some of Lazard's results the reader is referred to J. S. Wilson's book 'Profinite groups' (Wilson 1998); a fuller treatment is given in the chapter by Symonds and Weigel in [DSS].) Exactly which aspects of the theory of Lie groups over 'nonclassical' local fields are due to Lazard, Serre or Bourbaki, respectively, we do not know; but the central topic of this book, namely how the group-theoretic properties of a pro-p group reflect its status as a p-adic analytic group, is entirely the brainchild of Lazard.

Dependence of chapters

				13
				9
		12		8
5	10	11	7	
		4		6
		3		
		2		
		1		

Notation

$\mathbb{C}$	complex numbers
$\mathbb{R}$	real numbers
$\mathbb{Q}$	rational numbers
$\mathbb{Z}$	integers
$\mathbb{N}$	non-negative integers (*including* 0)
$\mathbb{Z}_p$	p-adic integers
$\mathbb{Q}_p$	p-adic numbers
$\mathbb{F}_q$	finite field of size q
$v(x) = v_p(x)$	p-adic valuation of x
$\|x\| = \|x\|_p$	p-adic absolute value of x
$[x]$	greatest integer $\leq x$
$\lceil x \rceil$	least integer $\geq x$
$\lambda(r) = \lceil \log_2(r) \rceil$	so $2^{\lambda(r)-1} < r \leq 2^{\lambda(r)}$
GL_n	$n \times n$ general linear group
SL_n	$n \times n$ special linear group
U_n	$n \times n$ upper uni-triangular matrix group
M_n	$n \times n$ matrix ring
1_n	$n \times n$ identity matrix
$\overline{X}$	closure of X
$\subseteq$	subset
$\subseteq_o, \subseteq_c$	open subset, closed subset
$\leq$	subgroup
$\leq_o, \leq_c$	open subgroup, closed subgroup
$<$	proper subgroup
$<_o, <_c$	open proper subgroup, closed proper subgroup
$\lhd$	normal subgroup
$\lhd_o, \lhd_c$	open normal subgroup, closed normal subgroup
$\langle X \rangle$	group generated by X
$\mathrm{Z}(G)$	centre of G
$\Phi(G)$	Frattini subgroup of G
$\mathrm{Aut}(G)$	automorphism group of G
$K[G]$	group ring of G over K
$\mathbb{Z}_p[[G]], \mathbb{F}_p[[G]]$	completed group algebras of G
$\mathrm{C}_G(X)$	centraliser of X in G

$A \times B$	direct product of A and B
$A \rtimes B$	semidirect product of A by B
$A^{(n)}$	n-fold direct power of A
$A \oplus B$	direct sum of A and B
A^n	direct sum of n copies of (additive group) A
$\mathrm{d}(G)$	minimal number of generators of G
$\mathrm{d}_p(G)$	minimal number of generators of $G/[G,G]G^p$
$\mathrm{rk}(G)$	rank of G
$\dim(V)$	dimension of V

$x^y = y^{-1}xy$

$[x,y] = x^{-1}x^y$, or $-x + xy$ if x is in a module acted on by y

$[x_1, \dots, x_n] = [[x_1, \dots, x_{n-1}], x_n]$

$[x,_n y] = [x, y, \dots, y]$, with n occurrences of y

$[A, B] = \langle \{[x,y] \mid x \in A, \ y \in B\} \rangle$

$[A_1, \dots, A_n] = [[A_1, \dots, A_{n-1}], A_n]$

$[A,_n B] = [A, B, \dots, B]$, with n occurrences of B

$X^{\{n\}} = \{x^n \mid x \in X\}$

$G^n = \langle G^{\{n\}} \rangle$ (if G is a multiplicative group)

$\gamma_1(G) = G, \quad \gamma_n(G) = [\gamma_{n-1}(G), G]$

$P_1(G) = G, \quad P_n(G) = \overline{[P_{n-1}(G), G]P_{n-1}(G)^p}$

$D_n(G)$ nth modular dimension subgroup of G

$\|\cdot\|$	norm
deg	degree
$W(\mathbf{X})$	set of words in non-commuting indeterminates $X_1, \dots, X_n$
$K\langle X_1, \dots, X_n \rangle = K\langle \mathbf{X} \rangle$	polynomial ring over K in non-commuting indeterminates $X_1, \dots, X_n$
$K\langle\langle X_1, \dots, X_n \rangle\rangle = K\langle\langle \mathbf{X} \rangle\rangle$	formal power series ring over K in non-commuting indeterminates $X_1, \dots, X_n$

$K[X_1, \ldots, X_n] = K[\mathbf{X}]$ polynomial ring over K in commuting indeterminates $X_1, \ldots, X_n$

$K[[X_1, \ldots, X_n]] = K[[\mathbf{X}]]$ formal power series ring over K in commuting indeterminates $X_1, \ldots, X_n$

$\langle \alpha \rangle = \alpha_1 + \cdots + \alpha_r$, where $\alpha = (\alpha_1, \ldots, \alpha_r)$

$\epsilon_i = (0, \ldots, 0, 1, 0, \ldots, 0)$ with 1 in the ith place

$\binom{X}{n} = X(X-1) \ldots (X - n + 1)/n!$

$(a, b) = ab - ba$

$(a_1, \ldots, a_n) = ((a_1, \ldots, a_{n-1}), a_n)$

$(a, b)_{\mathbf{e}} = (a, \underbrace{b, \ldots, b}_{e_1}, \underbrace{a, \ldots, a}_{e_2}, \ldots)$ where $\mathbf{e} = (e_1, e_2, \ldots)$

Prelude

A reminder

We have done our best to make this book reasonably self-contained. The intention is that a reader with no specialist knowledge of group theory, topology, number theory or Lie algebra theory should follow the main thread of the argument without undue difficulty. (This is less true of the 'Interludes', which touch on various topics, and of Chapter 13, which depends on a certain amount of commutative algebra.)

However, there are a number of elementary facts and concepts which are frequently used and can reasonably be classed under the heading of 'non-specialist knowledge'. These (apart from any that we may have missed) are collected together here, for the convenience of the reader.

0.1 Commutators

G denotes a group, x, y, z elements of G, and A, B, C subgroups of G.

$$x^y = y^{-1}xy, \quad [x,y] = x^{-1}x^y, \quad [x_1, \ldots, x_n] = [[x_1, \ldots, x_{n-1}], x_n].$$

$$[A, B] = \langle [a, b] \mid a \in A, b \in B \rangle$$

where $\langle X \rangle$ denotes the subgroup of G generated by a subset X of G.

$$[A, B, C] = [[A, B], C].$$

0.1. $[xy, z] = [x, z]^y[y, z]$, $[x, yz] = [x, z][x, y]^z$.

These are verified by inspection. Repeated applications of 0.1 give the first two claims of:

0.2. *For a positive integer n,*
  (i) $[x^n, y] = [x, y]^{x^{n-1}} \cdot [x, y]^{x^{n-2}} \ldots [x, y]^x \cdot [x, y]$;

1

(ii) $[x, y^n] = [x, y] \cdot [x, y]^y \dots [x, y]^{y^{n-1}}$;

(iii) $(xy)^n \equiv x^n y^n [y, x]^{n(n-1)/2} \pmod{\gamma_3(G)}$.

Part (iii) is easily proved by induction on n, using (i). It is the beginning of the *Hall–Petrescu formula*, proved in full in Appendix A.

0.3 'Three-subgroup lemma'. *If A, B and C are normal subgroups of G then*

$$[A, B, C] \leq [B, C, A][C, A, B].$$

Proof This follows from the Hall–Witt identity:

$$[a, b^{-1}, c]^b [b, c^{-1}, a]^c [c, a^{-1}, b]^a = 1, \qquad (*)$$

which is most quickly verified by putting $u = acb^a$, $v = bac^b$, $w = cba^c$ and noting that $[a, b^{-1}, c]^b = u^{-1} v$, etc.

0.2 Nilpotent groups

The terms of the *lower central series* of a group G are defined by $\gamma_1(G) = G$, $\gamma_{i+1}(G) = [\gamma_i(G), G]$ for $i \geq 1$. The group G is *nilpotent*, of class at most c, if $\gamma_{c+1}(G) = 1$. The centre of G is $Z(G) = \{x \in G \,|\, xg = gx$ for all $g \in G\}$.

G is a *finite p-group* (where p is a prime) if $|G| = p^n$ for some n.

0.4. *Let G be a finite group.*

(i) *G is nilpotent if and only if $Z(G/N) > 1$ for every proper normal subgroup N of G.*

(ii) *If G is a p-group then G is nilpotent.*

(iii) *If G is a p-group then every maximal proper subgroup of G is normal and has index p in G.*

(iv) *If G is nilpotent and $1 < N \lhd G$ then $[N, G] < N$ and $N \cap Z(G) > 1$.*

(v) *If G is nilpotent and $1 < N \lhd G$ then some maximal proper subgroup of N is normal in G.*

(vi) *If G is nilpotent then so are every quotient and every subgroup of G.*

(vii) *If G is nilpotent then the elements of finite order in G form a subgroup.*

Proof (i) *'If'*: induction on $|G|$. By hypothesis, $1 < Z(G) = Z$, say. Inductively, $\gamma_m(G/Z) = 1$ for some m. Then $\gamma_{m+1}(G) = 1$.

'Only if': Suppose $\gamma_{i+1}(G) = 1$. For some k, we then have $\gamma_{k+1}(G) \leq N$, $\gamma_k(G) \not\leq N$. This implies $1 < \gamma_k(G)N/N \leq Z(G/N)$.

(ii) By (i), it suffices to show that $Z(G) \neq 1$. Let $h_1, \ldots, h_r$ represent the non-central conjugacy classes of G. Then $|G : C_G(h_i)| = p^{e_i} > 1$ so

$$|Z(G)| = |G| - \sum_{i=1}^{r} |G : C_G(h_i)| \equiv 0 \pmod{p}.$$

Since $|Z(G)| \geq 1$ it follows that $|Z(G)| \geq p$.

(iii) Let M be a maximal proper subgroup of G. By (ii), $Z(G)$ contains an element z of order p. If $z \in M$, $M/\langle z \rangle$ is a maximal proper subgroup of $G/\langle z \rangle$ and we argue by induction. If $z \notin M$ then $M\langle z \rangle = G$. In this case $M \lhd G$ and $|G : M| = p$.

(iv) For some k, $\gamma_k(G) \cap N \neq 1$ and $\gamma_{k+1}(G) \cap N = 1$. Then $\gamma_k(G) \cap N \leq Z(G) \cap N$, so $Z(G) \cap N \neq 1$. If $[N, G] = 1$ then $[N, G] < N$. If $[N, G] > 1$, then (since $[N, G] \lhd G$) we have $1 < [N, G] \cap Z(G) = K$, say. Then $1 < N/K \lhd G/K$ and we argue by induction.

(v) By (iv), $[N, G] < N$. Any maximal proper subgroup of N containing $[N, G]$ is necessarily normal in G.

(vi) Clear.

(vii) Let $x, y \in G$ have finite order; it will suffice to prove that the subgroup $H = \langle x, y \rangle$ that they generate is finite. Now $H/\gamma_2(H)$ is finite. Suppose that $\gamma_{i-1}(H)/\gamma_i(H)$ is finite for some $i \geq 2$. There is a well-defined bilinear mapping from $H/\gamma_2(H) \times \gamma_{i-1}(H)/\gamma_i(H)$ into $\gamma_i(H)/\gamma_{i+1}(H)$ given by

$$(a\gamma_2(H), b\gamma_i(H)) \mapsto [a, b]\gamma_{i+1}(H),$$

which induces an epimorphism $H/\gamma_2(H) \otimes \gamma_{i-1}(H)/\gamma_i(H) \to \gamma_i(H)/\gamma_{i+1}(H)$. Hence $\gamma_i(H)/\gamma_{i+1}(H)$ is finite. It follows by induction that $\gamma_i(H)/\gamma_{i+1}(H)$ is finite for every i; but $\gamma_{c+1}(H) = 1$ where c is the nilpotency class of G.

0.5. *Let G be any group and α an automorphism of G which induces the identity on $G/\gamma_2(G)$. Then α induces the identity on $\gamma_i(G)/\gamma_{i+1}(G)$ for every i.*

Proof There is an $\langle \alpha \rangle$-module epimorphism from

$$G/\gamma_2(G) \otimes \gamma_{i-1}(G)/\gamma_i(G)$$

onto $\gamma_i(G)/\gamma_{i+1}(G)$, given by

$$x\gamma_2(G) \otimes y\gamma_i(G) \mapsto [x, y]\gamma_{i+1}(G).$$

The result follows by induction on i.

0.3 Stability group theory

Suppose that H is a group acting faithfully by automorphisms on a group G.

0.6. If $N \triangleleft G$ and H induces the trivial action on both N and G/N, then H can be embedded in a Cartesian product of copies of $Z(N)$.

Proof Let X be a set of generators for G. The embedding is given by the map $H \longrightarrow \prod_{x \in X} Z(N)$,

$$h \mapsto (x^{-1}x^h)_{x \in X};$$

though not obvious, it is easily verified that $x^{-1}x^h$ does indeed lie in $Z(N)$ for all $x \in G$.

If, in 0.6, the centre of N is torsion-free, or has a given finite exponent, etc., it follows that H has the same property.

0.7. *Let $G = G_0 \geq G_1 \geq \cdots \geq G_k = 1$ be a series of normal subgroups of G, and suppose that H fixes each G_i, and induces the trivial action on each factor G_i/G_{i+1}. Then*
   (i) $\gamma_k(H) = 1$;
   (ii) *if each G_i/G_{i+1} has exponent dividing m (respectively: is torsion-free), then H has exponent dividing m^{k-1} (respectively: is torsion-free).*

Proof Induction on k. Put $K = C_H(G_1) \cap C_H(G/G_{k-1})$. By the inductive hypothesis, $H/C_H(G_1)$ and $H/C_H(G/G_{k-1})$ satisfy (i) and (ii), with $k - 1$ replacing k. Therefore so does H/K. Taking $a \in K$, $b \in H$ and $c \in G$ in the Hall–Witt identity $(*)$, we see that K is contained in the centre of H (apply $(*)$ to the semidirect product $G \rtimes H$). Therefore

$$\gamma_k(H) = [\gamma_{k-1}(H), H] \leq [K, H] = 1.$$

Also, 0.6 shows that in case (ii), K has exponent dividing m (respectively, K is torsion-free); hence (ii) follows.

Applying 0.7 to the conjugation action of G on $\gamma_i(G)$, we deduce that

$$[\gamma_i(G), \gamma_k(G)] \leq \gamma_{i+k}(G),$$

for any group G and all i and k (take $H = G/\mathrm{C}_G(\gamma_i(G)/\gamma_{i+k}(G))$ and replace G by $\gamma_i(G)/\gamma_{i+k}(G)$).

0.4 Unipotent groups

Let k be a finite field of characteristic p and $V = k^n$. The group $\mathrm{GL}_n(k)$ of all invertible $n \times n$ matrices over k may be identified with the group $\mathrm{GL}(V)$ of all k-linear automorphisms of V. We denote by $\mathrm{U}_n(k)$ the subgroup consisting of upper uni-triangular matrices. An automorphism g of V is *unipotent* if $(g-1)^n$ is the zero endomorphism. A *subgroup* H of $\mathrm{GL}_n(K)$ is said to be unipotent if each of its elements is unipotent.

0.8. *Let $H \leq \mathrm{GL}_n(k)$. The following are equivalent:*
(i) *H is unipotent.*
(ii) *The semidirect product $V \rtimes H$ is a nilpotent group.*
(iii) *There is a chain of H-invariant k-subspaces*

$$V = V_n > V_{n-1} > \cdots > V_1 > V_0 = 0$$

such that H induces the trivial action on each factor V_i/V_{i-1}.
(iv) *$(h_1 - 1)(h_2 - 1) \cdots (h_n - 1) = 0$ for all $h_1, \ldots, h_n \in H$.*
(v) *There exists $g \in \mathrm{GL}_n(k)$ such that $g^{-1}Hg \leq \mathrm{U}_n(k)$.*
(vi) *H is a finite p-group.*

Proof If h is unipotent then

$$h^{p^e} - 1 = (h-1)^{p^e} = 0$$

whenever $p^e \geq n$. Therefore (i) implies (vi). If H is a finite p-group then so is $V \rtimes H$, so (vi) implies (ii). It is easy to see that (ii)$\Rightarrow$(iii)$\Rightarrow$(iv)$\Rightarrow$(i) and that (iii)$\Leftrightarrow$(v).

0.5 Frattini subgroup

In this section, G denotes a finite p-group. The *Frattini subgroup* of G, denoted $\Phi(G)$, is the intersection of all maximal proper subgroups of G.

0.9. (i) $\Phi(G) = [G, G]G^p$.
(ii) *If $X \subseteq G$ and $X\Phi(G)$ generates $G/\Phi(G)$ then X generates G.*

(iii) $G/\Phi(G) \cong \mathbb{F}_p^d$ where d is the minimal cardinality of any generating set for G.

Proof (i) $G/[G,G]G^p$ is an elementary abelian p-group, so its maximal proper subgroups intersect in the identity. Therefore $\Phi(G) \le [G,G]G^p$. The reverse inclusion follows from 0.4 (iii).

(ii) is clear, since if $\langle X \rangle < G$ then $\langle X \rangle \, \Phi(G)$ lies inside some maximal proper subgroup of G.

(iii) This now follows from (i) and (ii).

0.10. *Let H be the set of all automorphisms of G which induce the identity on $G/\Phi(G)$. Then H is a finite p-group.*

Proof It is enough to show that if $\alpha \in H$ has prime order q, then $q = p$. Let $\{x_1, \ldots, x_d\}$ be a generating set for G, and put

$$\Omega = \{(u_1 x_1, \ldots, u_d x_d) \mid u_1, \ldots, u_d \in \Phi(G)\},$$

a subset of $G \times \cdots \times G$ (with d factors). Then α permutes Ω, and has no fixed points, in view of 0.9 (ii). Therefore each orbit has length q and so $q \mid |\Omega| = |\Phi(G)|^d$. It follows that $q = p$.

0.6 Group algebras

Let G be a group and K a commutative ring. The group algebra of G over K, denoted $K[G]$, is defined to be the free K-module on the basis G, endowed with a product which extends simultaneously the group operation on G and the ring multiplication in K. Thus the elements of $K[G]$ are sums of the form $\sum_{g \in G} a_g g$, with each $a_g \in K$ and $a_g = 0$ for all but finitely many $g \in G$, and

$$\left(\sum_{g \in G} a_g g \right)\left(\sum_{g \in G} b_g g \right) = \sum_{g \in G} c_g g$$

where

$$c_g = \sum_{x \in G} a_x b_{x^{-1} g}.$$

It is easily verified that $K[G]$ is a ring. One usually identifies K with the subring $K \cdot 1_G$ of $K[G]$, and G with the subgroup $1_K \cdot G$ of the group of units of $K[G]$. There is a homomorphism $\varepsilon : K[G] \to K$ given by

$$\sum_{g \in G} a_g g \mapsto \sum_{g \in G} a_g;$$

this is called the *augmentation*. Its kernel I is the *augmentation ideal* of $K[G]$; it is easy to see that

$$I = \{ \sum a_g g \mid \sum a_g = 0 \} = \sum_{g \in G \backslash 1} (g-1)K[G] = \bigoplus_{g \in G \backslash 1} K(g-1).$$

More generally, for any normal subgroup N of G there is a natural epimorphism $K[G] \to K[G/N]$, whose kernel is the right (or equivalently left) ideal generated by the set $\{ g - 1 \mid g \in N \}$.

0.7 Topology

A topological space X is *Hausdorff* if distinct points of X have disjoint neighbourhoods; by a neighbourhood of x we mean any subset of X which contains an open set U with $x \in U$. A topological space X is *compact* if for any covering of X by open sets

$$X = \bigcup_{\alpha \in A} U_\alpha$$

there is a finite subset $\{ \alpha_1, \dots, \alpha_r \}$ of A such that

$$X = \bigcup_{i=1}^{r} U_{\alpha_i}.$$

0.11. *A space X is compact if and only if for each family $(Y_\alpha)_{\alpha \in A}$ of closed subsets of X with*

$$\bigcap_{\alpha \in A} Y_\alpha = \varnothing,$$

there exists a finite subset $\{ \alpha_1, \dots, \alpha_r \}$ of A such that

$$\bigcap_{i=1}^{r} Y_{\alpha_r} = \varnothing.$$

0.12. *If $f : X \longrightarrow Y$ is continuous and X is compact then $f(X)$ is compact.*

0.13. *Let X be a Hausdorff space.*
   (i) *Every compact subspace of X is closed in X.*
   (ii) *If X is compact, then every closed subset of X is compact (with the subspace topology).*

(iii) *If X is compact, then every infinite subset of X has a limit point in X.*

The proofs of 0.11–0.13 are simple exercises.

0.14. *If A and B are disjoint compact subsets of a Hausdorff space X, then there exist disjoint open subsets U and V of X with $A \subseteq U$ and $B \subseteq V$.*

Proof For each $a \in A$ and $b \in B$ there exist open sets $U(a,b)$, $V(a,b)$ with $a \in U(a,b)$, $b \in V(a,b)$ and $U(a,b) \cap V(a,b) = \varnothing$. Fix $a \in A$. Since B is compact, there exist $b_1, \ldots, b_r \in B$ such that $B \subseteq \bigcup_{i=1}^{r} V(a,b_i) = V(a)$, say. Put $U(a) = \bigcap_{i=1}^{r} U(a,b_i)$. Then the compactness of A gives $a_1, \ldots, a_s \in A$ such that $A \subseteq \bigcup_{i=1}^{s} U(a_j)$. Now let $U = \bigcup_{i=1}^{s} U(a_j)$, $V = \bigcap_{i=1}^{s} V(a_j)$.

0.15. *Let $f : X \longrightarrow Y$ be a continuous bijection, where X is compact and Y is Hausdorff. Then f is a homeomorphism.*

Proof We have to show that $f^{-1} : Y \longrightarrow X$ is continuous, i.e. that $(f^{-1})^{-1}(U)$ is open in Y for all U open in X. Now

$$(f^{-1})^{-1}(U) = f(U) = Y \setminus f(X \setminus U)$$

and $X \setminus U$ is compact. Therefore $f(X \setminus U)$ is compact, hence closed in Y, giving the result.

0.16. (Tychonoff's Theorem) *The product of any family of compact spaces is compact.*

For the proof see for example Higgins (1974), Chapter 1 (or any introductory topology textbook).

A *topological group* is a group G which is also a topological space, such that the maps

$$g \mapsto g^{-1} : G \to G$$
$$(g,h) \mapsto gh : G \times G \to G$$

are both continuous. Some less trivial results about topological groups are given in Appendix B; for most purposes, the following will suffice:

0.17. *Let G be a topological group.*

(i) *For each $g \in G$, the maps $x \mapsto xg$, $x \mapsto gx$, and $x \mapsto x^{-1}$ are homeomorphisms of G.*

(ii) *If H is a subgroup of G and H is open (respectively, closed), then every coset of H is open (respectively, closed).*

(iii) *Every open subgroup of G is closed.*

(iv) *G is Hausdorff if and only if {1} is a closed subset of G.*

(v) *If N is a closed normal subgroup of G and G is Hausdorff, then G/N is Hausdorff (with the quotient topology).*

(vi) *If H is a subgroup of G and H contains a non-empty open subset U of G then H is open in G.*

Proof (i), (ii), (iii) are easy exercises. (iv): In a Hausdorff space, singleton subsets are closed (easy). Conversely, suppose $\{1\}$ is closed. Then every singleton is closed, by (i). Let $x \neq y$ be elements of G. Then $U = G \setminus \{xy^{-1}\}$ is open. Since the map $(a, b) \mapsto a^{-1}b$ is continuous, and $1 \in U$, there exist open neighbourhoods V_1 and V_2 of 1 such that $V_1^{-1} \cdot V_2 \subseteq U$. Then $V_1 x$ and $V_2 y$ are disjoint neighbourhoods of x, y. (v) follows from (iv). (vi): note that $H = \bigcup_{h \in H} Uh$.

0.8 Lie algebras

Let k be a commutative ring. A *Lie algebra* over k is a k-module L with a binary operation, (occasionally called the *Lie bracket*)

$$(\cdot, \cdot) : L \times L \to L$$

that is *k-bilinear* and satisfies

$$(a, a) = 0$$
$$((a, b), c) + ((b, c), a) + ((c, a), b) = 0$$

for all $a, b, c \in L$. The first condition implies

$$(a, b) = -(b, a)$$

for all a and b, and is equivalent to it unless 2 is a zero-divisor on L. The second condition is known as the *Jacobi identity*.

If A is any associative algebra over k, we may define a new binary operation on A, called *commutation*, as follows:

$$(a, b) = ab - ba.$$

It is easy to verify that with this operation A becomes a Lie algebra, the *commutation Lie algebra* on A.

A Lie algebra over $\mathbb{Z}$ is sometimes called a *Lie ring*.

Lie algebras over $\mathbb{R}$ appeared originally in the guise of 'infinitesimal Lie groups'; that is, as a sort of linearised approximation to a (real) Lie group. As we shall see, Lie algebras over the p-adic numbers play the same role relative to p-adic Lie groups.

0.9 p-adic numbers

p will denote an arbitrary, but fixed, prime number. Each rational number $x \neq 0$ can be written uniquely as

$$x = p^n \cdot \frac{a}{b}$$

with $n, a, b \in \mathbb{Z}$, $b > 0$, $\gcd(a, b) = 1$ and $p \nmid ab$. We put

$$v_p(x) = n, \ |x|_p = p^{-n};$$

here $|\cdot|_p$ is the *p-adic absolute value* on $\mathbb{Q}$. This absolute value induces a metric on $\mathbb{Q}$, and the completion of $\mathbb{Q}$ with respect to this metric is *the p-adic field* $\mathbb{Q}_p$. Each element α of $\mathbb{Q}_p$ is thus the limit of a Cauchy sequence

$$\alpha = \lim_{i \to \infty} x_i$$

with $x_i \in \mathbb{Q}$ for each i, and the absolute value is extended to $\mathbb{Q}_p$ by setting

$$|\alpha|_p = \lim_{i \to \infty} |x_i|_p .$$

The 'valuation ring' in $\mathbb{Q}_p$ is the subring of *p-adic integers*

$$\mathbb{Z}_p = \left\{ \alpha \in \mathbb{Q}_p \mid |\alpha|_p \leq 1 \right\}.$$

Each element of $\mathbb{Z}_p$ is the limit of a Cauchy sequence in $\mathbb{Q}$ whose terms all lie in $\mathbb{Z}$. It follows that each p-adic integer is the sum of a series

$$\sum_{n=0}^{\infty} a_n p^n \tag{†}$$

with each $a_n \in \mathbb{Z}$; moreover, each a_n may be chosen to lie in the set $\{0, 1, \dots, p-1\}$, in which case the expression (†) is uniquely determined.

An alternative, equivalent, definition of $\mathbb{Z}_p$ is as the inverse limit of the system of rings

$$(\mathbb{Z}/p^n\mathbb{Z})_{n \in \mathbb{N}} ;$$

this construction is discussed in detail in Chapter 1.

The additive group of $\mathbb{Z}_p$ is a compact Hausdorff topological group, having $\mathbb{Z}$ as a dense subgroup. Intuitively, one should think of doing ring-theoretic operations in $\mathbb{Z}_p$ as follows: do the operations in the integers modulo p^n, then let n tend to ∞.

Our convention throughout the book will be that p denotes an arbitrary, but fixed, prime; and we shall use $|\cdot|$, without the subscript, to stand for $|\cdot|_p$. (The usual real or complex absolute value will scarcely appear; when it does, the reader will forgive us.)

Part I
Pro-p groups

1
Profinite groups and pro-p groups

Most of this book is about pro-p groups. The purpose of this introductory chapter is to explain what these are and where they come from, and to establish their basic properties. Many of these properties belong to the larger class of *profinite groups*, and we begin by discussing these.

Notation   If G is a topological group and X is a subset of G, we write $\overline{X}$ to denote the closure of X in G, and denote by $\langle X \rangle$ the subgroup of G generated as an abstract group by X. We write $X \leq_o G$, $X \vartriangleleft_o G$, $X \leq_c G$, $X \vartriangleleft_c G$ to denote: X is an open subgroup, open normal subgroup, closed subgroup, closed normal subgroup of G, respectively.

1.1 Profinite groups

1.1 Definition A *profinite group* is a compact Hausdorff topological group whose open subgroups form a base for the neighbourhoods of the identity.

Thus a discrete group is profinite if and only if it is finite. Since in a topological group any subgroup containing a non-empty open set is itself open, we see that the second part of the definition comes down to: *every open set containing 1 contains an open subgroup*. There are several equivalent definitions, some of which are discussed in the exercises; the most important one is given in Proposition 1.3, below. First we list some elementary consequences of Definition 1.1.

1.2 Proposition *Let G be a profinite group.*
  (i) *Every open subgroup of G is closed, has finite index in G, and contains an open normal subgroup of G. A closed subgroup of G is*

15

open if and only if it has finite index. The family of all open subgroups of G intersects in $\{1\}$.

(ii) *A subset of G is open if and only if it is a union of cosets of open normal subgroups.*

(iii) *For any subset X of G,*

$$\overline{X} = \bigcap_{N \triangleleft_o G} XN.$$

If X is a subgroup of G then

$$\overline{X} = \bigcap \{K \mid X \leq K \leq_o G\}.$$

(iv) *If X and Y are closed subsets of G then so is the set $XY = \{xy \mid x \in X,\, y \in Y\}$. If X is closed and n is an integer then the set $\{x^n \mid x \in X\}$ is closed.*

(v) *Let H be a closed subgroup of G. Then H (with the induced topology) is a profinite group. Every open subgroup of H is of the form $H \cap K$ with $K \leq_o G$.*

(vi) *Let N be a closed normal subgroup of G. Then G/N (with the quotient topology) is a profinite group, and the natural homomorphism $G \to G/N$ is an open and closed continuous mapping.*

(vii) *A sequence (g_i) in G converges if and only if it is a Cauchy sequence: i.e. for each $N \triangleleft_o G$ there exists $n = n(N)$ such that $g_i^{-1} g_j \in N$ for all $i \geq n$ and $j \geq n$.*

Proof   Most of this can be safely left as an exercise. The essential points to note are that every coset of an open subgroup is open, and that G is compact. We prove the 'if' part of (vii). If the Cauchy sequence (g_i) contains only finitely many distinct terms, it is easy to see that it must be eventually constant. If not, then since G is compact and Hausdorff the infinite set $\{g_i \mid i \in \mathbb{N}\}$ has a limit point $g \in G$. Now let $N \triangleleft_o G$. The neighbourhood gN of g then contains infinitely many of the g_i, so there exists $i \geq n(N)$ with $g_i \in gN$. For each $j \geq n(N)$ we then have $g_j \in g_j N = g_i N = gN$, showing that the sequence (g_j) converges to g.

A second definition of profinite groups is based on the concept of an *inverse limit*. We recall briefly what this is. A *directed set* is a non-empty partially ordered set $(\Lambda, \leq)$ with the property that for every $\lambda, \mu \in \Lambda$ there exists $\nu \in \Lambda$ with $\nu \geq \lambda$ and $\nu \geq \mu$. An *inverse system* of sets (or groups, rings or topological spaces) over Λ is a family of sets (or groups, etc.) $(G_\lambda)_{\lambda \in \Lambda}$, with maps (respectively homomorphisms, continuous maps) $\pi_{\lambda\mu} : G_\lambda \to G_\mu$ whenever $\lambda \geq \mu$, satisfying the natural compatibility conditions

$$\pi_{\lambda\lambda} = \mathrm{Id}_{G_\lambda}, \quad \pi_{\lambda\mu}\pi_{\mu\nu} = \pi_{\lambda\nu}$$

whenever $\lambda \geq \mu \geq \nu$. The *inverse limit*

$$\varprojlim G_\lambda = \varprojlim (G_\lambda)_{\lambda\in\Lambda}$$

is the subset (or subgroup, etc.) of the Cartesian product $\prod_{\lambda\in\Lambda} G_\lambda$ consisting of all 'vectors' (g_λ) such that $g_\lambda \pi_{\lambda\mu} = g_\mu$ whenever $\lambda \geq \mu$. It is easily verified (see Exercise 1) that this is the (unique) solution to an appropriate universal problem (thus it is a limit in the sense of category theory).

If the G_λ are finite groups, we give each of them the discrete topology, and $\prod G_\lambda$ the product topology. In this way, $\varprojlim G_\lambda$, with the induced topology, becomes a topological group.

If Λ is a family of normal subgroups of a given group G, we may order Λ by reverse inclusion and obtain an inverse system $(G/N)_{N\in\Lambda}$, the maps being the natural epimorphisms $G/N \to G/M$ for $N \leq M$. We are now ready for

1.3 Proposition *If G is a profinite group then G is (topologically) isomorphic to $\varprojlim(G/N)_{N \vartriangleleft_o G}$. Conversely, the inverse limit of any inverse system of finite groups is a profinite group.*

Proof Write $\widehat{G} = \varprojlim(G/N)_{N \vartriangleleft_o G}$. There is a natural homomorphism

$$\iota : G \to \prod G/N,$$

given by $g\iota = (gN)_{N \vartriangleleft_o G}$. Since $\bigcap_{N \vartriangleleft_o G} N = 1$, ι is injective; and it is clear that $G\iota \leq \widehat{G}$. To see that ι is surjective, let $(g_N N) \in \widehat{G}$. Then every finite collection of cosets $g_N N$ has non-empty intersection; since these cosets are closed subsets of the compact space G, it follows that $\bigcap_{N \vartriangleleft_o G} g_N N$ is non-empty. Choosing g to lie in this intersection we get $g\iota = (g_N N)$.

Since a continuous isomorphism between compact Hausdorff groups is a topological isomorphism, it only remains to show that ι is continuous. For $M \vartriangleleft_o G$ let

$$U(M) = \prod_{N \not\geq M} G/N \times \prod_{N \geq M} \{1\} \leq \prod_{N \vartriangleleft_o G} G/N.$$

The subgroups $U(M) \cap \widehat{G}$ form a base for the neighbourhoods of 1 in $\widehat{G}$, and for each M we have $U(M)\iota^{-1} = M$ which is open in G. So ι is continuous.

For the converse, we consider an inverse system of finite groups G_λ ($\lambda \in \Lambda$), each with the discrete topology. Then $\prod_{\lambda \in \Lambda} G_\lambda$ is Hausdorff, and Tychonoff's theorem shows that it is compact. It follows from the definition of the product topology that every neighbourhood of 1 contains a subgroup of the form $U(S) = \prod_{\lambda \notin S} G_\lambda \times \prod_{\lambda \in S}\{1\}$ for some finite subset S of Λ. Thus $\prod G_\lambda$ is a profinite group, and we only have to show that $\varprojlim G_\lambda = \widehat{G}$, say, is a closed subgroup. Now let $\widehat{g} = (g_\lambda) \in \prod G_\lambda \setminus \widehat{G}$. Then there exist $\nu > \mu$ in Λ such that $g_\nu \pi_{\nu\mu} \neq g_\mu$. Now $\widehat{g}U(\nu, \mu)$ is an open neighbourhood of $\widehat{g}$ in $\prod G_\lambda$, and $\widehat{g}U(\nu, \mu) \cap \widehat{G} = \varnothing$. This shows that $\prod G_\lambda \setminus \widehat{G}$ is open in $\prod G_\lambda$, and the result follows.

Examples of profinite groups

(1) Profinite groups arise as the Galois groups of algebraic field extensions: as such they form an essential component in the modern formulation of class field theory (see Cassells and Fröhlich (1967), Fried and Jarden (1986)), and the Galois cohomology of algebraic groups (Serre (1997), Platonov and Rapinchuk (1994), Chapter 6). Two chapters of [DSS] are devoted to recent applications of (specifically) pro-p Galois groups in number theory and the theory of local fields.

(2) Let Γ be a group and Λ a family of normal subgroups of finite index in Γ, directed by reverse inclusion. The family of quotients $(\Gamma/N)_{N \in \Lambda}$ forms an inverse system of finite groups, whose inverse limit

$$\widehat{\Gamma}_\Lambda = \varprojlim(\Gamma/N)_{N \in \Lambda}$$

is a profinite group by Proposition 1.3. The natural homomorphism $\Gamma \to \widehat{\Gamma}_\Lambda$ has kernel equal to $\bigcap_{N \in \Lambda} N = K$, say, and it embeds Γ/K as a dense subgroup in $\widehat{\Gamma}_\Lambda$. The group $\widehat{\Gamma}_\Lambda$ is called *a profinite completion* of Γ. (If we define Cauchy sequences in Γ by taking the elements of Λ as a base for the neighbourhoods of 1 in Γ, we can identify $\widehat{\Gamma}_\Lambda$ with the set of all Cauchy sequences modulo the 'null sequences', so $\widehat{\Gamma}_\Lambda$ is a completion in the usual sense.) When Λ consists of all the normal subgroups of finite index, $\widehat{\Gamma}_\Lambda$ is *the* profinite completion of Γ, usually denoted simply $\widehat{\Gamma}$. The kernel of $\Gamma \to \widehat{\Gamma}$ is the *finite residual* of Γ. Later, we shall be particularly concerned with the case where Λ consists of all the normal subgroups whose index is a power of some (fixed) prime p: in this case,

$\hat{\Gamma}_\Lambda = \hat{\Gamma}_p$ is the *pro-p completion* of Γ, and Γ embeds into $\hat{\Gamma}_p$ if and only if Γ is residually a finite p-group.

(3) Let R be a commutative ring with identity, and let Λ be a set of ideals of finite index in R, directed by reverse inclusion. Fix a positive integer n. For each $I \in \Lambda$ let $K(I)$ be the principal congruence subgroup modulo I in $\mathrm{SL}_n(R)$, i.e. the kernel of the natural homomorphism $\mathrm{SL}_n(R) \to \mathrm{SL}_n(R/I)$. The family $(K(I))_{I \in \Lambda}$ is a directed set of normal subgroups of finite index in $\mathrm{SL}_n(R)$, and we obtain a corresponding profinite completion $\widehat{\mathrm{SL}_n(R)}_\Lambda$. On the other hand, we can form the ring $\hat{R}_\Lambda = \varprojlim (R/I)_{I \in \Lambda}$; it is not hard to see that $\widehat{\mathrm{SL}_n(R)}_\Lambda$ may be identified with a subgroup of $\mathrm{SL}_n(\hat{R}_\Lambda)$. If $R = \mathbb{Z}$ and $\Lambda = \{p^i \mathbb{Z} \mid i \geq 0\}$, where p is a prime, $\hat{R}_\Lambda$ is the ring $\mathbb{Z}_p$ of p-adic integers; in this case, $\widehat{\mathrm{SL}_n(\mathbb{Z})}_\Lambda$ can be identified with $\mathrm{SL}_n(\mathbb{Z}_p)$ (see Exercise 9). If $R = \mathbb{Z}$ and Λ consists of all ideals of finite index in $\mathbb{Z}$, we can identify $\widehat{\mathrm{SL}_n(\mathbb{Z})}_\Lambda$ with $\mathrm{SL}_n(\hat{\mathbb{Z}})$ where $\hat{\mathbb{Z}}$ denotes the profinite completion of $\mathbb{Z}$. The relationship between $\mathrm{SL}_n(\hat{\mathbb{Z}})$ and *the* profinite completion of $\mathrm{SL}_n(\mathbb{Z})$ is a deep matter (the 'congruence subgroup problem'); see Platonov and Rapinchuk (1994), §9.5.

Next, we state a fundamental set-theoretic principle, which is extremely useful when one is trying to deduce properties of a profinite group from properties of its finite quotient groups.

1.4 Proposition *Let $(X_\lambda; \pi_{\lambda\mu})$ be an inverse system of non-empty compact spaces over a directed set Λ. Then $\varprojlim X_\lambda$ is non-empty.*

Proof For each subset $S \subseteq \Lambda$ let $L(S)$ be the subset of $P = \prod_{\lambda \in \Lambda} X_\lambda$ consisting of 'vectors' (x_λ) satisfying $x_\lambda \pi_{\lambda\mu} = x_\mu$ whenever $\lambda \in S$, $\mu \in S$ and $\lambda \geq \mu$. If S is finite then $L(S)$ is closed in P (we give P the product topology), and $L(S) \neq \varnothing$: for there exists $\nu \in \Lambda$ with $\nu \geq \lambda$ for each $\lambda \in S$, and choosing $x_\nu \in X_\nu$, $x_\lambda = x_\nu \pi_{\nu\lambda}$ for each $\lambda \in S$, and x_λ an arbitrary element of X_λ for $\lambda \notin S \cup \{\nu\}$, we get $(x_\lambda) \in L(S)$. Since P is compact, by Tychonoff's theorem, it follows that

$$\bigcap \{L(S) \mid S \text{ a finite subset of } \Lambda\} = L, \text{ say,}$$

is non-empty. But L is exactly $\varprojlim X_\lambda$.

Proposition 1.4 will mostly be applied to inverse systems of *finite* sets: these are compact when endowed with the discrete topology.

A subset X of a topological group G *generates* G *topologically* if $\overline{\langle X \rangle} = G$. The topological group G is *finitely generated* if it is generated topologically by a finite subset.

1.5 Proposition *Let G be a profinite group and let H be a closed subgroup.*

(i) *Let $X \subseteq H$. Then X generates H topologically if and only if XN/N generates HN/N for every $N \lhd_o G$.*

(ii) *Let d be a positive integer. If HN/N can be generated by d elements for every $N \lhd_o G$, then H can be generated topologically by a d-element subset.*

Proof (i) Follows from Proposition 1.2 (iii). (ii) For each $N \lhd_o G$, let Y_N be the set of all d-tuples of elements of G/N which generate HN/N. Each set Y_N is finite and non-empty, and if $\pi_{MN} : G/M \to G/N$ is the natural projection for $M \leq N$ (where $M, N \lhd_o G$) then $Y_M \pi_{MN} \subseteq Y_N$. So the Y_N (with $N \lhd_o G$) form an inverse system. By Proposition 1.4, the inverse limit of this inverse system is non-empty; let $(X_N) \in \varprojlim Y_N$. Then there exist $x_1, \ldots, x_d \in G$ such that for each $N \lhd_o G$, $X_N = (x_1 N, \ldots, x_d N)$ (see the proof of Proposition 1.3), and part (i) shows that $\{x_1, \ldots, x_d\}$ generates H topologically.

1.6 Proposition *If G is a finitely generated profinite group and m is a positive integer then G has only finitely many open subgroups of index m, and every open subgroup contains an open topologically characteristic subgroup.*

(A subgroup of G is *topologically characteristic* if it is invariant under all continuous automorphisms of G.)

Proof Suppose G is topologically generated by a d-element subset. Then there are at most $(m!)^d$ continuous homomorphisms of G into the symmetric group S_m (where S_m has the discrete topology). Now if $H \leq_o G$ and $|G : H| = m$ then the permutation representation of G on the right cosets of H is continuous, since the inverse image of a given permutation is the intersection of m open sets of the form $x^{-1}Hy$; and H is exactly the inverse image in G of a one-point stabiliser in S_m. Hence there are no more than $m \cdot (m!)^d$ possibilities for H.

The second claim follows at once, since if $H \leq_o G$ then $|G : H^\alpha| = |G : H|$ for each automorphism α of G, and H^α is open if α is continuous, so the topologically characteristic subgroup $\bigcap \{H^\alpha \mid \alpha \in \mathrm{Aut}\,(G),\, \alpha$ continuous$\}$ is again open in G.

1.7 Proposition *If G is a finitely generated profinite group then every open subgroup of G is finitely generated.*

Proof Let X be a finite (topological) generating set for G, and assume without loss of generality that $X^{-1} = X$. Let H be an open subgroup of G, and let T be a transversal to the right cosets of H in G, with $1 \in T$. Note that T is finite. For each $x \in X$ and $t \in T$ there exists $s = s(t, x) \in T$ such that $Htx = Hs$. We put

$$Y = \{tx \cdot s(t, x)^{-1} \mid t \in T, x \in X\},$$

and claim that Y generates H topologically.

Consider the subgroup $M = \overline{\langle Y \rangle}$ of G. If $a \in M$, $t \in T$ and $x \in X$, then

$$at \cdot x = atxs(t, x)^{-1} \cdot s(t, x) \in MT;$$

so $MTX = MT$. Since $1 \in MT$ and $X = X^{-1}$, it follows that $MT \supseteq \langle X \rangle$; as T is finite, MT is closed, and therefore $MT = G$. But clearly $M \leq H$, and so

$$H = MT \cap H = M(T \cap H) = M$$

as claimed.

1.8 Definition Let G be a profinite group. The *Frattini subgroup* of G is

$$\Phi(G) = \bigcap \{M \mid M \text{ is a maximal proper open subgroup of } G\}.$$

1.9 Proposition *Let G be a profinite group.*
(i) $\Phi(G) \lhd_c G$.
(ii) *If $K \lhd_c G$ and $K \leq \Phi(G)$ then $\Phi(G/K) = \Phi(G)/K$.*
(iii) *For a subset X of G the following are equivalent:*
  (a) *X generates G topologically;*
  (b) *$X \cup \Phi(G)$ generates G topologically;*
  (c) *$X\Phi(G)/\Phi(G)$ generates $G/\Phi(G)$ topologically.*

Proof (i) and (ii) are immediate from the definition. In (iii), it is clear that (a) implies (b) and that (b) implies (c). Suppose that (c) holds, and let K be an open subgroup of G containing X. If $K \neq G$, then $K \leq M$ for some maximal proper open subgroup M of G, and then

$$\overline{\langle X \rangle}\Phi(G)/\Phi(G) \leq M/\Phi(G) \neq G/\Phi(G),$$

contradicting (c). Hence $K = G$, and it follows by Proposition 1.2 (iii) that $\overline{\langle X \rangle} = G$. Thus (c) implies (a).

1.2 Pro-p groups

Henceforth, we adopt the convention that p stands for an arbitrary, but fixed, prime.

1.10 Definition A *pro-p group* is a profinite group in which every open normal subgroup has index equal to some power of p.

Thus a finite group is pro-p if and only if its order is a power of p. Note that in a pro-p group, *every* open subgroup has p-power index, since it contains an open normal subgroup. The following is immediate from Proposition 1.2, (v) and (vi):

1.11 Proposition *Let G be a profinite group.*
(i) *If G is pro-p and $H \leq_c G$ then H is pro-p.*
(ii) *Let $K \lhd_c G$. Then G is pro-p if and only if both K and G/K are pro-p groups.*

1.12 Proposition *A topological group G is a pro-p group if and only if G is topologically isomorphic to an inverse limit of finite p-groups.*

Proof   If G is a pro-p group, then

$$G \cong \varprojlim (G/N)_{N \lhd_o G}$$

by Proposition 1.3; and each G/N is a finite p-group. For the converse, suppose $G = \varprojlim (G_\lambda)_{\lambda \in \Lambda}$ where each G_λ is a finite p-group. Then G is profinite, by Proposition 1.3. Also every open subgroup of G contains a subgroup

$$G(S) = G \cap \left(\prod_{\lambda \notin S} G_\lambda \times \prod_{\lambda \in S} \{1\} \right)$$

for some finite subset S of Λ. Since

$$|G : G(S)| \mid \prod_{\lambda \in S} |G_\lambda|$$

it follows that every open subgroup of G has index equal to a power of p.

Examples of pro-p groups

(1) The additive group $\mathbb{Z}_p$ of p-adic integers. This is the prototype of all pro-p groups. As well as being historically the origin of the subject, it plays a rôle in pro-p groups analogous to that of the cyclic groups in abstract group theory; a major theme of the later chapters will be to show how our principal objects of study, the analytic pro-p groups, are built up in a simple way from finitely many copies of $\mathbb{Z}_p$.

(2) The 'Sylow subgroups' of an arbitrary profinite group: see the exercises, below.

(3) The principal congruence subgroups in $\mathrm{GL}_n(\mathbb{Z}_p)$: see the exercises.

(4) More generally, we shall see in Part II that every *p-adic analytic group* is 'locally' a pro-p group.

(5) As mentioned in §1.1, the *pro-p completion* of any (abstract) group is a pro-p group (this follows from Proposition 1.12). As we shall see in various Interludes, this fact is the basis whereby the theory of pro-p groups can be usefully applied to problems in (abstract) group theory.

In the study of pro-p groups the Frattini subgroup plays a particularly useful rôle.

1.13 Proposition *If G is a pro-p group then*

$$\Phi(G) = \overline{G^p[G,G]} \,.$$

Here, $[G,G]$ is the derived group and $G^p = \langle g^p \mid g \in G \rangle$.

Proof In a finite p-group, each maximal proper subgroup is normal and has index p. Hence if M is a maximal proper open subgroup of G, we can find $N \lhd_o G$ with $N \le M$, observe that M/N is a maximal subgroup of the finite p-group G/N, and conclude that $M \lhd G$ and $|G : M| = p$. It follows that $G^p[G,G] \le M$. This shows that

$$\Phi(G) = \bigcap M \ge G^p[G,G]$$

and since $\Phi(G)$ is closed we obtain $\Phi(G) \ge \overline{G^p[G,G]}$.

Now consider the group $Q = G/\overline{G^p[G,G]}$. This is a pro-$p$ group, so its open normal subgroups intersect in the identity. If $N \lhd_o Q$ then Q/N is a finite elementary abelian p-group, and so $\Phi(Q/N) = 1$. Therefore $\Phi(Q) \le \bigcap_{N \lhd_o Q} N = 1$, and it follows by Proposition 1.9(ii) and the first part that

$$\Phi(G)/\overline{G^p[G,G]} = \Phi(Q) = 1 \,.$$

1.14 Proposition *Let G be a pro-p group. Then G is finitely generated if and only if $\Phi(G)$ is open in G.*

Proof   If $\Phi(G)$ is open then $G/\Phi(G)$ is finite. Thus there is a finite subset X of G such that $G = X\Phi(G)$, and then X generates G topologically by Proposition 1.9.

For the converse, suppose $G = \overline{\langle X \rangle}$ where $|X| = d$ is finite. If $\Phi(G) \leq N \lhd_o G$ then G/N is an elementary abelian p-group, by Proposition 1.13, and can be generated by d elements; consequently $|G : N| \leq p^d$. Among all such subgroups N we may therefore choose one, N_0 say, whose index in G is as large as possible. Then $N_0 \leq N$ whenever $\Phi(G) \leq N \lhd_o G$. Since $\Phi(G)$ is both closed and normal in G, it follows that

$$\Phi(G) = \bigcap \{N \mid \Phi(G) \leq N \lhd_o G\} = N_0 \,.$$

Thus $\Phi(G)$ is open in G.

Next, we introduce an important series of (topologically) characteristic subgroups, the *lower p-series*:

1.15 Definition *Let G be a pro-p group. Then $P_1(G) = G$, and for $i \geq 1$*

$$P_{i+1}(G) = \overline{P_i(G)^p[P_i(G), G]} \,.$$

Thus $P_2(G) = \Phi(G)$ by Proposition 1.13. Note that $P_{i+1}(G) \geq \Phi(P_i(G))$ for each i. (Pro-p groups G in which this inclusion is an equality will be of particular interest to us later on.)

1.16 Proposition *Let G be a pro-p group.*
  (i) *$P_i(G/K) = P_i(G)K/K$ for all $K \lhd_c G$ and all i.*
  (ii) *$[P_i(G), P_j(G)] \leq P_{i+j}(G)$ for all i and j.*
  (iii) *If G is finitely generated then $P_i(G)$ is open in G for each i, and the set $\{P_i(G) \mid i \geq 1\}$ is a base for the neighbourhoods of 1 in G.*

Proof   Write $G_i = P_i(G)$ for each i. For (i), let $K \lhd_c G$. Then $(P_i(G/K))$ is the fastest-descending series of closed normal subgroups of G/K such that each factor is central and of exponent dividing p. Since (G_iK/K) is a series with these properties, it follows that $P_i(G/K) \leq G_iK/K$ for all i. Now suppose that, for some n, $P_n(G/K) = G_nK/K$.

Put $M/K = P_{n+1}(G/K)$. Then M is closed in G and $M \geq G_n^p[G_n, G]K$, so $M \geq G_{n+1}K$. Hence $M = G_{n+1}K$, and the result follows by induction.

(ii) Certainly $[G_i, G_1] \leq G_{i+1}$ for all i. Let $n \geq 2$ and suppose inductively that $[G_i, G_{n-1}] \leq G_{i+n-1}$ for all i. Now we fix $m \geq 1$, and want to show that $[G_m, G_n] \leq G_{m+n}$. Since G_{m+n} is closed, it will suffice to show that $[G_m, G_n] \leq N$ whenever $G_{m+n} \leq N \lhd_o G$. Thus, in view of (i), we may replace G by the finite p-group G/N, and assume further that $G_{m+n} = 1$. Then $[G_m, G_{n-1}] \leq G_{m+n-1}$ is central and has exponent dividing p. If $g \in G_m$ and $x \in G_{n-1}$ we have

$$[g, x^p] = [g, x]^p = 1 \, ;$$

so $[G_m, G_{n-1}^p] = 1$. Also

$$[G_m, [G_{n-1}, G]] \leq [G, [G_m, G_{n-1}]][G_{n-1}, [G, G_m]]$$
$$\leq [G, G_{m+n-1}][G_{n-1}, G_{m+1}]$$
$$\leq G_{m+n} = 1 \, ,$$

by the three-subgroup lemma and the inductive hypothesis. It follows that

$$[G_m, G_{n-1}^p[G_{n-1}, G]] = 1 \, .$$

Since G is finite, this is the same as $[G_m, G_n] = 1$, which is what we had to show.

(iii) Now we assume that G is finitely generated. Certainly $G_1 = G$ is finitely generated and open in G. Let $n \geq 1$ and suppose inductively that G_n is finitely generated and open in G. Then Proposition 1.14 shows that $\Phi(G_n)$ is open in G_n. Since $\Phi(G_n) \leq G_{n+1} \leq G_n$ it follows that G_{n+1} is open in G_n, hence also in G, and Proposition 1.7 shows that G_{n+1} is finitely generated. The first claim follows by induction.

To show that $\{G_i \mid i \geq 1\}$ is a base for the neighbourhoods of 1 in G, it now suffices to show that every open normal subgroup of G contains G_i for some i. This follows from (i), since if $N \lhd_o G$ then G/N is a finite p-group and so $P_i(G/N) = 1$ for sufficiently large i.

It is a remarkable feature of finitely generated pro-p groups that the topology is completely determined by the group structure. The fundamental theorem is

1.17 Theorem *If G is a finitely generated pro-p group then every subgroup of finite index in G is open.*

This depends on two further facts, which we shall establish below:

1.18 Lemma *If G is a pro-p group and K is a subgroup of finite index in G then $|G:K|$ is a power of p.*

1.19 Proposition *If G is a finitely generated pro-p group then the derived group $[G,G]$ is closed in G.*

Proof of Theorem 1.17 G is a finitely generated pro-p group. Write

$$G^{\{p\}} = \{g^p \mid g \in G\}\,.$$

This set is compact, hence closed in G (being the image of the continuous mapping $g \mapsto g^p$ of G into G). Since $G/[G,G]$ is abelian, $G^p[G,G] = G^{\{p\}}[G,G]$, and so by Proposition 1.19 we see that $G^p[G,G]$ is also closed; hence it is equal to $\Phi(G)$, and therefore is open in G by Proposition 1.14.

Now let K be a proper normal subgroup of finite index in G. Arguing by induction, we may assume that K is open in M whenever M is a finitely generated pro-p group with $K \le M < G$. Take $M = G^p[G,G]K$. By Lemma 1.18, G/K is a finite p-group; consequently $M < G$, and the result of the first paragraph shows that M is open in G. Therefore M is a finitely generated pro-p group, by Proposition 1.7, and our inductive hypothesis shows that K is open in M. Thus K is open in G. Since every subgroup of finite index in G contains a normal subgroup of finite index, this proves the theorem.

In the course of the above proof, we saw that $\Phi(G) = G^p[G,G]$. This important observation means that, in the case of a *finitely generated* pro-p group, we can retrospectively simplify Definition 1.15 by removing the 'bars':

1.20 Corollary. *If G is a finitely generated pro-p group, then $\Phi(G) = G^p[G,G]$ and $P_{i+1}(G) = P_i(G)^p[P_i(G),G]$ for each i.*

Proof We have already established the first claim. Put $G_i = P_i(G)$ for each i. Propositions 1.16 and 1.14 show that each G_i is a finitely generated pro-p group, and that $\Phi(G_i)$ is open in G_i. By the first part, applied with G_i in place of G, we have

$$\Phi(G_i) = [G_i, G_i]G_i^p \le [G_i, G]G_i^p\,.$$

It follows that $[G_i, G]G_i^p$ is open (in G_i, hence in G), therefore closed, and therefore equal to $P_{i+1}(G)$ as claimed.

It is an interesting open problem whether the conclusion of Theorem 1.17 holds for all finitely generated profinite groups. The theorem has some striking consequences.

1.21 Corollary (i) *Every (abstract) homomorphism from a finitely generated pro-p group to a profinite group is continuous.*

(ii) *The topology of a finitely generated pro-p group is determined by its group structure.*

Proof (i) Suppose $\theta : G \to H$ is a homomorphism, where H is a profinite group and G is a finitely generated pro-p group. If $K \leq_o H$ then $K\theta^{-1}$ is a subgroup of finite index (at most $|H : K|$) in G, so $K\theta^{-1}$ is open in G. Since subgroups like K form a base for the neighbourhoods of 1 in H it follows that θ is continuous.

(ii) Now take $H = G$ (with possibly a different topology), and let θ be the identity map.

As a consequence of (i), we have

1.22 Corollary *If G is a finitely generated pro- p group then every automorphism of G (as an abstract group) is a topological automorphism, and every topologically characteristic subgroup of G is characteristic.*

Proof of Lemma 1.18 G is a pro-p group and K is a subgroup of finite index, which we may as well take to be normal in G. Say

$$|G : K| = m = p^r q$$

with $p \nmid q$, and put $X = \{g^m \mid g \in G\}$. Then $X \subseteq K$, and X is a closed subset of G.

Now let $g \in G$ and let $N \lhd_o G$. Then $g^{p^e} \in N$ for some e, and we may suppose that $e \geq r$. There exist integers a and b such that $am + bp^e = p^r$, and

$$g^{p^r} = (g^a)^m (g^{p^e})^b \in XN .$$

Since r is independent of N and X is closed this shows that $g^{p^r} \in X \subseteq K$. Thus G/K is a p-group.

The proof of Proposition 1.19 is more delicate; it depends on the following group-theoretic result:

1.23 Lemma *If $H = \langle a_1, \ldots, a_d \rangle$ is a nilpotent group then every element of $[H, H]$ is equal to a product of the form $[x_1, a_1] \ldots [x_d, a_d]$ with $x_1, \ldots, x_d \in H$.*

Proof   This is by induction on the class c of H, which we may take to be ≥ 2. Note first that if $u \in \gamma_{c-1}(H)$ then

$$[u, a_1^{e_1} \ldots a_d^{e_d}] = [u^{e_1}, a_1] \ldots [u^{e_d}, a_d]$$

and that if $u_1, \ldots, u_d, v_1, \ldots, v_d \in \gamma_{c-1}(H)$ then

$$\prod [u_i, a_i] \cdot \prod [v_i, a_i] = \prod [u_i v_i, a_i] \, .$$

These follow from the fact that $\gamma_c(H)$ is central in H, and together they imply that every element of $\gamma_c(H)$ can be written in the form

$$w = [w_1, a_1] \ldots [w_d, a_d] \qquad\qquad (*)$$

with $w_1, \ldots, w_d \in \gamma_{c-1}(H)$.

Now to prove the lemma, let $g \in [H, H]$. By the inductive hypothesis, we have

$$g = [y_1, a_1] \ldots [y_d, a_d] w$$

with $y_1, \ldots, y_d \in H$ and $w \in \gamma_c(H)$. From $(*)$ this gives

$$g = \prod [y_i, a_i] \prod [w_i, a_i] = \prod [w_i y_i, a_i] \, ,$$

using the fact that each $[w_i, a_i]$ is central.

Proof of Proposition 1.19   Suppose the pro-p group G is generated topologically by $\{a_1, \ldots, a_d\}$. Let

$$X = \{ [g_1, a_1] \ldots [g_d, a_d] \mid g_1, \ldots, g_d \in G \} \, .$$

Then X is closed in G, being the image of $G \times \ldots \times G$ under the continuous map $(g_1, \ldots, g_d) \mapsto \prod [g_i, a_i]$. Now let $N \lhd_o G$. Then G/N is a finite p-group, so nilpotent; and $G/N = \langle a_1 N, \ldots, a_d N \rangle$. By Lemma 1.23, $[G/N, G/N] = XN/N$, and so

$$[G, G]N = XN \, .$$

It follows that

$$[G, G] \subseteq \bigcap_{N \lhd_o G} XN = \overline{X} = X \, .$$

But plainly $X \subseteq [G, G]$, so we have equality, and $[G, G]$ is closed as claimed.

1.3 Procyclic groups

We have mentioned the special rôle of the p-adic integers in the theory of pro-p groups. This rests on the possibility of defining p-adic powers in a pro-p group.

1.24 Lemma *Let G be a pro-p group, let g be an element of G, and let $(a_i), (b_i)$ be p-adically convergent sequences of integers tending to the same limit in $\mathbb{Z}_p$. Then the sequences (g^{a_i}) and (g^{b_i}) both converge in G, and their limits are equal.*

Proof Let $N \lhd_o G$. Then $|G/N| = p^j$ for some j. For all sufficiently large i and k we have $a_i \equiv a_k \pmod{p^j}$, and then $g^{a_i} \equiv g^{a_k} \pmod{N}$. Thus (g^{a_i}) is a Cauchy sequence in G, and hence converges to some element $g_1 \in G$, by Proposition 1.2. Similarly, the sequence (g^{b_i}) converges to some element $g_2 \in G$. Now let N be as above. For sufficiently large k, we have $b_k \equiv a_k \pmod{p^j}$, $g^{b_k} \equiv g_2 \pmod{N}$, and $g^{a_k} \equiv g_1 \pmod{N}$. This gives

$$g_1 g_2^{-1} \equiv g^{a_k - b_k} \equiv 1 \pmod{N}.$$

Since N was arbitrary it follows that $g_1 = g_2$.

In view of Lemma 1.24, we can make the following definition without ambiguity:

1.25 Definition Let G be a pro-p group, $g \in G$ and $\lambda \in \mathbb{Z}_p$. Then

$$g^\lambda = \lim_{n \to \infty} g^{a_n}$$

where (a_n) is a sequence of integers with $\lim_{n \to \infty} a_n = \lambda$.

The operation of 'p-adic exponentiation' is very well behaved:

1.26 Proposition *Let G be a pro-p group, let g and h be elements of G, and let $\lambda, \mu \in \mathbb{Z}_p$.*
  (i) *$g^{\lambda + \mu} = g^\lambda g^\mu$ and $g^{\lambda \mu} = (g^\lambda)^\mu$.*
  (ii) *If $gh = hg$ then $(gh)^\lambda = g^\lambda h^\lambda$.*
  (iii) *The map $\nu \mapsto g^\nu$ defines a continuous homomorphism of $\mathbb{Z}_p$ into G. Its image $g^{\mathbb{Z}_p}$ is the closure in G of $\langle g \rangle$.*

Proof   (i) and (ii) are true 'modulo N' for every open normal subgroup N of G, by the ordinary laws of exponents. Hence they are true in G. The mapping $\nu \mapsto g^\nu$ is a group homomorphism of $\mathbb{Z}_p$ in G, by (i). It is continuous, by Corollary 1.21, so its image is a compact, hence closed, subgroup of G. Clearly $g^{\mathbb{Z}_p}$ contains $\langle g \rangle$; since each element of $g^{\mathbb{Z}_p}$ is the limit of a sequence of elements of $\langle g \rangle$ it follows that $g^{\mathbb{Z}_p} \leq \overline{\langle g \rangle}$. Therefore $g^{\mathbb{Z}_p} = \overline{\langle g \rangle}$.

1.27 Definition A group G is *procyclic* if G is profinite and G/N is a cyclic group for every open normal subgroup N of G.

1.28 Proposition *Let G be a pro-p group. Then the following are equivalent.*

(a) *G is procyclic;*

(b) *G can be topologically generated by a one-element subset;*

(c) *$G = g^{\mathbb{Z}_p}$ for some $g \in G$;*

(d) *G is either finite and cyclic or else is topologically isomorphic to $\mathbb{Z}_p$.*

Proof   Assume that G is procyclic, and suppose that G has two distinct maximal proper open subgroups M and N. Then $M \cap N \geq \Phi(G) \geq G^p[G,G]$, so M and N are normal subgroups of index p in G and $G/(M \cap N)$ is elementary abelian of order p^2. Such a group is not cyclic. Therefore either $G = 1$ or G has a unique maximal proper open subgroup; in either case $\Phi(G)$ is open in G and $G/\Phi(G)$ is cyclic. Hence G can be generated topologically by a single element, by Proposition 1.9, and we see that (a) implies (b). Proposition 1.26 (iii) shows that (b) is equivalent to (c). Suppose that (c) holds, and let K be the kernel of the homomorphism $\theta : \mathbb{Z}_p \to G$ given by $\lambda\theta = g^\lambda$. By hypothesis, θ is surjective; and θ is continuous by Corollary 1.21. Since both $\mathbb{Z}_p/K$ and G are compact Hausdorff groups it follows that G is topologically isomorphic to $\mathbb{Z}_p/K$; this implies (d), since every proper quotient group of $\mathbb{Z}_p$ is cyclic. Thus (c) implies (d), and the last fact just mentioned shows that (d) implies (a).

Notes

This is mostly 'standard' material; the substantial Theorem 1.17 is due to Serre. Introductory accounts of profinite and pro-p groups can be found in Serre (1997), Chapter I §1, Fried and Jarden (1986), Chapter 1, and Wilson (1998).

Exercises

1. Let G be a Hausdorff topological group. Show that G is profinite if and only if (a) there is a base for the neighbourhoods of 1 consisting of subgroups of finite index, and (b) G is *complete*, i.e. every Cauchy net in G converges. [A *net* is a family $(g_U)_{U \in \mathcal{B}}$, where $\mathcal{B}$ is a base for the neighbourhoods of 1, and each $g_U \in G$. It is *Cauchy* if for each $U \in \mathcal{B}$ there exists $V \in \mathcal{B}$ such that $g_S^{-1} g_T \in U$ whenever $S, T \in \mathcal{B}$ and $S \subseteq V$ and $T \subseteq V$. The net *converges* to an element $g \in G$ if for each $U \in \mathcal{B}$ there exists $V \in \mathcal{B}$ such that $g_S^{-1} g \in U$ whenever $S \in \mathcal{B}$ and $S \subseteq V$. These concepts generalise the corresponding ones for sequences, to the case where a countable base for the neighbourhoods of 1 may fail to exist.]

2. Let G be a compact Hausdorff topological group. Show that G is profinite if and only if G is *totally disconnected*, i.e. each connected component has just one element. (Though elementary, this is quite tricky. See Appendix B, or Higgins (1974), Chapter 2 §9.)

3. (i) Formulate and prove the appropriate universal property of $\varprojlim G_\lambda$. (ii) Let H be an abstract group and let $\theta : H \to G$ be a homomorphism, where G is a profinite (or pro-p) group. Show that θ extends uniquely to a continuous homomorphism $\hat{\theta} : \widehat{H} \to G$ (respectively, $\hat{\theta} : \widehat{H}_p \to G$): i.e. that there exists a unique $\hat{\theta}$ making the diagram

$$
\begin{array}{ccc}
H & \xrightarrow{\;\theta\;} & G \\
{\scriptstyle j}\searrow & & \nearrow{\scriptstyle \hat{\theta}} \\
& \widehat{H} \text{ or } \widehat{H}_p &
\end{array}
$$

commute, where j is the natural map $H \to \widehat{H}$ (or $H \to \widehat{H}_p$). Deduce that if H is a dense subgroup of G then G is a (continuous) epimorphic image of $\widehat{H}$ (respectively $\widehat{H}_p$). (iii) Give an example of a finitely generated pro-p group G and a dense subgroup H of G, with H finitely generated as an abstract group, such that $\widehat{H}_p \not\cong G$.

4. Let Γ be a group, Λ a family of normal subgroups of finite index in Γ, directed with respect to reverse inclusion, and put $G = \hat{\Gamma}_\Lambda$. Write $\theta : \Gamma \to G$ for the natural homomorphism.
(i) Show that θ induces an isomorphism $\Gamma/N \to G/\overline{N\theta}$ for each $N \in \Lambda$.
(ii) Show that every open subgroup of G is of the form $\overline{H\theta}$ where $H \leq \Gamma$ and $H > N$ for some $N \in \Lambda$.

(iii) Let Γ and Δ be finitely generated groups. Show that Γ and Δ have the same class of finite groups as their finite homomorphic images if and only if $\hat{\Gamma}$ and $\hat{\Delta}$ are isomorphic (as profinite groups).
(iv) Is (iii) still true without assuming finite generation?

5. (i) Show that every closed subgroup of a procyclic group is procyclic.
(ii) More generally, show that if G is a d-generator abelian profinite group then every closed subgroup of G can be generated by d elements.
(iii) Let G be a d-generator pro-p group. Show that every generating set for G contains a subset of size at most d that generates G [*Hint*: consider $G/\Phi(G)$].

6. Let G be a profinite group, and let $w(X_1, \ldots, X_n)$ be a group word.
(i) Show that the set $w(G) = \{w(x_1, \ldots, x_n) \mid x_1, \ldots, x_n \in G\}$ is closed in G. (ii) Deduce that if $g \in G$ and if for every $N \lhd_o G$ there exist $x_1(N), \ldots, x_n(N) \in G$ with $g \equiv w(x_1(N), \ldots, x_n(N)) \, (\mathrm{mod} N)$, then there exist $x_1, \ldots, x_n \in G$ such that $g = w(x_1, \ldots, x_n)$. [An important special case: *if $m \in N$ and g is congruent to an mth power modulo N for every $N \lhd_o G$, then g is an mth power in G.*]

7. Let G be a profinite group and A a subgroup of G. (i) Show that $C_G(A)$ is closed. (ii) Show that if A is abelian (or nilpotent of class c, or soluble of derived length d), then $\overline{A}$ has the same property.

8. Give an example of a finitely generated profinite group G such that $\Phi(G)$ is not open.

9. Show that the natural map $\mathrm{SL}_n(\mathbb{Z}) \to \mathrm{SL}_n(\mathbb{Z}/m\mathbb{Z})$ is surjective, for all m and n. Denoting its kernel by $K_n(m)$, show that

$$\varprojlim(\mathrm{SL}_n(\mathbb{Z})/K_n(p^i))_{i \in \mathbb{N}} \cong \mathrm{SL}_n(\mathbb{Z}_p)$$

$$\varprojlim(\mathrm{SL}_n(\mathbb{Z})/K_n(m))_{m \in \mathbb{N}} \cong \mathrm{SL}_n(\widehat{\mathbb{Z}}).$$

[*Hint for first part*: find a simple generating set of $\mathrm{SL}_n(\mathbb{Z}/m\mathbb{Z})$. The fact that $\widehat{\mathrm{SL}_n(\mathbb{Z})} \cong \mathrm{SL}_n(\widehat{\mathbb{Z}})$ if $n \geq 3$ lies much deeper; see Platonov and Rapinchuk (1994), §9.5]

10. Fix a prime p and a positive integer n. For each j put

$$\Gamma_j = \{g \in \mathrm{SL}_n(\mathbb{Z}_p) \mid g \equiv 1_n \, (\mathrm{mod}\, p^j)\}.$$

(i) Show that Γ_1 is a pro-p group (with the subspace topology induced by the p-adic topology on $\mathrm{M}_n(\mathbb{Z}_p)$.

(ii) Show that Γ_1 is finitely generated (topologically). Deduce that every subgroup of finite index in $\mathrm{SL}_n(\mathbb{Z}_p)$ contains Γ_j for some j. [Thus $\mathrm{SL}_n(\mathbb{Z}_p)$ has the *congruence subgroup property*: for more on this, see Chapter 5. *Hint for* (ii): use Proposition 1.14.]

11. Let G be a profinite group and let p be a prime. For $N \lhd_o G$ denote by $\mathcal{P}(N)$ the set of Sylow p-subgroups of G/N. By applying Proposition 1.4 to a suitable inverse system of finite sets, show that G has a closed subgroup P such that $PN/N \in \mathcal{P}(N)$ for every $N \lhd_o G$. Show that P is a maximal pro-p subgroup of G.

12. Let G be a profinite group and let p be a prime. A *Sylow pro-p subgroup* of G is a maximal pro-p subgroup. Show that every pro-p subgroup of G is contained in a Sylow pro-p subgroup, and that the Sylow pro-p subgroups of G are all conjugate.

13. A profinite group G is *pronilpotent* if G/N is nilpotent for every $N \lhd_o G$. Show that G is pronilpotent if and only if G has a unique Sylow pro-p subgroup for each prime p, and that this holds if and only if G is isomorphic to a Cartesian product of pro-p groups for various primes p.

14. Let G be a pro-p group. Show that G satisfies the ascending chain condition for closed subgroups (i.e. every such chain becomes stationary after finitely many steps) if and only if every closed subgroup of G is topologically finitely generated. Give an example of a profinite group having the latter property but not the former. [*Hint for the first part*: use Proposition 1.14.]

15. For a pro-p group G, define $\pi_0(G) = G$ and, for $i \geq 0$, $\pi_{i+1}(G) = \overline{\pi_i(G)^p}$. Now Kostrikin's celebrated contribution to the 'restricted Burnside problem' is the **Theorem**: *For each positive integer d, there is a positive integer $f(d,p)$ such that the order of every finite d-generator group of exponent p is bounded above by $f(d,p)$* (see Vaughan-Lee 1993). Show that Kostrikin's theorem is equivalent to the statement: *if G is any finitely generated pro-p group then $\pi_k(G)$ is open in G for every k.*

16. Zel'manov's solution of the restricted Burnside problem shows that *for each pair of positive integers d, k, there is a positive integer $f(d, p^k)$ such that the order of every finite d-generator group of exponent p^k is bounded above by $f(d, p^k)$* (see Vaughan-Lee (1993)). Show that Zel'manov's theorem is equivalent to the statement: *if G is any finitely generated pro-p group then $\overline{G^{p^k}}$ is open in* G.

17. (i) Let $H = \langle a_1, \ldots, a_d \rangle$ be a nilpotent group, and let $m \geq 2$. Show that each element of $\gamma_m(H)$ is equal to a product of the form $[x_1, a_1] \ldots [x_d, a_d]$ with $x_1, \ldots, x_d \in \gamma_{m-1}(H)$.

(ii) Let G be a finitely generated pronilpotent group. Show that for each $m \geq 1$ the subgroup $\gamma_m(G)$ is closed.

18. Let G be a profinite group and S a closed subset. Write $S^{(n)} = \{x_1 \ldots x_n \mid x_1, \ldots, x_n \in S\}$ for each $n \geq 1$, and $S^\infty = \bigcup_{n=1}^{\infty} S^{(n)}$.

(i) Show that S^∞ is closed if and only if $S^\infty = S^{(n)}$ for some finite n.

(ii) Show that $S^\infty = S^{(n)}$ if and only if $NS^\infty = NS^{(n)}$ for every $N \lhd_o G$.

(iii) What is the significance of $S^\infty = S^{(n)}$ when $S = w(G) \cup w(G)^{-1}$, with $w(G)$ as in Exercise 6?

[*Hint for the 'only if' in (i)*: use Baire's Category Theorem, see Ex. 3.6.]

19. Prove the following theorem (Martinez 1994): *If G is a finitely generated pro-p group then for every k the subgroup G^{p^k} is closed in G.*

[*Step 1*: Show that it suffices to prove the following: *if G is a d-generator finite p-group and $S = \{x^{p^k} \mid x \in G\}$ then $G^{p^k} = S^{(n)}$ for some n that depends only on p^k and d.*

Step 2: Let G be a d-generator finite p-group and $H = G^{p^k}$. Prove that $H = \left\langle x_1^{p^k}, \ldots, x_r^{p^k} \right\rangle$ for some $x_1, \ldots, x_n \in G$, where $r \leq 2df(d, p^k)$ and f is Zel'manov's function (see Exercise 16). [*Hint*: Use the proof of Proposition 1.7, Exercise 5(iii), and Proposition 1.9.]

Step 3: Let X denote the set of all conjugates of the elements $x_1^{p^k}, \ldots, x_r^{p^k}$ in G. Show that $[H, H] \subseteq X^{(2r)}$. Deduce that $[H, H]H^{p^k} \subseteq S^{(2r+1)}$, where S is as defined above. [*Hint for the first claim*: use Lemma 1.23.]

Step 4: Deduce that $H = S^{(n)}$ where $n = r(p^k + 1) + 1$. [*Hint*: each element of H is congruent modulo $[H, H]H^{p^k}$ to a product of low powers of $x_1^{p^k}, \ldots, x_r^{p^k}$, in that order.]

20. **Free pro-p groups** . Let X be a finite set and let $F = F(X)$ denote the free group on X. (i) Show that the pro-p completion $\widehat{F}_p$ is 'free as a pro-p group': namely, every mapping ϕ from X into a pro-p group

G extends to a continuous homomorphism $\overline{\phi} : \widehat{F}_p \to G$. (ii) Suppose $Y \subseteq X$. Show that the inclusion mapping $F(Y) \to F(X)$ induces an isomorphism of $\widehat{F(Y)}_p$ onto the closed subgroup of $\widehat{F(X)}_p$ generated by the image of Y. [*Hint*: prove that if $N \lhd F(Y)$ then $N = F(Y) \cap M$ for a suitable $M \lhd F(x)$.] (iii) Suppose that $Y \subseteq X$. Put $N = \langle Y^{\widehat{F(X)}_p} \rangle$, the normal closure of Y in $\widehat{F(X)}_p$. Prove that $\widehat{F(X)}_p / N$ is a free pro-p group on $|X| - |Y|$ generators. (The 'free pro-p group' $\widehat{F(X)}_p$ is useful when one comes to discuss presentations of pro-p groups by generators and relations: see §4.5 and §5.5. For infinite sets X, the definition of the free pro-p group on X needs to be modified slightly: see Fried and Jarden (1986), §15.5.)

21. In this exercise, $\widehat{\Gamma}$ denotes *either* the profinite completion *or* the pro-p completion of a group Γ.

(i) Let $f : \Delta \to \Gamma$ be a homomorphism of groups. Show that f induces a continuous homomorphism $\widehat{f} : \widehat{\Delta} \to \widehat{\Gamma}$, and that this makes 'profinite completion' and 'pro-p completion' into functors.

(ii) Let $\Delta \leq \Gamma$ with $i : \Delta \to \Gamma$ the inclusion mapping, and write $\theta : \Gamma \to \widehat{\Gamma}$ for the natural mapping. Show that $\widehat{\Delta i} = \overline{\Delta \theta}$, the closure of $\Delta \theta$ in $\widehat{\Gamma}$. Deduce that $\widehat{i}$ is an isomorphism of $\widehat{\Delta}$ onto $\overline{\Delta \theta}$ if and only if the following holds: for every normal subgroup K of finite (respectively: p-power) index in Δ there exists a normal subgroup N of finite (respectively: p-power) index in Γ such that $K = N \cap \Delta$; that is, if and only if the profinite (resp. pro-p) topology on Γ induces the profinite (resp. pro-p) topology on the subspace Δ of Γ.

(iii) Now suppose that $\Delta \lhd \Gamma$, with $\pi : \Delta \to \Gamma/\Delta$ the natural epimorphism. Prove that the sequence

$$\widehat{\Delta} \xrightarrow{\widehat{i}} \widehat{\Gamma} \xrightarrow{\widehat{\pi}} \widehat{\Gamma/\Delta} \to 1$$

is exact.

(iv) Show that 'profinite completion' is an exact functor on the category of polycyclic-by-finite groups. [*Hint*: In such a group, every subgroup is closed in the profinite topology; see Segal (1983), Chapter 1.]

22. Let Γ be a torsion-free finitely generated nilpotent group, with upper central series $1 = Z_0 < \ldots < Z_c = \Gamma$. Then $Z_i/Z_{i-1} \cong \mathbb{Z}^{r_i}$ for certain positive integers $r_1, \ldots, r_c$. Show that the upper central series of $\widehat{\Gamma}$ is $1 = \overline{Z_0} < \ldots < \overline{Z_c} = \Gamma$, and that for each i we have $\overline{Z_i} \cong \widehat{Z_i}$ and $\overline{Z_i}/\overline{Z_{i-1}} \cong \widehat{Z_i/Z_{i-1}}$, where $\widehat{\ }$ denotes either profinite completion or pro-p completion.

[*Hint*: If $\Delta \lhd \Gamma$ and Γ/Δ is torsion-free, then the conditions of Exercise 21(ii) are satisfied; see Segal (1983), Chapter 1.]

23. Show that a finitely generated abelian pro-p group is isomorphic to $\mathbb{Z}_p^d \times F$ for some d and some finite group F.

[*Hint:* Show that the group is a finitely generated module for $\mathbb{Z}_p$, using Proposition 1.26(iii), and note that $\mathbb{Z}_p$ is a PID.]

24. Let G be a finitely generated pro-p group and M, N open normal subgroups of G. Prove that $[M, N]$ is closed in G.

[*Hint*: Put $D = M \cap N$ and show that

$$[M, N] = [D, D] \prod_{x \in X} [M, x] \prod_{y \in Y} [y, D]$$

where X and Y are suitable finite subsets of N and M respectively.]

2

Powerful p-groups

In this chapter we restrict attention to finite p-groups. It turns out that the key to understanding the structure of analytic pro-p groups lies in the properties of a special class of finite groups.

2.1 Definition (i) A finite p-group G is *powerful* if p is odd and G/G^p is abelian, or $p = 2$ and G/G^4 is abelian.
(ii) A subgroup N of a finite p-group G is *powerfully embedded* in G, written N p.e. G, if p is odd and $[N, G] \leq N^p$, or $p = 2$ and $[N, G] \leq N^4$.

Thus G is powerful if and only if G p.e. G; and if N p.e. G then $N \lhd G$ and N is powerful. When p is odd, G is powerful if and only if $G^p = \Phi(G)$. One should think of 'powerful' as a generalization of 'abelian'. We shall see that powerful p-groups (and, later, pro-p groups) share many of the simple structural features of abelian groups.

2.2 Lemma. *Let G be a finite p-group and let N, K and W be normal subgroups of G with $N \leq W$.*
(i) *If N p.e. G then NK/K p.e. G/K.*
(ii) *If p is odd and $K \leq N^p$, or if $p = 2$ and $K \leq N^4$, then N p.e. G if and only if N/K p.e. G/K.*
(iii) *If N p.e. G and $x \in G$ then $\langle N, x \rangle$ is powerful.*
(iv) *If N is not powerfully embedded in W, then there exists a normal subgroup J of G such that*

- *if p is odd,*

$$N^p[N, W, W] \leq J < N^p[N, W] \quad and \quad |N^p[N, W] : J| = p;$$

37

• *if* $p = 2$,

$$N^4[N,W]^2[N,W,W] \le J < N^4[N,W] \quad and \quad |N^4[N,W] : J| = 2.$$

Proof   Parts (i) and (ii) are obvious from the definition. To prove (iii), put $H = \langle N, x \rangle$. Then $[H, H] = [N, H]$ since $N \lhd H$, so if N p.e. G then $[H, H] \le N^p \le H^p$ (respectively, $[H, H] \le H^4$ if $p = 2$). For part (iv), suppose that p is odd and that $[N, W] \not\le N^p$. Then $N^p < N^p[N, W] = M$, say. Since G is a p-group and M and N are normal in G, there exists $J \lhd G$ such that $N^p \le J < M$ and $|M : J| = p$. Then M/J is central in G/J and the result follows. A similar argument deals with the case where $p = 2$.

The point of part (iv) is that in order to establish that N p.e. W, where $N \le W$ are normal subgroups of a p-group G, we can factor out a suitable J and thereby reduce to the case where $N^p = 1$ (if p is odd) or $N^4 = 1$ (if $p = 2$), and $[N, W]$ has order p (and $[N, W]$ is central in G). This technique is illustrated in the proof of the following important result.

2.3 Proposition. *Let G be a finite p-group and $N \le G$. If N p.e. G then N^p p.e. G.*

Proof Case 1: where p is odd. It is given that $[N, G] \le N^p$, and we may assume that $(N^p)^p = 1 = [N^p, G, G]$. Then $[N, G, G] \le Z(G)$, and it follows that for any given $x \in N$ and $g \in G$, the map $w \mapsto [x, g, w]$ is a homomorphism from G into $Z(G)$. Then

$$\prod_{j=0}^{p-1}[x,g,x^j] = \prod_{j=0}^{p-1}[x,g,x]^j = [x,g,x]^{p(p-1)/2} .$$

Hence

$$[x^p, g] = [x,g]^{x^{p-1}}[x,g]^{x^{p-2}} \ldots [x,g]$$

$$= \prod_{j=p-1}^{0} [x,g][x,g,x^j]$$

$$= [x,g]^p \prod_{j=0}^{p-1}[x,g,x^j] \qquad \text{since } [x,g,x^j] \in Z(G) \text{ for each } j$$

$$= [x,g]^p[x,g,x]^{p(p-1)/2} = 1$$

since $[N, G]^p = 1$. Thus $[N^p, G] = 1$, giving the result.

Case 2: $p = 2$. We may now assume that $[N, G] \leq N^4$ and that

$$[N^2, G, G] = [N^2, G]^2 = (N^2)^4 = 1 .$$

For $x \in N$ and $g \in G$ we have

$$[x^4, g] = [x^2, g][x^2, g, x^2][x^2, g] = [x^2, g]^2 = 1 ,$$

so $N^4 \leq Z(G)$. Since N has exponent dividing 8, N^4 is generated by elements of order 2, hence $(N^4)^2 = 1$. Then with x and g as above we have

$$[x^2, g] = [x, g][x, g, x][x, g] = [x, g]^2 = 1 ,$$

since $[x, g, x] \in [N, G, G] \leq [N^4, G] = 1$ and $[x, g] \in [N, G] \leq N^4$. Thus $[N^2, G] = 1$ and the result follows.

Now recall Definition 1.15. When G is a finite p-group, this becomes

$$P_1(G) = G, \quad P_{i+1}(G) = P_i(G)^p [P_i(G), G] \quad \text{for } i \geq 1 .$$

For the rest of this chapter, we simplify the notation by writing

$$G_i = P_i(G) .$$

2.4 Lemma. *Let G be a powerful p-group.*
   (i) *For each i, G_i p.e. G and $G_{i+1} = G_i^p = \Phi(G_i)$.*
   (ii) *For each i, the map $x \mapsto x^p$ induces a homomorphism from G_i/G_{i+1} onto G_{i+1}/G_{i+2}.*

Proof (i) Since $G = G_1$ is powerful, G_1 p.e. G. Suppose G_i p.e. G for some $i \geq 1$. Then $G_{i+1} = G_i^p[G_i, G] = G_i^p$, and Proposition 2.3 shows that G_{i+1} p.e. G. Since $G_i^p \leq \Phi(G_i) = G_i^p[G_i, G_i] \leq G_{i+1}$, this implies also that $G_{i+1} = \Phi(G_i)$. The result follows by induction.

   (ii) Part (i) shows that G_i is powerful, $G_{i+1} = P_2(G_i)$ and $G_{i+2} = P_3(G_i)$. So, changing notation, we may assume that $i = 1$; and then replacing G by G/G_3, we may assume that $G_3 = 1$. Then $[G, G] \leq G_2 \leq Z(G)$, so for $x, y \in G$ we have

$$(xy)^p = x^p y^p [y, x]^{p(p-1)/2} .$$

If p is odd then $p \mid p(p-1)/2$, so

$$[y, x]^{p(p-1)/2} \in G_2^p = G_3 = 1 .$$

If $p = 2$ then $[G, G] \leq G^4 \leq G_3 = 1$. Thus in either case we have

$(xy)^p = x^p y^p$. Since $G_2^p = G_3 = 1$ and $G^p = G_2$, this shows that $x \mapsto x^p$ induces a homomorphism from G/G_2 onto G_2/G_3, and completes the proof.

2.5 Lemma. *If $G = \langle a_1, \ldots, a_d \rangle$ is a powerful p-group, then $G^p = \langle a_1^p, \ldots, a_d^p \rangle$.*

Proof Let $\theta : G/G_2 \to G_2/G_3$ be the homomorphism given in the preceding lemma. Then G_2/G_3 is generated by $\{(a_1 G_2)\theta, \ldots, (a_d G_2)\theta\}$, so $G_2 = \langle a_1^p, \ldots, a_d^p \rangle G_3$. Since $G_3 = \Phi(G_2)$ and $G_2 = G^p$, by Lemma 2.4, this gives the result.

2.6 Proposition. *If G is a powerful p-group then every element of G^p is a pth power in G.*

Proof We argue by induction on $|G|$. Let $g \in G^p$. By Lemma 2.4, there exist $x \in G$ and $y \in G_3$ such that $g = x^p y$. Put $H = \langle G^p, x \rangle$. Since $G^p = G_2$ p.e. G by Lemma 2.4, Lemma 2.2 (iii) shows that H is powerful. Also $g \in H^p$, since $y \in G_3 = G_2^p$. If $H \neq G$ then the inductive hypothesis gives that g is a pth power in H. If $H = G$, then $G = \langle x \rangle$ is cyclic, since now $G = \langle G^p, x \rangle = \Phi(G)\langle x \rangle$; and in this case the result is trivial.

We can now summarise the main features of the lower p-series in a powerful p-group:

2.7 Theorem. *Let $G = \langle a_1, \ldots, a_d \rangle$ be a powerful p-group, and put $G_i = P_i(G)$ for each i.*
  (i) *G_i p.e. G;*
  (ii) *$G_{i+k} = P_{k+1}(G_i) = G_i^{p^k}$ for each $k \geq 0$;*
  (iii) *$G_i = G^{p^{i-1}} = \{x^{p^{i-1}} \mid x \in G\} = \langle a_1^{p^{i-1}}, \ldots, a_d^{p^{i-1}} \rangle$;*
  (iv) *the map $x \mapsto x^{p^k}$ induces a homomorphism from G_i/G_{i+1} onto G_{i+k}/G_{i+k+1}, for each i and k.*

Proof We have already established (i), and observed that $G_{i+1} = G_i^p = P_2(G_i)$ for each i. It follows from Proposition 2.6 that $G_{i+1} = \{x^p \mid x \in G_i\}$, and then by induction that $G_i = \{x^{p^{i-1}} \mid x \in G\}$. Since G_i is a subgroup this implies that $G_i = G^{p^{i-1}}$. Similarly, repeated applications of Lemma 2.5 show that $G_i = \langle a_1^{p^{i-1}}, \ldots, a_d^{p^{i-1}} \rangle$. Thus we have (iii).

Part (iv) follows from Lemma 2.4 (ii). Finally, taking G_i in place of G and $k + 1$ in place of i, in (iii), we get

$$P_{k+1}(G_i) = G_i^{p^k} = \{x^{p^k} \mid x \in G_i\}$$
$$= \{y^{p^{i-1+k}} \mid y \in G\} = G_{i+k},$$

giving (ii).

2.8 Corollary. *If* $G = \langle a_1, \ldots, a_d \rangle$ *is a powerful p-group then* $G = \langle a_1 \rangle \ldots \langle a_d \rangle$, *i.e.* G *is the product of its cyclic subgroups* $\langle a_i \rangle$.

Proof Say $G_e > G_{e+1} = 1$. Arguing by induction on e, we may suppose that $G = \langle a_1 \rangle \ldots \langle a_d \rangle G_e$. But $G_e = \langle a_1^{p^{e-1}}, \ldots, a_d^{p^{e-1}} \rangle$ and G_e is central in G, so the result follows. (For a partial converse, see Exercise 8.)

The two major results of this chapter, Theorems 2.9 and 2.13, relate the property of being powerful to the existence of 'small' generating sets for subgroups of a p-group. For a finite p-group G, we denote by d(G) the minimal cardinality of a set of generators for G. Thus d(G) is also the dimension of $G/\Phi(G)$ as a vector space over $\mathbb{F}_p$.

2.9 Theorem. *If* G *is a powerful p-group and* $H \leq G$ *then* d(H) $\leq$ d(G).

Proof The proof is by induction on $|G|$. Let $d = $ d(G) and put $m = $ d(G_2). Lemma 2.4 (i) shows that G_2 is powerful, so by the inductive hypothesis we may suppose that the subgroup $K = H \cap G_2$ satisfies d(K) $\leq m$.

Now the map $\pi : G/G_2 \to G_2/G_3$ given by $x \mapsto x^p$ is an epimorphism (by Lemma 2.4 (ii)), and dim(ker π) $= d - m$ (where dim denotes dimension as an $\mathbb{F}_p$-vector space). So dim(ker$\pi \cap HG_2/G_2$) $\leq d - m$, whence

$$\dim\left((HG_2/G_2)\pi\right) \geq \dim(HG_2/G_2) - (d - m) = m - (d - e)$$

where $e = \dim(HG_2/G_2)$. Let $h_1, \ldots, h_e$ be elements of H such that $HG_2 = \langle h_1, \ldots, h_e \rangle G_2$. Since $\Phi(K) \leq K^p \leq G_3$, the subspace of $K/\Phi(K)$ spanned by the cosets of $h_1^p, \ldots, h_e^p$ has dimension at least $\dim((HG_2/G_2)\pi) \geq m - (d - e)$. Since d($K$) $\leq m$, we can find $d - e$ elements $y_1, \ldots, y_{d-e}$ of K such that

$$K = \langle h_1^p, \ldots, h_e^p, y_1, \ldots, y_{d-e} \rangle \Phi(K).$$

Then $K = \langle h_1^p, \ldots, h_e^p, y_1, \ldots, y_{d-e}\rangle$ and so

$$H = H \cap \langle h_1, \ldots, h_e\rangle G_2 = \langle h_1, \ldots, h_e\rangle K = \langle h_1, \ldots, h_e, y_1, \ldots, y_{d-e}\rangle.$$

Thus $d(H) \leq d$ as required.

The *rank* of a finite group G is defined to be

$$\mathrm{rk}(G) = \sup\{d(H) \mid H \leq G\}.$$

Theorem 2.9 can be re-stated succinctly: *if G is a powerful p-group then* $\mathrm{rk}(G) = d(G)$. The exact converse of this statement is false (see Exercise 3); our second major theorem is nevertheless a sort of converse: it shows that in any finite p-group G, there is a powerful normal subgroup of index bounded by a function of $\mathrm{rk}(G)$. The proof requires some preparation.

2.10 Definition For a finite p-group G and a positive integer r, $V(G,r)$ denotes the intersection of the kernels of all homomorphisms of G into $\mathrm{GL}_r(\mathbb{F}_p)$.

Since the image of any homomorphism of a p-group G into $\mathrm{GL}_r(\mathbb{F}_p)$ is a p-group, and every p-subgroup of $\mathrm{GL}_r(\mathbb{F}_p)$ is conjugate to a subgroup of the lower uni-triangular group $U_r(\mathbb{F}_p)$, we could equally well define $V(G,r)$ as the intersection of the kernels of all homomorphisms of G into $U_r(\mathbb{F}_p)$. Note that an element g of G belongs to $V(G,r)$ if and only if g acts trivially in every linear representation of G on any $\mathbb{F}_p$-vector space of dimension at most r.

For $r \in \mathbb{N}$, define the integer $\lambda(r)$ by

$$2^{\lambda(r)-1} < r \leq 2^{\lambda(r)}.$$

2.11 Lemma. (i) *The group $U_r(F_p)$ has a series, of length $\lambda(r)$, of normal subgroups, with elementary abelian factors.*

(ii) *If G is a finite p-group, then $G/V(G,r)$ has a series with these properties.*

Proof (ii) follows from (i), since $G/V(G,r)$ is isomorphic to a subgroup of the direct product of finitely many copies of $U_r(\mathbb{F}_p)$. To prove (i), note that the result is trivial if $r = 1$. If $r \geq 2$, put $s = [r/2]$. Then the elements of $U_r(\mathbb{F}_p)$ have the form

$$x = \begin{pmatrix} A & 0 \\ B & C \end{pmatrix}$$

with $A \in U_s(\mathbb{F}_p)$ and $C \in U_{r-s}(\mathbb{F}_p)$. The mapping which sends x to (A, C) is a homomorphism from $U_r(\mathbb{F}_p)$ into $U_s(\mathbb{F}_p) \times U_{r-s}(\mathbb{F}_p)$, and its kernel is easily seen to be an elementary abelian p-group. The result follows by an inductive argument.

2.12 Proposition. *Let G be a finite p-group and r a positive integer. Put $V = V(G,r)$ and let $W = V$ if p is odd, $W = V^2$ if $p = 2$. If $N \lhd G$, $d(N) \le r$, and $N \le W$, then N p.e. W.*

Proof The proof is by induction on $|N|$. Suppose first that p is odd and that $[N, V] \not\le N^p$. In view of Lemma 2.2 (iv), we may assume that $N^p = 1$ and $|[N, V]| = p$. Since G is a p-group, there exists $M \lhd G$ with $[N, V] \le M < N$ and $|N : M| = p$. Since $N/[N, V]$ is elementary abelian, we have $d(M/[N, V]) = d(N/[N, V]) - 1 \le r - 1$; as $[N, V]$ is cyclic it follows that $d(M) \le r$. Hence, by the inductive hypothesis, $[M, V] \le M^p = 1$. Thus M is central in N, and as N/M is cyclic it follows that N is abelian. Then N is an $\mathbb{F}_p$-vector space of dimension at most r, so the conjugation action of V on N must be trivial. Thus $[N, V] = 1$, in contradiction to the initial assumption.

Now suppose $p = 2$. As above, we reduce to the case where $N^4 = 1$ and $|[N, W]| = 2$. Since any product of squares in N is congruent to a square modulo $[N, W]$, it follows that $(N^2)^2 = 1$. Also, if $a, b \in N$ then

$$[a^2, b] = [a, b]^2 \in [N, W]^2 = 1,$$

so $N^2 \le Z(N)$. Now N/N^2 is an $\mathbb{F}_2$-vector space of dimension at most r, so $[N, V] \le N^2$. Hence for $a \in N$ and $v \in V$ we have

$$(a^2)^v = (a^v)^2 = (ba)^2 \quad \text{with } b \in N^2$$
$$= a^2.$$

Thus $[N^2, V] = 1$. Therefore $[N, V, V] = 1$ and so

$$[N, W] = [N, V^2] \le [N, V]^2[N, V, V] = 1,$$

contrary to assumption.

Remark In the proof of Proposition 2.12, the defining property of $V(G, r)$ was only brought into play with regard to linear representations of G arising from the conjugation action of G on elementary abelian sections of G. Hence the result remains valid if we replace V, in the statement, by the (possibly larger) subgroup

$$V^* = \bigcap C_G(A/B),$$

where (A, B) runs over all pairs of normal subgroups of G with $B < A$ and A/B elementary abelian of rank at most r.

2.13 Theorem. *Let G be a finite p-group of rank r. Then G has a powerful characteristic subgroup of index at most $p^{r\lambda(r)}$ if p is odd, $2^{r+r\lambda(r)}$ if $p = 2$.*

Proof Put $V = V(G, r)$. By Lemma 2.11, there is a series of normal subgroups running from G to V, of length at most $\lambda(r)$, with each factor elementary abelian. Since G has rank r, each of these factors has order at most p^r, so $|G : V| \le p^{r\lambda(r)}$. If p is odd, Proposition 2.12 shows that V is powerful. If $p = 2$, we know by Proposition 2.12 that V^2 is powerful; and since $|V/V^2| \le 2^r$ we have $|G : V^2| \le 2^{r+\lambda(r)}$. This completes the proof.

Remark   It is clear from the proof that we could take $r = \sup\{\mathrm{d}(K) \mid K \lhd G\}$ in Theorem 2.13, giving a sharper result. With a little more care, r can be reduced further: see Exercise 6. For an interesting application see Exercise 7.

Notes

The material of this chapter is all from Lubotzky and Mann (1987a). The proof of Theorem 2.9 is due to A. Caranti.

Exercises

G always denotes a finite p-group.

1. Let N p.e. G. Prove: (a) If $H \leq G$ and H is powerful then NH is powerful. (b) If $S \subseteq G$ and $N = \langle S^G \rangle$ then $N = \langle S \rangle$. (c) $[N, G]$ p.e. G. (d) If M p.e. G then NM and $[N, M]$ are both powerfully embedded in G. (e) G has a unique maximal powerfully embedded subgroup.

2. Let $M, N \lhd G$. Prove the following:
  (i) if N p.e. G then N^{p^i} p.e. G for all i;
  (ii) if $[N, M]$ p.e. N then $[N^p, M] = [N, M]^p$;
  (iii) (Shalev 1993b) if N p.e. G and M p.e. G then $[N^{p^i}, M^{p^j}] = [N, M]^{p^{i+j}}$ for all i and j.
[*Hints: for (i)*, induction, using Proposition 2.3 and Theorem 2.7. *For (ii)*, examine the proof of Proposition 2.3. *For (iii)*, induction on $i + j$.]

3. (a) Give an example of a p-group G such that $G^{p^2} \neq (G^p)^p$. [*Hint:* There exists a group with $\mathrm{d}(G) = 2$, $G^4 = 1$ and $|G| = 2^{12}$.] (b) Give an example of a p-group G with $\mathrm{d}(G) = \mathrm{rk}(G)$ such that G is not powerful.

4. (i) Show that a 2-group G is powerful if and only if $G/(G^2)^2$ is abelian.
  (ii) Show that if G is a powerful 2-group then $[P_i(G), P_j(G)] \leq P_{i+j+1}(G)$ for all i and j.

5. Suppose G is powerful and has exponent p^e. For each k, let $E_k = \{x \in G \mid x^{p^k} = 1\}$. (a) Show that the mapping $x \mapsto x^{p^{e-1}}$ is an endomorphism of G. (b) Show that E_{e-1} is a subgroup of G, of order equal to $|G : P_e(G)|$. (c) Show by an example that E_{e-1} need not be powerful. (d) Show that if all characteristic subgroups of G are powerful then E_k is a subgroup of G for each k.

6. Define $\Phi^0(G) = G$ and, for $i \geq 0$, $\Phi^{i+1}(G) = \Phi(\Phi^i(G))$. Let $s = \sup_{i \geq 0} \mathrm{d}(\Phi^i(G))$ and put $W = \Phi^{\lambda(s)+\epsilon}(G)$, where $\epsilon = 0$ if p is odd, $\epsilon = 1$ if $p = 2$. Show that W is powerful and that $|G : W| \leq p^{s(\lambda(s)+\epsilon)}$.

7. Let s and ϵ be as in Exercise 6. Show that $\mathrm{rk}(G) \leq s(\lambda(s) + 1 + \epsilon)$.
[Thus a bound for $\mathrm{d}(K)$ as K runs over the *characteristic* subgroups of a p-group gives a (slightly larger) bound for $\mathrm{d}(H)$ as H runs over *all* subgroups.]

8. (*a*) Suppose p is odd. Show that G is powerful if (and only if) G is the product of $\mathrm{d}(G)$ cyclic subgroups. (*b*) Show by example that this fails if $p = 2$.

9. A d-tuple $(x_1, \ldots, x_d)$ of elements of G is called a *basis* if each element of G can be expressed uniquely in the form $x_1^{a_1} \ldots x_d^{a_d}$. This is equivalent to the stipulation that $G = \langle x_1 \rangle \ldots \langle x_d \rangle$ and $|G| = |\langle x_1 \rangle| \ldots |\langle x_d \rangle|$. Show that if G is powerful then G has a basis of cardinality $\mathrm{d}(G)$.

[*Hint:* Put $G_i = P_i(G)$ for each i, and suppose that $G_{e+1} = 1 \neq G_e$. Let $(x_1 G_e, \ldots, x_d G_e)$ be a basis for G/G_e, found inductively, and suppose that $(x_1^{p^{e-1}}, \ldots, x_s^{p^{e-1}})$ is a basis for G_e. Let $s < k \leq d$, and suppose that $x_k G_e$ has order p^ℓ in G/G_e. Using the fact that $G_{e-\ell}\langle x_k \rangle$ is powerful (why?), show that for some $w \in \langle x_1, \ldots, x_s \rangle$ the element $x_k' = w x_k$ has order p^ℓ. Show that then $(x_1, \ldots, x_s, x_{s+1}', \ldots, x_d')$ is a basis for G.]

10. Let $M = G^{p^n}$ if p is odd, $M = (G^{2^n})^2$ if $p = 2$. Suppose that M is not powerful. Show that $\mathrm{d}(M) > p^n$. Deduce that there exist normal subgroups $B < A$ of G and an element $x \in G$ such that (*i*) A/B is elementary abelian of rank $> p^n$, and (*ii*) $[A, x^{p^n}] \not\leq B$.
[*Hint:* For the first part, use Proposition 2.12 and consider the exponent of $\mathrm{U}_r(\mathbb{F}_p)$. Then use the *Remark* following Proposition 2.12.]

11. The *wreath product* $C_p \wr X$, for a finite group X, is the semi-direct product of the group algebra $\mathbb{F}_p[X]$ by X, with X acting by right multiplication. Now let H be a finite p-group containing an elementary abelian normal subgroup A and an element x such that $[A, x^{p^n}] \neq 1$. Show that there exist $a \in A$ such that $[a, x^{p^n}] = 1$ and $\langle a, x \rangle / \langle x^{p^n} \rangle \cong C_p \wr C_{p^n}$.

12. Prove the following theorem, due to Shalev (1992a): *if G does not involve $C_p \wr C_{p^n}$ then G^{p^n} is powerful if p is odd, and $(G^{2^n})^2$ is powerful if $p = 2$.*
[*Hint:* Use Exercises 10 and 11. (We say that G *involves* W if there exist $K \lhd H \leq G$ such that $H/K \cong W$.)]

13. A powerful 2-generator p-group is metacyclic (i.e. has a cyclic normal subgroup with cyclic quotient).
[*Hint:* Use Exercise 1 (b) to show that the derived group is cyclic.]

14. If p is odd then every metacyclic p-group is powerful.

15. Show that if p is odd, a p-group which is the product of two cyclic subgroups is metacyclic.

16. Let $H = \langle x_1, \ldots, x_d \rangle$ be the free-nilpotent group of class c on the given generators (so $H = F/\gamma_{c+1}(F)$ where F is free on $\{x_1, \ldots, x_d\}$). Put $r = \mathrm{rk}(H)$, $\ell = \lambda(r)$ if p is odd, $\ell = 1 + \lambda(r)$ if $p = 2$. (i) Show that H^{p^ℓ}/H^{p^n} is a powerful p-group for every $n \geq \ell$. (ii) Show that the map $\mu : x_i \mapsto x_i^{p^\ell}$, $(i = 1, \ldots, d)$, defines an injective endomorphism of H, with $|H : H\mu| = p^m$ for some m. (iii) Deduce that *every finite p-group is involved in a powerful p-group, of the same nilpotency class.*

[*Hint:* For (i), use Lemma 2.11 and Proposition 2.12. For (ii): the fact that $|H : H\mu|$ is a power of p follows from Proposition 3 in Segal (1983), Chapter 6. The injectivity of μ depends on a Hirsch length argument together with the fact that H is torsion-free (see Hall (1969), Theorem 5.6, Corollary).]

3

Pro-p groups of finite rank

We begin this chapter by developing the theory of *powerful pro-p groups*. This parallels the theory of Chapter 2, which can indeed be considered a special case of it: however we feel that the structure of the whole theory emerges more clearly when the arguments which belong essentially to finite p-groups are presented separately.

The heart of the chapter is in the second section, where our first major theorem on pro-p groups appears: this characterises the pro-p groups of finite rank as exactly those which contain a finitely generated powerful open subgroup. Various alternative characterisations of this class of pro-p groups are derived in §3.3. Further characterisations appear in the exercises, and later throughout the book; they are summarised in Interlude A.

3.1 Powerful pro-p groups

3.1 Definition Let G be a pro-p group

(i) G is *powerful* if p is odd and $G/\overline{G^p}$ is abelian, or if $p = 2$ and $G/\overline{G^4}$ is abelian

(ii) Let $N \leq_o G$. Then N is *powerfully embedded in* G, written N p.e. G, if p is odd and $[N, G] \leq \overline{N^p}$, or if $p = 2$ and $[N, G] \leq \overline{N^4}$.

Note that if N p.e. G then $N \lhd_o G$ and N is powerful. Using the fact that $\overline{N^p}$ (resp. $\overline{N^4}$) is the intersection of the open normal subgroups of G containing it, we deduce the following important criterion:

3.2 Proposition *Let G be a pro-p group and $N \leq_o G$. Then N p.e. G if and only if NK/K p.e. G/K for every $K \lhd_o G$.*

3.3 Corollary. *A topological group G is a powerful pro-p group if and only if G is the inverse limit of an inverse system of powerful finite p-groups in which all the maps are surjective.*

Proof Suppose G is a powerful pro-p group. Then $G \cong \varprojlim G/N$ where N runs over the open normal subgroups of G, and each G/N is a powerful finite p-group. Conversely, suppose $G = \varprojlim G_\lambda$ where each G_λ is a powerful finite p-group, and $G_\mu \to G_\lambda$ is surjective whenever $\mu \geq \lambda$. Then G is a pro-p group, and if $K \triangleleft_o G$ then G/K is a quotient of some G_λ, hence powerful; hence G is powerful by Proposition 3.2.

The way is now clear for us to carry over the results of Chapter 2 from finite p-groups to pro-p groups. However, it is necessary to restrict attention to *finitely generated* pro-p groups, in which the lower p-series is well-behaved, i.e. consists of *open* subgroups: see Proposition 1.16.

3.4 Lemma *Let G be a powerful finitely generated pro-p group. Then every element of G^p is a pth power in G, and $G^p = \Phi(G)$ is open in G. If $p = 2$, then G^4 is open in G.*

Proof Let $g \in \overline{G^p}$. Then $gN \in (G/N)^p$ for each $N \triangleleft_o G$, hence by Proposition 2.6 we see that gN is a pth power in G/N for each such N. It follows that g is a pth power in G (see Exercise 1.6). Hence $\overline{G^p} \leq G^p$, and so $G^p = \overline{G^p}$ consists of pth powers. Since $[G,G] \leq \overline{G^p}$, this shows that $G^p = \Phi(G) = P_2(G)$, which is open by Proposition 1.16. If $p = 2$, a similar argument shows that $G^4 = \overline{G^4} \geq P_3(G)$, giving that G^4 is open.

3.5 Corollary. *Let G be as in Lemma 3.4. Then for each i we have*

$$G^{p^i} = (G^{p^{i-1}})^p = \{x^{p^i} \mid x \in G\} \ p.e. \ G^{p^{i-1}}. \qquad (*)$$

Proof It follows from Proposition 2.3 and Proposition 3.2 that $\overline{G^p}$ p.e. G. Thus the case $i = 1$ of $(*)$ reduces to Lemma 3.4. The general case follows by induction, on replacing G by $G^{p^{i-1}}$.

3.6 Theorem. *Let $G = \overline{\langle a_1, \ldots, a_d \rangle}$ be a finitely generated powerful pro-p group, and put $G_i = P_i(G)$ for each i.*

(i) G_i p.e. G;

(ii) $G_{i+k} = P_{k+1}(G_i) = G_i^{p^k}$ *for each $k \geq 0$, and in particular $G_{i+1} = \Phi(G_i)$;*

(iii) $G_i = G^{p^{i-1}} = \{x^{p^{i-1}} \mid x \in G\} = \overline{\langle a_1^{p^{i-1}}, \ldots, a_d^{p^{i-1}} \rangle}$;

(iv) *the map* $x \mapsto x^{p^k}$ *induces a homomorphism from* G_i/G_{i+1} *onto* G_{i+k}/G_{i+k+1}, *for each* i *and* k.

Proof The second equality in (iii) follows from Corollary 3.5. Everything else follows from Theorem 2.7, applied to the finite p-groups G/G^{p^n} for sufficiently large n; the subgroups G^{p^n} being open by Corollary 3.5 (indeed they form a base for the neighbourhoods of 1 in G, by Proposition 1.16 (iv), since $G^{p^n} \le P_{n+1}(G)$ for each n).

3.7 Proposition. *If* $G = \overline{\langle a_1, \ldots, a_d \rangle}$ *is a powerful pro-p group, then* $G = \overline{\langle a_1 \rangle} \ldots \overline{\langle a_d \rangle}$, *i.e.* G *is the product of its procyclic subgroups* $\overline{\langle a_1 \rangle}, \ldots, \overline{\langle a_d \rangle}$.

Proof Let $A = \overline{\langle a_1 \rangle} \ldots \overline{\langle a_d \rangle}$. As a product of finitely many closed, hence compact, subsets of G, A is a closed subset of G. So $A = \bigcap_{N \lhd_o G} AN$. But Corollary 2.8 shows that $AN/N = G/N$ for each $N \lhd_o G$; consequently $A = G$.

For any topological group G, $\mathrm{d}(G)$ denotes the minimal cardinality of a topological generating set for G. If G is a finitely generated pro-p group, we thus have

$$\mathrm{d}(G) = \dim_{\mathbb{F}_p}(G/\Phi(G)).$$

Combining Theorem 2.9 and Proposition 1.5, we obtain

3.8 Theorem. *Let* G *be a powerful finitely generated pro-p group and* H *a closed subgroup. Then* $\mathrm{d}(H) \le \mathrm{d}(G)$.

We now extend Definition 2.10 to the case where G is a finitely generated pro-p group: thus $V(G, r)$ will denote the intersection of the kernels of all homomorphisms of G into $\mathrm{GL}_r(\mathbb{F}_p)$. Since G is finitely generated and $\mathrm{GL}_r(\mathbb{F}_p)$ is finite, these homomorphisms are all continuous, by Corollary 1.21, and so there are only finitely many of them: hence $V(G, r)$ is open in G, as well as being (obviously) characteristic. (When $p = 2$, we shall also need to know that $V(G, r)^2$ is open in G; this follows from Proposition 1.7, Corollary 1.20 and Proposition 1.14.)

With the help of Proposition 3.2, Proposition 2.12 translates into

3.9 Proposition. *Let G be a finitely generated pro-p group and r a positive integer. Put $V = V(G, r)$. Let $N \lhd_o G$ satisfy $d(N) \le r$, and $N \le V$ if p is odd, $N \le V^2$ if $p = 2$. Then N p.e. V if p is odd, N p.e. V^2 if $p = 2$.*

We can now carry over the proof of Theorem 2.13, virtually word for word, to obtain

3.10 Theorem. *Let G be a finitely generated pro-p group, and suppose that $r = \sup_{N \lhd_o G} d(N)$ is finite. Then G has a powerful characteristic open subgroup of index at most $p^{r\lambda(r)}$ if p is odd, $2^{r+r\lambda(r)}$ if $p = 2$.*

(Here, $\lambda(r)$ is the integer defined, as in Chapter 2, by $2^{\lambda(r)-1} < r \le 2^{\lambda(r)}$.)

3.2 Pro-p groups of finite rank

The next thing to do is to clarify the various possible definitions of *rank:*

3.11 Proposition. *Let G be a profinite group, and put*

$$r_1 = \sup\{d(H) \mid H \le_c G\}$$
$$r_2 = \sup\{d(H) \mid H \le_c G \text{ and } d(H) < \infty\}$$
$$r_3 = \sup\{d(H) \mid H \le_o G\}$$
$$r_4 = \sup\{rk(G/N) \mid N \lhd_o G\}.$$

Then $r_1 = r_2 = r_3 = r_4$.

Proof Obviously $r_2 \le r_1$ and $r_3 \le r_1$. If $N \lhd_o G$ and $M/N \le G/N$, then $d(M/N) \le d(M) \le r_3$; so $r_4 \le r_3$. Also for such M and N we have $M = NX$, where X is a finite subset of G; putting $H = \overline{\langle X \rangle}$ we see that $d(M/N) - d(HN/N) \le d(H) \le r_2$, giving that $r_4 \le r_2$. Finally, let $H \le_c G$. Then $d(H) = \sup\{d(HN/N) \mid N \lhd_o G\} \le r_4$, by Proposition 1.5; hence $r_1 \le r_4$.

3.12 Definition Let G be a profinite group. The *rank* $rk(G)$ of G is the common value of $r_1, \ldots, r_4$ given in Proposition 3.11.

Note that a profinite group of finite rank is, by definition, finitely generated. If G is a finitely generated powerful pro-p group, then Theorem 3.8 shows that $rk(G) = d(G)$, so G has finite rank. More generally, if

G is finitely generated and has a powerful open subgroup, then G has finite rank: for it is easy to see that a profinite group which has an open subgroup of finite rank has finite rank itself (cf. Exercise 1). Combining this fact with Theorem 3.10, we obtain the main result of this chapter:

3.13 Theorem. *Let G be a pro-p group. Then G has finite rank if and only if G is finitely generated and G has a powerful open subgroup; in that case, G has a powerful open characteristic subgroup.*

3.14 Corollary. *Let G be a pro-p group and r a positive integer. Suppose that every open subgroup of G contains an open normal subgroup N of G with $d(N) \leq r$. Then G has finite rank.*

Proof The hypothesis implies that G is finitely generated. Now let $W = V(G, r)$ if p is odd, $W = V(G, r)^2$ if $p = 2$. Then W contains an r-generator open normal subgroup N of G, and Proposition 3.9 shows that N is powerful.

Similarly (see Exercises 2.6 and 2.7) it is easy to derive

3.15 Corollary. *Let G be a pro-p group and r a positive integer. If $d(K) \leq r$ for every topologically characteristic subgroup K of G, then $\mathrm{rk}(G) \leq r(\lambda(r) + 1 + \epsilon)$ where $\epsilon = 0$ if p is odd, $\epsilon = 1$ if $p = 2$.*

3.3 Characterisations of finite rank

In this final section, we explore various conditions on a pro-p group that are necessary and/or sufficient for it to have finite rank; many more will appear in later chapters. They are summarised in Interlude A.

3.16 Theorem. *Let G be a pro-p group. Then the following are equivalent:*

(a) *there exist $s \in \mathbb{N}$ and $c > 0$ such that $|G : \overline{G^{p^k}}| \leq cp^{ks}$ for all k;*

(b) *there exist $s \in \mathbb{N}$ and $c > 0$ such that $|G : G^{p^k}| \leq cp^{ks}$ for all k;*

(c) *G has finite rank.*

Moreover, if in (c) G has rank r then we can take $s = r$ in (a) and (b); and given s as in (a), G has an open normal subgroup K with $\mathrm{rk}(K) \leq s$.

Proof Suppose that G has finite rank r. By Theorem 3.10, G has a powerful open normal subgroup H, say. Put $H_i = P_i(H)$ for each i.

Then $|H : H_2| \le p^r$, and it then follows from Theorem 3.6 (iv) that $|H : H_{k+1}| \le p^{kr}$ for each k. Now $H_{k+1} = H^{p^k}$ by Theorem 3.6 (iii), so we have

$$|G : G^{p^k}| \le |G : H^{p^k}| \le |G : H|p^{kr}.$$

Thus (c) implies (b). Obviously (b) implies (a). Finally, suppose (a) holds. Then $|G : \Phi(G)| \le |G : \overline{G^p}| \le cp^s$ is finite, so G is finitely generated. Put $W = V(G, s)$ if p is odd, $W = V(G, s)^2$ if $p = 2$, and write $G_i = \overline{G^{p^i}}$ for each i. Since $W \lhd_o G$, there exists m such that $G_m \le W$. Now our hypothesis implies that for some $k \ge m$, $|G_k : G_{k+1}| \le p^s$: for if not, we would have, for sufficiently large n,

$$|G : G_{m+n}| \ge |G_m : G_{m+n}| \ge p^{(s+1)n} > cp^{ms}p^{ns} = cp^{(m+n)s},$$

contradicting the hypothesis. Choose such a k and put $K = G_k$. Then $\Phi(K) \ge \overline{K^p} \ge G_{k+1}$, so $|K/\Phi(K)| \le p^s$ and so $d(K) \le s$. Now Proposition 3.9 shows that K is powerful, and Theorem 3.8 then shows that $\mathrm{rk}(K) = d(K) \le s$. This proves the final statement of the theorem, and shows that (a) implies (c).

3.17 Theorem *Let G be a pro-p group. Then the following are equivalent:*

(a) *G is the product of finitely many procyclic subgroups;*

(b) *G is the product of finitely many closed subgroups of finite rank;*

(c) *G has finite rank;*

(d) *G is finitely generated as a '$\mathbb{Z}_p$-powered group', i.e. G has a finite subset X such that every element of G is equal to a product of the form $x_1^{\lambda_1} \ldots x_s^{\lambda_s}$ with $x_j \in X$ and $\lambda_j \in \mathbb{Z}_p$.*

(e) *G is countably generated as a '$\mathbb{Z}_p$-powered group'.*

Proof Suppose G has finite rank. By Theorem 3.10, G has a powerful open normal subgroup, K say. Also K is finitely generated, so Proposition 3.7 shows that $K = C_1 \ldots C_d$ for some procyclic subgroups C_j of K. Now let $\{x_1, \ldots, x_m\}$ be a transversal to the cosets of K in G. Then $G = C_1 \ldots C_d \langle x_1 \rangle \ldots \overline{\langle x_m \rangle}$. Thus (c) implies (a).

Trivially (a) implies (b). Now suppose (b) holds. Thus $G = H_1 \ldots H_t$ where each H_j is a closed subgroup of finite rank. By Theorem 3.16, there exist $c > 0$ and $s \in \mathbb{N}$ such that $|H_j : H_j^{p^k}| \le cp^{ks}$ for each j and all k. But G/G^{p^k} is the product of its subgroups $H_j G^{p^k}/G^{p^k}$ ($j = 1, \ldots, t$), so $|G : G^{p^k}| \le (cp^{ks})^t = c^t p^{kst}$. Using Theorem 3.16 in the reverse direction now shows that G has finite rank. Thus (b) implies (c).

Now Proposition 1.28 shows that (a) implies (d). Clearly (d) implies (e). Finally, to show that (e) implies (a), suppose X is a countable subset of G which generates G as a '$\mathbb{Z}_p$-powered group'. For each finite sequence $\mathbf{x} = (x_1, \ldots, x_s)$ of elements of X, put

$$M(\mathbf{x}) = x_1^{\mathbb{Z}_p} \ldots x_s^{\mathbb{Z}_p} \,.$$

Then each of the sets $M(\mathbf{x})$ is closed in G, and by hypothesis G is the union of the $M(\mathbf{x})$ as $\mathbf{x}$ ranges over all finite sequences in X. Now, as X is countable, the set of all such sequences is countable; it follows by the Baire Category Theorem (see Exercise 6) that for some $\mathbf{x}$, $M(\mathbf{x})$ contains a non-empty open subset of G. Thus there exist $w \in G$ and $N \lhd_o G$ with $wN \subseteq M(\mathbf{x}) = x_1^{\mathbb{Z}_p} \ldots x_s^{\mathbb{Z}_p}$. Let $\{t_1, \ldots, t_m\}$ be a set of representatives for the cosets of N in G. Then we have

$$G = \overline{\langle t_1 \rangle} \ldots \overline{\langle t_m \rangle} \cdot \overline{\langle w \rangle} \cdot \overline{\langle x_1 \rangle} \ldots \overline{\langle x_s \rangle} \,;$$

thus (a) follows.

We conclude the chapter by looking at two kinds of *growth condition*: 'subgroup growth' in Theorem 3.19 and 'word growth' in Theorem 3.20. Both theorems will be refined later (as will Theorem 3.16), using more sophisticated machinery (see §11.1 and Interlude E). Their proofs illustrate the following useful principle:

3.18 Lemma *Let G be a finitely generated pro-p group and r a positive integer. Let N be an open normal subgroup of G maximal with respect to the property $\mathrm{d}(N) \geq r$. Then*
(i) $N = \mathrm{C}_G(N/\Phi(N))$, *and*
(ii) $|G : N| \leq p^{(r-1)\lambda(d)}$ *where* $d = \mathrm{d}(N)$.

Proof (i) Let $C = \mathrm{C}_G(N/\Phi(N))$. Then $N \leq C \lhd G$. Suppose $N < C$. Then there exists an element Nx of order p in $C/N \cap \mathrm{Z}(G/N)$. Putting $M = \langle N, x \rangle$, we have $N < M \lhd_o G$. But $M/\Phi(N)$ is abelian and contains $N/\Phi(N)$, so $\mathrm{d}(M) \geq \mathrm{d}(M/\Phi(N)) \geq \mathrm{d}(N/\Phi(N)) = \mathrm{d}(N) \geq r$, contradicting the maximality of N. Hence $N = C$ as claimed.

(ii) By (i), G/N acts faithfully by conjugation on $N/\Phi(N) \cong \mathbb{F}_p$, so G/N is isomorphic to a subgroup of $\mathrm{U}_d(\mathbb{F}_p)$. Since every normal subgroup of G/N can be generated by $r - 1$ elements, Lemma 2.11 shows that $|G/N| \leq p^{(r-1)\lambda(d)}$.

3.19 Theorem *Let G be a pro-p group, and for each n let σ_n denote the number of open subgroups of index at most p^n in G. Then the following are equivalent:*

(a) *there exist $c > 0$ and $s \in \mathbb{N}$ such that $\sigma_n \leq cp^{ns}$ for all n;*

(b) *G has finite rank.*

Proof Suppose first that G has finite rank r. If $H \leq_o G$ then $H/\Phi(H)$ is elementary abelian of rank at most r; since every open subgroup of index p in H contains $\Phi(H)$, it follows that H contains no more than p^r open subgroups of index p. Now every open subgroup of index p^{n+1} in G is contained in at least one open subgroup of index p^n. Hence $\sigma_{n+1} \leq \sigma_n + p^r \sigma_n \leq p^{r+1}\sigma_n$, and it follows by induction that $\sigma_n \leq p^{n(r+1)}$ for all n. Thus (b) implies (a).

Now suppose (a) holds. Since $\Phi(G)$ is the intersection of open subgroups of index p in G, and there are at most cp^s of these, we see that $\Phi(G)$ is open in G. Hence G is finitely generated. Our aim now is to show that there is a finite upper bound for the numbers $d(N)$ as N ranges over all open normal subgroups of G. Once this is established, it will follow by Corollary 3.14 that G has finite rank.

So let r be a positive integer, and suppose that the set of $N \lhd_o G$ such that $d(N) \geq r$ is non-empty. Choose N to be a maximal member of this set, and put $d = d(N)$. By Lemma 3.18, we then have

$$|G : N| \leq p^{(r-1)\lambda(d)} .$$

On the other hand, $N/\Phi(N)$ is a d-dimensional vector space over $\mathbb{F}_p$, and so contains at least $p^{(d-1)^2/4}$ subspaces of codimension $[d/2]$. Thus G contains at least $p^{(d-1)^2/4}$ open subgroups of index $\leq p^{(r-1)\lambda(d)+[d/2]}$, and so

$$cp^{s((r-1)\lambda(d)+[d/2])} \geq p^{(d-1)^2/4} .$$

Since $r \leq d$ and $\lambda(d) \leq 1 + \log_2 d$, this implies that $d < 4s\log_2 d + c^*$, where $c^* = 6s + 4\log_p c$, and hence that d is bounded above by a function of c and s. Thus we have the required upper bound for r.

Condition (a) in Theorem 3.19 says that G has *polynomial subgroup growth*. For a finite subset X of G, the *word growth* of G (relative to X) is the function f_X given by $f_X(n) = |W_n(X)|$ where

$$W_n(X) = \{x_1 x_2 \dots x_n \mid x_i \in X \cup X^{-1} \cup \{1\} \text{ for each } i\} \subseteq G$$

denotes the set of all (values in G of) group words of length at most n on X.

3.20 Theorem *Let G be a pro-p group and X a finite topological generating set for G. If there exist $c, s > 0$ such that $f_X(n) \leq cn^s$ for all n, then G has finite rank.*

Thus having 'polynomial word growth' is a sufficient condition for a finitely generated pro-p group to have finite rank; unlike the corresponding subgroup growth condition, however, it is certainly not necessary: see Exercise 13 and Interlude E.

Proof Let $r \geq 2$ be a positive integer, and suppose that the set of open normal subgroups N of G with $\mathrm{d}(N) \geq r$ is non-empty. We shall show that r is bounded above by a function of c and s only; that G has finite rank then follows by Corollary 3.14.

Choose N to be a maximal such subgroup and put $d = \mathrm{d}(N)$, so $d \geq r$. Then G/N acts faithfully by conjugation on $N/\Phi(N) \cong \mathbb{F}_p^d$, by Lemma 3.18. It follows by Lemma 2.11 that G has a series of normal subgroups

$$G = N_0 \geq N_1 \geq \cdots \geq N_k = N > N_{k+1} = \Phi(N)$$

with each factor N_{i-1}/N_i elementary abelian and $k = \lambda(d) \leq 1 + \log d$.

Now let $0 \leq i \leq k$ and suppose that N_i is generated by a subset of $W_\ell(X)$, for some $\ell \geq 1$. If $N_i > N$ then $\mathrm{d}(N_i) < r$, by the choice of N, while if $N_i = N$ then $\mathrm{d}(N_i) = d$; so in any case, $\mathrm{d}(N_i) \leq d$, and it follows that every generating set for N_i contains one of cardinality at most d (Exercise 1.5). Hence there exist $y_1, \ldots, y_d \in W_\ell(X)$ such that $N_i = \overline{\langle y_1, \ldots, y_d \rangle}$. As N_i/N_{i+1} is elementary abelian (and of course N_{i+1} is open in N_i), we can find a transversal T_i to the cosets of N_{i+1} in N_i consisting of elements of the form

$$t(\mathbf{a}) = y_1^{a_1} \ldots y_d^{a_d},$$

with $0 \leq a_j < p$ for each j; each such element evidently lies in $W_{(p-1)d\ell}(X)$.

It follows now that N_{i+1} is generated by a subset Z, say, consisting of elements of the form

$$t(\mathbf{a}) y_j^{\pm 1} t(\mathbf{b})^{-1}$$

(see the proof of Proposition 1.7). Thus $Z \subseteq W_{\ell^*}(X)$ where $\ell^* = (2(p-1)d + 1)\ell < 2pd\ell$.

Since $N_0 = G$ is generated by $X \subseteq W_1(X)$, we see by induction that $N = N_k$ can be generated by a subset of $W_{(2pd)^k}(X)$. Taking $\ell = (2pd)^k$ in the above discussion, we thus obtain a transversal T_k to

$N_k/N_{k+1} = N/\Phi(N)$ with $T_k \subseteq W_n(X)$ where $n = (p-1)d\ell < (2pd)^{k+1}$. By hypothesis, $|W_n(X)| = f_X(n) \le cn^s$; consequently

$$p^d = |N/\Phi(N)| = |T_k| \le cn^s < c(2pd)^{(k+1)s}.$$

Thus

$$d < \log_p c + (k+1)s(\log_p 2 + 1 + \log_p d)$$
$$\le \log c + s(2 + \log d)^2,$$

since $p \ge 2$ and $k \le 1 + \log d$. It follows that d is bounded above by a function of c and s, giving the result since $r \le d$.

Notes

The theory of §3.1 is from Lubotzky and Mann (1987b). Lemma 3.18 and Theorems 3.19 and 3.20 are from Lubotzky and Mann (1991). The other results are new in the stated form; many of them appeared in different versions in [L].

Warning: most authors use the word 'rank' to mean the *minimal size of a generating set* for a profinite group (what we denote by $d(G)$).

Exercises

1. Let G be a profinite group and let $N \lhd_c G$. Show that

$$\max\{\mathrm{rk}(N), \mathrm{rk}(G/N)\} \leq \mathrm{rk}(G) \leq \mathrm{rk}(N) + \mathrm{rk}(G/N).$$

Deduce that if $H \leq_o G$ and $\mathrm{rk}(H)$ is finite then $\mathrm{rk}(G)$ is finite.

2. Fix a prime p, and for $k \in \mathbb{N}$ put

$$P(k) = (1 - p^{-k})(1 - p^{1-k}) \ldots (1 - p^{-1}) S(k) = \sum_{j=1}^{k} P(j)^{-1}.$$

(a) Let H be a finite p-group, and put $\mathrm{d}(H) = k$. Show that the number of ordered k-element generating sets for H is $P(k)|H|^k$. [*Hint:* Consider $H/\Phi(H)$.]

(b) Let G be a finite p-group, and denote the number of subgroups H of index p^n in G with $\mathrm{d}(H) = k$ by $a_n(k)$. Show that $a_n(k) \leq P(k)^{-1}p^{nk}$. Deduce that the number of k-generator subgroups of index p^n in G is at most $S(k)p^{nk}$. [*Hint:* Count ordered k-tuples in G, and use (a).]

(c) Now let G be a finitely generated pro-p group. Show that the results of (b) are still valid. [*Hint:* Consider a suitable finite quotient of G.] (d) Let G be a pro-p group of finite rank r, and define σ_n as in Theorem 3.19. Show that $\sigma_n < 2S(r)p^{nr}$ for all $n \geq 1$.

3. Let G be a pro-p group, denote the number of open subgroups of index p^n in G by a_n, and the number of open k-generator subgroups of index p^n in G by $b_n(k)$ (so $b_n(k) = \sum_{j=0}^{k} a_n(j)$ in the notation of Exercise 2). The 'lower density' of the family of k-generator subgroups of G is then defined to be

$$\underline{\delta}(k) = \liminf_{n \to \infty} b_n(k)/a_n.$$

Prove: *If $\underline{\delta}(k) > 0$ for some finite k then G has finite rank.* [*Hint:* Use Exercise 2 and Theorem 3.19.]

4. (Shalev 1992a) Let G be a finitely generated pro-p group. Show that if G does not have finite rank then G involves $C_p \wr C_{p^n}$ for every n (i.e. for each n, there exist subgroups $K \lhd H$ of G such that $H/K \cong C_p \wr C_{p^n}$). [For the definition of $C_p \wr C_{p^n}$, see Exercise 2.11. *Hint:* Note that $\overline{G^{p^n}}$ is open in G for each n (Exercise 1.14). Then use Exercise 2.12.]

5. Show that for each n there is an epimorphism of $C_p \wr C_{p^{n+1}}$ onto $C_p \wr C_{p^n}$. Hence construct a 2-generator pro-p group of infinite rank.

6. Prove the following special case of the Baire Category Theorem: *Let G be a compact Hausdorff space, in which every non-empty open set contains a non-empty compact open subset. If $G = \bigcup_{i=1}^{\infty} M_i$ and each subset M_i is closed in G, then for some n, M_n contains a non-empty open subset of G.* [*Hint:* Suppose false. Put $Y_n = G \setminus \bigcup_{i=1}^{n} M_i$, show that $\overline{Y_n} = G$, and hence find a sequence of non-empty open compact sets (T_n) such that $T_{n+1} \subseteq T_n \cap Y_{n+1}$ for each n. Deduce that $\bigcap Y_n \supseteq \bigcap T_n \neq \varnothing$.]

7. Let G be a pro-p group of finite rank, and let $m \geq 0$. Show that for all sufficiently large k, $\gamma_{p^{k-m}}(G) \leq G^{p^{2k}}$.
[*Hint :* Use Theorem 3.16 .]

8. Let G be a pro-p group. Put $\epsilon = 0$ if p is odd, $\epsilon = 1$ if $p = 2$. Suppose that for some k, $\gamma_{p^{k-\epsilon}}(G) \leq \overline{G^{p^{2k}}}$. Prove that $\overline{G^{p^n}}$ is powerful where $n = 2k - \epsilon - 1$.
[*Hint:* This depends on the following commutator formula, valid in all groups:

$$[G^{p^n}, G^{p^n}] \leq \prod \gamma_{i+j}(G)^{f(i,j)}$$

where $f(i,j) = \binom{p^n}{i}\binom{p^n}{j}$ and the product runs over $1 \leq i \leq p^n$ and $1 \leq j \leq p^n$. It is a special case of a result due to Rex Dark, see Passman (1977), Chapter 11, Theorem 1.16. Show that for each relevant pair (i,j), either $i + j > p^{k-\epsilon}$ or $p^{n+1+\epsilon} \mid f(i,j)$.]

9. (A. Shalev) Let G be a finitely generated pro-p group, and let ϵ be as in Exercise 8. Prove that the following are equivalent:
 (a) G has finite rank;
 (b) for some $k \geq 1$, $\gamma_{p^{k-\epsilon}}(G) \leq \overline{G^{p^{2k}}}$;
 (c) for all sufficiently large k, $\gamma_{p^{k-\epsilon}}(G) \leq \overline{G^{p^{2k}}}$.
[*Hint:* Show that (b) implies that $\overline{G^{p^{2k}}}$ is open in G. Then use Exercise 7, Exercise 8 and Theorem 3.13. In Chapter 11 we shall show that G is virtually powerful if and only if, for some n and h with $n < p^h$, every n-fold commutator in G is a p^hth power in G.]

10. A pro-p group is *meta-procyclic* if it has a procyclic normal subgroup with procyclic quotient. Show that a pro-p group is meta-procyclic if and only if it is an inverse limit of metacyclic p-groups.

11. Show that a powerful pro-p group which can be generated (topologically) by 2 elements is meta-procyclic. Show also that such a group either has an open normal procyclic subgroup or else is torsion-free.

12. Let G be a finitely generated powerful pro-p group, $G_i = P_i(G)$ and $d_i = d(G_i)$.
   (i) Show that $d_{i+1} \leq d_i$ for each i.
   (ii) Suppose that G_{i+1} is (topologically) generated by $\{a_1, \ldots, a_{d_{i+1}}\}$; show that G_i has a topological generating set $\{b_1, \ldots, b_{d_i}\}$ such that $a_i = b_i^p$ for $i = 1, \ldots, d_{i+1}$.
   (iii) Deduce that G has a topological generating set $\{x_1, \ldots, x_{d_1}\}$ such that for every $i \geq 1$, G_i is (topologically) generated by $x_1^{p^{i-1}}, \ldots, x_{d_i}^{p^{i-1}}$.
   [*Hint: For (ii)*, having found $b_1, \ldots, b_{d_{i+1}}$, show that their images in G_i/G_{i+1} form part of a basis. *For (iii)*, start with a generating set for G_k where d_k is minimal, and work backwards to $G_1 = G$.]

13. Milnor (1968) and Wolf (1968) have proved that a soluble group having polynomial word growth relative to some finite generating set must be virtually nilpotent. Deduce from this that if a soluble pro-p group has polynomial word growth relative to some finite topological generating set, then it is virtually nilpotent.

4
Uniformly powerful groups

In the previous chapter, we saw that every pro-p group of finite rank has an open normal subgroup which is powerful. In this chapter we show that this subgroup may be chosen so as to satisfy a slightly stronger condition, that of being 'uniformly powerful'. We then show that uniformly powerful groups have a remarkable property: the group operation can be 'smoothed out', to give a new, abelian, group structure, and this new abelian group is in a natural way a finitely generated free $\mathbb{Z}_p$-module.

When, in Part II, we come to consider a pro-p group of finite rank as an analytic group, we shall see that this $\mathbb{Z}_p$-module structure provides a natural co-ordinate system on the group. More immediately, we obtain, free of charge, a faithful linear representation for the automorphism group of any uniformly powerful pro-p group.

4.1 Uniform groups

4.1 Definition A pro-p group G is *uniformly powerful* if
(i) G is finitely generated,
(ii) G is powerful, and
(iii) for all i, $|P_i(G) : P_{i+1}(G)| = |G : P_2(G)|$.

We shall usually abbreviate 'uniformly powerful' to '*uniform*'.

If G is a pro-p group satisfying (i) and (ii) of this definition, then we know from Theorem 3.6 that the pth power map $x \mapsto x^p$ induces an epimorphism $f_i : P_i(G)/P_{i+1}(G) \to P_{i+1}(G)/P_{i+2}(G)$, for each i; condition (iii) of Definition 4.1 is clearly equivalent to:

(iii)$'$ *for each $i \geq 1$, the map f_i is an isomorphism.*

61

The following is now almost obvious.

4.2 Theorem. *Let G be a finitely generated powerful pro-p group. Then $P_k(G)$ is uniform for all sufficiently large k.*

Proof Write $G_i = P_i(G)$, and suppose $|G_i : G_{i+1}| = p^{d_i}$. By Theorem 3.6 (iv) we have $d_1 \geq d_2 \geq \ldots \geq d_i \geq d_{i+1} \geq \ldots$, so there exists m such that $d_k = d_m$ for all $k \geq m$. Now Theorem 3.6 (ii) shows that $P_i(G_k) = G_{k+i-1}$ for all i and k, and Theorem 3.6 (i) shows that G_k is powerful.

If G is a characteristic open subgroup in a pro-p group H then $P_k(G)$ is also open and characteristic in H, so from Theorem 3.10 we deduce

4.3 Corollary. *A pro-p group of finite rank has a characteristic open uniform subgroup.*

Now let G be a powerful pro-p group with $\mathrm{d}(G) = d$ finite. Write $G_i = P_i(G)$ for each i. Then $G_{i+1} = \Phi(G_i)$ by Theorem 3.6, so

$$\mathrm{d}(G_i) = \mathrm{d}(G_i/G_{i+1}) \leq \mathrm{rk}(G) = d\,,$$

by Theorem 3.8. If $H \leq_o G$ and H happens to be powerful also, then $H \geq G_i$ for some i, and we have, similarly,

$$\mathrm{d}(G_i) \leq \mathrm{d}(H) \leq \mathrm{rk}(G) = d\,.$$

Now G is uniform if and only if $\mathrm{d}(G_i/G_{i+1}) = \mathrm{d}(G_1/G_2) = d$ for all i; so we have

4.4 Proposition. *Let G be a powerful finitely generated pro-p group. Then the following are equivalent:*
(a) *G is uniform;*
(b) *$\mathrm{d}(P_i(G)) = \mathrm{d}(G)$ for all $i \geq 1$;*
(c) *$\mathrm{d}(H) = \mathrm{d}(G)$ for every powerful open subgroup H of G.*

The most useful characterisation of uniform groups is simply stated:

4.5 Theorem. *A powerful finitely generated pro-p group is uniform if and only if it is torsion-free.*

Proof Let G be a finitely generated powerful pro-p group, and write $G_i = P_i(G)$ for each i. Suppose first that G is not torsion-free. Then G contains an element x of order p (an element of finite order coprime to p would have to lie in G_i for every i, and hence be 1). Say $x \in G_i \setminus G_{i+1}$. Then $1 \neq xG_{i+1} \in G_i/G_{i+1}$ and $1 = x^p G_{i+2} \in G_{i+1}/G_{i+2}$, so the map $f_i : G_i/G_{i+1} \to G_{i+1}/G_{i+2}$ is not injective. It follows that G is not uniform.

For the converse, suppose that G is not uniform. Then for some i, the epimorphism $f_i : G_i/G_{i+1} \to G_{i+1}/G_{i+2}$ is not injective, so there exists $x \in G_i \setminus G_{i+1}$ such that $x^p \in G_{i+2}$. Put $x_2 = x$, and suppose that for some $n \geq 2$ we have found $x_2, \dots, x_n$ satisfying $x_j^p \in G_{i+j}$ and $x_j \equiv x_{j-1} \pmod{G_{i+j-2}}$ for $2 < j \leq n$. There exists $z \in G_{i+n-1}$ such that $z^p = x_n^p$; put $x_{n+1} = z^{-1} x_n$. Then $x_{n+1} \equiv x_n \pmod{G_{i+n-1}}$. Also $x_{n+1}^p \in G_{i+n+1}$: for if p is odd we have

$$x_{n+1}^p = (z^{-1} x_n)^p \equiv z^{-p} x_n^p [x_n, z^{-1}]^{p(p-1)/2} \equiv 1 \pmod{G_{i+n+1}}$$

since $[G_{i+n-1}, G, G][G_{i+n-1}, G]^p \leq G_{i+n+1}$; while if $p = 2$ we have

$$x_{n+1}^p = z^{-2}[z^{-1}, x_n^{-1}] x_n^2 \equiv z^{-2} x_n^2 = 1 \pmod{G_{i+n+1}},$$

because $[G_{i+n-1}, G] \leq G_{i+n-1}^4 = G_{i+n+1}$ since G_{i+n-1} p.e. G.

Thus the sequence $x_2, \dots, x_n, \dots$ can be constructed recursively; it is a Cauchy sequence and therefore converges to an element $x_\infty \in G$, say. Then $x_\infty \equiv x \not\equiv 1 \pmod{G_{i+1}}$; and $x_\infty^p \equiv x_n^p \equiv 1 \pmod{G_{i+n-1}}$ for all n, so $x_\infty^p = 1$. Thus G is not torsion-free.

We conclude this section with an important definition; it depends on

4.6 Lemma. *If A and B are open uniform subgroups of some pro-p group then $\mathrm{d}(A) = \mathrm{d}(B)$.*

Proof For large enough i we have $P_i(A) \leq A \cap B \leq B$. Then $\mathrm{d}(B) = \mathrm{d}(P_i(A)) = \mathrm{d}(A)$ by Proposition 4.4.

4.7 Definition Let G be a pro-p group of finite rank. The *dimension* of G is

$$\dim(G) = \mathrm{d}(H)$$

where H is any open uniform subgroup of G.

The preceding lemma shows that this is unambiguous; the following result shows that it is reasonable:

4.8 Theorem *Let G be a pro-p group of finite rank and N a closed normal subgroup of G. Then*

$$\dim(G) = \dim(N) + \dim(G/N).$$

(This makes sense because both N and G/N have finite rank, by Exercise 3.1.)

Proof Suppose first of all that each of the groups G, N and G/N is uniform. Lemma 3.4 shows that $\Phi(G) = \{g^p \mid g \in G\}$ and $\Phi(N) = \{g^p \mid g \in N\}$. Since G/N is torsion-free, it follows that $\Phi(G) \cap N = \Phi(N)$. As $\Phi(G/N) = \Phi(G)N/N$, this gives

$$\mathrm{d}(G) = \dim_{\mathbb{F}_p}(G/\Phi(G))$$
$$= \dim_{\mathbb{F}_p}(N/\Phi(N)) + \dim_{\mathbb{F}_p}((G/N)/\Phi(G/N)) = \mathrm{d}(N) + \mathrm{d}(G/N),$$

and the theorem follows in this case.

To prove the theorem in general, it will now suffice to find a uniform open subgroup H in G such that $H \cap N$ and $H/(H \cap N)$ are also both uniform. Say $\mathrm{rk}(G) = r$, and put $G_0 = V(G, r)$ if p is odd, $G_0 = V(G; r)^2$ if $p = 2$ (see §3.1). Proposition 3.9 shows that every open normal subgroup of G contained in G_0 is powerful. Let k as in Theorem 4.2 be sufficiently large so that $P_k(G_0) = G_1$, say, is uniform; it follows by Theorem 4.5 that then every open normal subgroup of G contained in G_1 is uniform. Similarly, N has a characteristic open subgroup N_1 such that every open normal subgroup of N contained in N_1 is uniform. Finally, for the same reason, $G_1/(G_1 \cap N_1)$ has a uniform open normal subgroup $H/(G_1 \cap N_1)$.

Now $H/(G_1 \cap N_1)$ is torsion-free while $N/(G_1 \cap N_1)$ is finite; consequently $H \cap N = G_1 \cap N_1$. Thus $H/(H \cap N)$ is uniform, and $H \cap N$ is uniform by the choice of N_1. As H is uniform, by the choice of G_1, this completes the proof.

In Chapter 8, it will emerge that $\dim(G)$ is indeed the dimension of G as a p-adic analytic group. In this chapter, we shall content ourselves with setting up two (in general distinct) homeomorphisms between a uniform group of dimension d and the space $\mathbb{Z}_p^d$.

4.2 Multiplicative structure

In this section and the next, G will denote a uniform pro-p group, with $\mathrm{d}(G) = d$. For each n we write $G_n = P_n(G)$.

Let $\{a_1, \ldots, a_d\}$ be a topological generating set for G. Then Proposition 3.7 shows that $G = \overline{\langle a_1 \rangle} \ldots \overline{\langle a_d \rangle}$; thus each element $a \in G$ can be expressed in the form

$$a = a_1^{\lambda_1} \ldots a_d^{\lambda_d} \tag{1}$$

with $\lambda_1, \ldots, \lambda_d \in \mathbb{Z}_p$. Now fix a positive integer k, and consider the finite group G/G_{k+1}. This has order exactly p^{kd}, and is equal to the product of its d cyclic subgroups $\langle a_1 G_{k+1} \rangle, \ldots, \langle a_d G_{k+1} \rangle$, each of which has order at most p^k. It follows that these cyclic groups each have order exactly p^k, and hence that each element of G/G_{k+1} can be expressed in the form $a_1^{e_1} \ldots a_d^{e_d} G_{k+1}$ where the integers $e_1, \ldots e_d$ are uniquely determined modulo p^k. This implies now that, in the expression (1), the p-adic integers $\lambda_1, \ldots, \lambda_d$ are uniquely determined modulo p^k. As this holds for every k, it follows that $\lambda_1, \ldots, \lambda_d$ are uniquely determined p-adic integers. Thus we obtain a bijective mapping $\theta : G \to \mathbb{Z}_p$, where $a\theta = (\lambda_1, \ldots, \lambda_d)$. Let $\psi : \mathbb{Z}_p^d \to G$ be the inverse bijection, so $(\lambda_1, \ldots, \lambda_d)\psi = a_1^{\lambda_1} \ldots a_d^{\lambda_d}$. Since multiplication in G is continuous, Corollary 1.21 shows that ψ is continuous; and as G and $\mathbb{Z}_p^d$ are both compact Hausdorff spaces it follows that ψ is a homeomorphism.

Thus we have

4.9 Theorem. *Let G be a uniform pro-p group and $\{a_1, \ldots, a_d\}$ a topological generating set for G, where $d = \mathrm{d}(G)$. Then the mapping*

$$(\lambda_1, \ldots, \lambda_d) \mapsto a_1^{\lambda_1} \ldots a_d^{\lambda_d}$$

from $\mathbb{Z}_p^d$ to G is a homeomorphism.

4.3 Additive structure

The homeomorphism $\theta : G \to \mathbb{Z}_p^d$ defined in the last section is a 'system of co ordinates of the second kind', in the language of Lie groups. We shall examine its analytic properties in Chapter 8. Its algebraic properties, however, are not particularly good. We turn now to the construction of another co-ordinate system; it takes more work to set up, but the effort will be amply justified. We keep the notation of the previous section.

4.10 Lemma *Let $n \in \mathbb{N}$. The mapping $x \mapsto x^{p^n}$ is a homeomorphism from G onto G_{n+1}. For each k and m, it restricts to a bijection $G_k \to G_{k+n}$ and induces a bijection $G_k/G_{k+m} \to G_{n+k}/G_{n+k+m}$.*

Proof Write $f(x) = x^{p^n}$. Theorem 3.6 shows that $f(G_k) = G_{n+k}$ and that $f(G_{k+m}) = G_{n+k+m}$; and repeated applications of Theorem 3.6 (iv) show that if $x \equiv y \pmod{G_{k+m}}$ then $f(x) \equiv f(y) \pmod{G_{n+k+m}}$. Thus f induces a surjection from G_k/G_{k+m} onto G_{n+k}/G_{n+k+m}. Since G is uniform, $|G_k/G_{k+m}| = |G_{n+k}/G_{n+k+m}|$; hence this surjection is a bijection. It follows that if $x, y \in G_k$ and $f(x) = f(y)$ then $x \equiv y \pmod{G_{k+m}}$ for all m. As $\bigcap_m G_{k+m} = 1$, this shows that $f|_{G_k}$ is injective. Finally, it is clear that f is continuous: as everything in sight is a compact Hausdorff space, $f|_{G_k}$ is a homeomorphism $G_k \to G_{k+n}$. The first statement is the case $k = 1$.

Lemma 4.10 shows that each element $x \in G_{n+1}$ has a unique p^nth root in G, which we shall denote $x^{p^{-n}}$. We can use this bijection between G and G_{n+1} to transfer the group operation from G_{n+1} to G, thereby defining a new group structure on G. For $x, y \in G$ we define

$$x +_n y = (x^{p^n} y^{p^n})^{p^{-n}} \; ;$$

thus the map $x \mapsto x^{p^{-n}}$ becomes an isomorphism from G_{n+1} onto the group $(G, +_n)$.

4.11 Lemma. *If $n > 1$, $x, y \in G$, and $u, v \in G_n$ then*

$$xu +_n yv \equiv x +_n y \equiv x +_{n-1} y \pmod{G_n},$$

and for all $m > n$

$$x +_m y \equiv x +_n y \pmod{G_{n+1}}.$$

Proof The final claim follows from the preceding one by induction on $m - n$ (where n is playing the role of $n - 1$). Now recall from Proposition 1.16 that $[G_n, G_n] \le G_{2n}$. This implies that

$$x^{p^n} y^{p^n} \equiv (x^{p^{n-1}} y^{p^{n-1}})^p \pmod{G_{2n}}$$
$$= (x +_{n-1} y)^{p^n}.$$

Extracting p^nth roots, and using Lemma 4.10, we infer that

$$x +_n y = (x^{p^n} y^{p^n})^{p^{-n}} \equiv x +_{n-1} y \pmod{G_n}.$$

Since $(xu)^{p^n} \equiv x^{p^n} \pmod{G_{2n}}$ and $(yv)^{p^n} \equiv y^{p^n} \pmod{G_{2n}}$, again by Lemma 4.10, the same argument also gives

$$xu +_n yv \equiv x +_n y \pmod{G_n}.$$

Thus for a given pair (x, y), the sequence $(x +_n y)$ is a Cauchy sequence, and we can make the following definition:

4.12 Definition For $x, y \in G$,

$$x + y = \lim_{n \to \infty} x +_n y .$$

It is clear from Lemma 4.11 that

$$x + y \equiv x +_n y \pmod{G_{n+1}} \tag{2}$$

and that if $u, v \in G_n$ then

$$xu + yv \equiv x + y \pmod{G_n} . \tag{3}$$

4.13 Proposition *The set G with the operation $+$ is an abelian group, with identity element 1 and inversion given by $x \mapsto x^{-1}$.*

Proof For each n, $x +_n 1 = x$ and $x +_n x^{-1} = 1$. Hence $x + 1 = x$ and $x + x^{-1} = 1$. To verify the associative law, let $x, y, z \in G$ and let $n > 1$. Then $x + y = (x +_n y)u$ for some $u \in G_{n+1}$, so

$$(x + y) + z \equiv (x +_n y) + z \pmod{G_{n+1}}$$
$$\equiv (x +_n y) +_n z \pmod{G_{n+1}} .$$

Similarly, $x + (y + z) \equiv x +_n (y +_n z) \pmod{G_{n+1}}$. Since the operation $+_n$ is associative, it follows that

$$(x + y) + z \equiv x + (y + z) \pmod{G_{n+1}} ;$$

and as n was arbitrary the associativity of $+$ follows. Finally, we verify that $+$ is commutative. Since $[x^{p^n}, y^{p^n}] \in [G_{n+1}, G_{n+1}] \leq G_{2n+2}$, we have $x^{p^n} y^{p^n} \equiv y^{p^n} x^{p^n} \pmod{G_{2n+2}}$. Extracting p^nth roots and using Theorem 4.9 we see that $x +_n y \equiv y +_n x \pmod{G_{n+2}}$. Thus $x + y \equiv y + x \pmod{G_{n+1}}$, and as this holds for each n the result follows.

Henceforth, we shall use 'additive' notation for the group operations in $(G, +)$, so we write 0 for 1, $-x$ for x^{-1}, $x - y$ for $x + (-y)$ and mx for $x + \ldots + x$ (m times) if m is positive, mx for $|m| \cdot (-x)$ if m is negative. Our next task is to elucidate the structure of this additive group.

4.14 Lemma (i) *If $xy = yx$ then $x + y = xy$.*
(ii) *For each integer m, $mx = x^m$.*

(iii) *For each $n \geq 1$, $p^{n-1}G = G_n$.*

(iv) *If $x, y \in G_n$ then $x + y \equiv xy \pmod{G_{n+1}}$.*

Proof (i) is immediate from the definition. (ii) follows, for positive m, by induction on m, and then for negative m from the fact that $-x = x^{-1}$. Part (iii) then follows from Theorem 3.6 (iii). To prove (iv), recall that the mapping $x \mapsto x^{p^n}$ induces a homomorphism from G_n/G_{n+1} into G_{2n}/G_{2n+1} (this follows from Theorem 2.7 (iv)). Thus for $x, y \in G_n$ we have

$$(xy)^{p^n} \equiv x^{p^n} y^{p^n} \pmod{G_{2n+1}}.$$

Extracting p^nth roots and using Lemma 4.10 we get

$$xy \equiv x +_n y \pmod{G_{n+1}},$$

and (iv) follows by (2).

4.15 Corollary. *For each n, G_n is an additive subgroup of G; the additive cosets of G_n in G are the same as the multiplicative cosets of G_n in G. Also the identity map $G_n/G_{n+1} \to G_n/G_{n+1}$ is an isomorphism of the additive group G_n/G_{n+1} onto the multiplicative group G_n/G_{n+1}, and the index of G_n in the additive group $(G, +)$ is equal to $|G : G_n|$.*

Proof Lemma 4.14 (iii) shows that $G_n = p^{n-1}G$ is an additive subgroup of $(G, +)$. Now let $a \in G$, $u \in G_n$. Then

$$a + u = a + 1 \cdot u \equiv a + 1 = a \pmod{G_n}$$

by (3), so $a + u \in aG_n$. Thus $a + G_n \subseteq aG_n$. On the other hand,

$$au - a = au + (-a) \equiv a + (-a) = 0 \pmod{G_n},$$

by (3) again, so $au - a \in G_n$ and $au \in a + G_n$. Thus $aG_n \subseteq a + G_n$. This shows that the additive cosets modulo G_n are the same as the multiplicative cosets. (Hence the notation G/G_n is unambiguous: we get the same quotient *set* whether we consider the additive group $(G, +)$ or the multiplicative group G). In particular, the index $|G : G_n|$ is the same when calculated in either group. Finally, Lemma 4.14 (iv) shows that the restriction of the identity map on G/G_{n+1} to G_n/G_{n+1} is an isomorphism between the additive and the multiplicative structures.

4.16 Proposition *With the original topology of G, $(G, +)$ is a uniform pro-p group of dimension $d = d(G)$. Moreover any set of topological generators for G is a set of topological generators for $(G, +)$.*

Proof G is a compact Hausdorff space. We already know that the map $x \mapsto -x = x^{-1}$ is continuous; and (3) shows that the map $(x, y) \mapsto x+y$, from $G \times G$ to G, is continuous. So $(G, +)$ is a topological group. We also know that the family $\{G_n\}_{n \in \mathbb{N}}$ is a base for the neighbourhoods of $0 = 1$ in G; as each G_n is a subgroup of p-power index in the additive group $(G, +)$, by Corollary 4.15, it follows that $(G, +)$ is a pro-p group. It is powerful by virtue of being abelian; and for the same reason, the subgroups $p^{n-1}G = G_n$ are exactly the terms of the lower p-series of $(G, +)$. Since $|G_n : G_{n+1}| = p^d$ for all n, it follows that $(G, +)$ is uniform of dimension d. Finally, suppose X is a topological generating set for G. Then $G/G_2 = \langle X \rangle G_2/G_2$ (as multiplicative groups). But Lemma 4.14 (iv) shows that the additive group G/G_2 is identical to the multiplicative group, so we have $(G, +)/G_2 = \langle X \rangle_+ + G_2/G_2$, where $\langle X \rangle_+$ denotes the additive subgroup generated by X. Since $G_2 = pG$ is the Frattini subgroup of $(G, +)$ it follows that X is a topological generating set for $(G, +)$.

As $(G, +)$ is a pro-p group, it admits a natural action by $\mathbb{Z}_p$ (see §1.3). Since $(G, +)$ is abelian, Proposition 1.26 shows that this makes it into a $\mathbb{Z}_p$-module. We are now ready for the main result, which gives the structure of this module:

4.17 Theorem *Let G be a uniform pro-p group of dimension d, and let $\{a_1, \ldots, a_d\}$ be a topological generating set for G. Then, with the operations defined above, $(G, +)$ is a free $\mathbb{Z}_p$-module on the basis $\{a_1, \ldots, a_d\}$.*

Proof By Proposition 4.16, the set $\{a_1, \ldots, a_d\}$ generates the uniform pro-p group $(G, +)$ topologically, and $\mathrm{d}(G, +) = d$. We now apply Theorem 4.9 to this group: in additive notation, this shows that each element of $(G, +)$ has a unique expression in the form

$$a = \lambda_1 a_1 + \cdots + \lambda_d a_d$$

with $\lambda_1, \ldots, \lambda_d \in \mathbb{Z}_p$ (note that for $x \in G$ and $\lambda \in \mathbb{Z}_p$, we have $x^\lambda = \lambda x$, as follows from Lemma 4.14 (ii) on taking limits). But this is exactly the statement of the theorem.

From the point of view of applications, the following corollaries are particularly useful.

4.18 Corollary. *Let G be a uniform pro-p group of dimension d. Then the action of $\operatorname{Aut}(G)$ on G is $\mathbb{Z}_p$-linear with respect to the $\mathbb{Z}_p$-module structure on $(G, +)$. Hence $\operatorname{Aut}(G)$ may be identified with a subgroup of $\operatorname{GL}_d(\mathbb{Z}_p)$.*

Proof Let α be an automorphism of G. Then α is continuous, by Corollary 1.22. For each n, α respects the operation of taking p^nth roots: this follows from the uniqueness of p^nth roots. It follows that α respects the operation $+_n$, for each n; and hence, by continuity, that α respects the operation $+$. Similarly, since the operation of $\mathbb{Z}_p$ is defined by taking limits of integral powers in G, it follows by continuity that α respects the operation of $\mathbb{Z}_p$.

4.19 Corollary. *Let G be a pro-p group of finite rank and dimension d. Then there is an exact sequence*

$$1 \to \mathbb{Z}_p^e \to G \to \operatorname{GL}_d(\mathbb{Z}_p) \times F$$

for some $e \le d$ and some finite p-group F.

Proof G has a uniform open normal subgroup H. Put $A = Z(H)$. Then A is closed in H, so A is a torsion-free abelian pro-p group of rank at most $\operatorname{rk}(H) = d$. It follows that $A \cong \mathbb{Z}_p^e$ for some $e \le d$ (e.g. by Theorem 4.8 and Theorem 4.9: but this can easily be seen more directly). Now for each $g \in G$ let g^* denote the automorphism of H induced by conjugation with g. We have a homomorphism $\theta : G \to \operatorname{Aut}(H) \times G/H$ given by $g\theta = (g^*, gH)$. Clearly $\ker \theta = A$. The result follows by Corollary 4.18.

4.4 On the structure of powerful pro-p groups

4.20 Theorem *Let G be a finitely generated powerful pro-p group. Then the elements of finite order in G form a characteristic subgroup T of G. Also T is a powerful finite p-group and G/T is uniform.*

The proof depends on the following lemma (which generalises a well-known result due to Minkowski). As before, we write $G_i = P_i(G)$ for each i.

4.21 Lemma *Let G be a uniform pro-p group, and for each i let Γ_i be the group of all automorphisms of G which induce the identity on G/G_i. Then Γ_2 is torsion-free if p is odd, Γ_3 is torsion-free if $p = 2$.*

Proof The map $x \mapsto x^{p^{j-1}}$ induces bijections from G/G_2 onto G_j/G_{j+1} and from G/G_3 onto G_j/G_{j+2}, by Lemma 4.10; it follows that Γ_2 acts trivially on G_j/G_{j+1} and that Γ_3 acts trivially on G_j/G_{j+2}, for each j. It follows (by stability group theory) that for each $i > 2$, Γ_2/Γ_i is a finite p-group. Since $\bigcap_{i=2}^{\infty} \Gamma_i = 1$, any element of finite order in Γ_2 must have p-power order. Thus it will suffice to show that Γ_2 (or Γ_3 if $p = 2$) has no elements of order p.

Now let γ satisfy $\gamma^p = 1$, where $\gamma \in \Gamma_2$ (and $\gamma \in \Gamma_3$ if $p = 2$) and suppose that for some i we have $[G, \gamma] \subseteq G_i$. Then for $g \in G$ we have

$$1 = [g, \gamma^p]$$
$$= [g, \gamma][g, \gamma]^{\gamma} \dots [g, \gamma]^{\gamma^{p-1}}$$
$$\equiv [g, \gamma]^p [g, \gamma, \gamma^{p(p-1)/2}] \pmod{G_{i+2}},$$

because $[g, \gamma, \gamma^n] \in G_{i+1}$ for each n, and G_{i+2} contains both $[G_{i+1}, G]$ and $[G_{i+1}, \langle \gamma \rangle]$.

If p is odd then $\gamma^{p(p-1)/2} = 1$, while if $p = 2$ and $\gamma \in \Gamma_3$ then $[g, \gamma, \gamma] \in [G_i, \Gamma_3] \subseteq G_{i+2}$. In either case, therefore, we may infer that $[g, \gamma]^p \in G_{i+2}$, and hence, by Lemma 4.10, that $[g, \gamma] \in G_{i+1}$. Thus $[G, \gamma] \subseteq G_{i+1}$.

It follows by induction that $[G, \gamma] \subseteq \bigcap_{i=1}^{\infty} G_i = 1$. So $\gamma = 1$ as required.

Proof of Theorem 4.20   Now G is finitely generated and powerful. For some m, G_m is uniform. Put $K = C_G(G_m)$. Then $Z(K) \geq G_m \cap K$, so $K/Z(K)$ is a finite p-group. Hence K is nilpotent, and so the elements of finite order in K form a subgroup T, say. Then $T \lhd G$ and K/T is torsion-free.

Now G/K acts faithfully on the uniform group G_m, by conjugation. Since G is powerful, G_m p.e. G, so G/K acts trivially on $G_m/P_i(G_m)$ where $i - 2$ if p is odd, $i = 3$ if $p = 2$. Thus Lemma 4.21 shows that G/K is torsion-free.

Therefore all elements of finite order in G lie in T. Theorem 4.8 shows that G/T is uniform. Finally, T is a finite p-group since $T \cap G_m = 1$; and T is powerful: for $[T, T] \leq T \cap [G, G]$, and if p is odd we have

$$T \cap [G, G] \leq T \cap \overline{G^p} = T \cap \{g^p \mid g \in G\} \subseteq T^p,$$

while if $p = 2$ we have $T \cap [G, G] \subseteq T^4$ in a similar way. This concludes the proof.

4.22 Corollary. *Let G be a finitely generated powerful pro-p group of dimension d. Then* $\mathrm{Aut}(G)$ *is isomorphic to a subgroup of* $\mathrm{GL}_d(\mathbb{Z}_p) \times F$ *for some finite group F. In particular* $\mathrm{Aut}(G)$ *is isomorphic to a linear group over* $\mathbb{Z}_p$.

Proof Let G_m and T be as in the above proof. Then $G_m \cap T = 1$ so $\mathrm{Aut}(G)$ is isomorphic to a subgroup of $\mathrm{Aut}(G/T) \times \mathrm{Aut}(G/G_m)$. The result follows by Corollary 4.18.

It follows from 4.3 and 4.5 that a pro-p group of finite rank has a torsion-free open normal subgroup, and hence that its finite subgroups have bounded order. On the other hand, 'Sylow's theorem' (Exercise 1.12) shows that if G is any profinite group containing an open pro-p subgroup, then the maximal q-subgroups of G are all conjugate, when q is a prime distinct from p. Here we establish a sort of common generalisation of these facts:

4.23 Theorem *Let G be a profinite group having an open normal subgroup that is a pro-p group of finite rank. Then the finite subgroups of G lie in finitely many conjugacy classes.*

This applies, for example, to the group $G = \mathrm{GL}_d(\mathbb{Z}_p)$ (see Theorem 5.2 in the next chapter).

Now let G be as in the theorem. By Corollary 4.3, G has an open normal subgroup H that is a uniform pro-p group. As G/H is finite, there are only finitely many possibilities for the subgroup HF as F ranges over all the finite subgroups of G. So the theorem will follow once we have proved

4.24 Proposition *Let H be a uniform pro-p group and F a finite group acting by automorphisms on H. Then the complements to H in the semi-direct product $H \rtimes F$ lie in finitely many conjugacy classes.*

The proof is an exercise in 'non-abelian cohomology'. For the time being, let H and F denote arbitrary groups, with a given action of F on H.

Definition A 1-*cocycle* $F \to H$ is a mapping $\delta : F \to H$ such that

$$\delta(xy) = \delta(x)^y \delta(y) \qquad \text{for all } x, y \in F.$$

Two 1-cocycles $\beta, \gamma : F \to H$ are *equivalent*, written $\beta \sim \gamma$, if there exists $v \in H$ such that

$$\gamma(x) = v^x \beta(x) v^{-1} \qquad \text{for all } x \in F.$$

It is clear that 'equivalence' is indeed an equivalence relation. A 1-cocycle is *trivial* if it is equivalent to the constant map $x \mapsto 1$. Note that every 1-cocycle δ satisfies $\delta(1) = 1$.

Now suppose that B is another complement to H in the group $H \rtimes F$. We may define a map $\delta_B : F \to H$ by putting $\delta_B(x)$ equal to the unique element $h \in H$ such that $xh \in B$. Then for $x, y \in F$ we have

$$(xy).\left(\delta_B(x)^y \delta_B(y)\right) = x\delta_B(x) \cdot y\delta_B(y) \in B,$$

so $\delta_B(x)^y \delta_B(y) = \delta_B(xy)$, and we see that δ_B is a 1-cocycle. It is equally easy to check that for any 1-cocycle $\delta : F \to H$, the set

$$B = \{x\delta(x) \mid x \in F\}$$

is a subgroup of $H \rtimes F$, that it complements H, and that it satisfies $\delta_B = \delta$. Thus the complements we are interested in correspond bijectively with 1-cocycles from F to H; moreover, we have

4.25 Lemma *Let B be a complement to H in $H \rtimes F$. Then B is conjugate to F if and only if the 1-cocycle δ_B is trivial.*

Proof Let $v \in H$. Then for all $x \in F$ we have

$$x \cdot v^x \delta_B(x) v^{-1} = v(x\delta_B(x))v^{-1} \in vBv^{-1}. \tag{4}$$

If $v^x \delta_B(x) v^{-1} = 1$ for each x, it follows that $F \leq vBv^{-1}$; as both groups complement H in $H \rtimes F$ this implies that $F = vBv^{-1}$. Thus if $\delta_B \sim 1$ then B is conjugate to F. Conversely, if B is conjugate to F then there exists $v \in H$ such that $vBv^{-1} = F$, and (4) then implies that $v^x \delta_B(x) v^{-1} = 1$ for each $x \in A$, showing that $\delta_B \sim 1$.

4.26 Lemma *Suppose that $|F| = mp^e$ where $p \nmid m$, and that H is an abelian p-group. If $\delta : F \to H$ is a 1-cocycle then the mapping δ^{p^e} given by $\delta^{p^e}(x) = \delta(x)^{p^e}$ for $x \in F$ is a trivial 1-cocycle from F to H.*

Proof Given that H is abelian, it is clear that δ^{p^e} is a 1-cocycle. Write H additively, and put

$$b = -\sum_{x \in F} \delta(x).$$

As H is a p-group, there exists $a \in H$ such that $ma = b$. Then for each $y \in F$ we have

$$m(a^y - a) = \sum_{x \in F} \delta(x) - \sum_{x \in F} \delta(x)^y$$

$$= \sum_{x \in F} \delta(x) - \sum_{x \in F} (\delta(xy) - \delta(y))$$

$$= |F|\delta(y) = mp^e\delta(y).$$

The result follows on cancelling m.

From now on, H is supposed to be a uniform pro-p group. We write $H_n = P_n(H)$ and $\pi_n : H \to H/H_n$ for the natural epimorphism. Let us prove

4.27 Proposition *Let F be a finite group acting on the uniform pro-p group H, and let $\delta : F \to H$ be a 1-cocycle. Then δ is trivial if and only if $\pi_{3e+1} \circ \delta : F \to H/H_{3e+1}$ is trivial, where p^e is the exact power of p dividing $|F|$.*

Proof Put $k = 3e + 1$. Suppose we can show that $\pi_n \circ \delta$ is trivial for all $n \geq k$. Then for each $n \geq k$ the set $V_n = \{v \in H \mid \delta(x) \equiv v^x v^{-1} (\text{mod } H_n)$ for all $x \in F\}$ is non-empty, and it is a union of cosets of H_n, so it is closed in H. Since, clearly, $V_n \supseteq V_{n+1}$ for all n, it follows that $\bigcap_{n \geq k} V_n \neq \varnothing$ (as H is compact). Any element v lying in this intersection then satisfies $\delta(x) = v^x v^{-1}$ for all $x \in F$, and we conclude that $\delta \sim 1$.

Now fix $n \geq k$, and suppose we have shown that $\pi_n \circ \delta \sim 1$. We want to deduce that $\pi_{n+1} \circ \delta \sim 1$. Replacing δ by an equivalent cocycle, we may assume that in fact $\pi_n \circ \delta$ is the constant mapping 1, i.e. that $\delta(F) \subseteq H_n$.

Now put $M = H_{n-e}/H_{2n-2e}$. Then M is abelian, and $\pi_{2n-2e} \circ \delta = \bar{\delta}$, say, is a 1-cocycle from F into $H_n/H_{2n-2e} = M^{p^e}$. By the preceding lemma, there exists $a \in M^{p^e}$ such that $\bar{\delta}(x)^{p^e} = a^x a^{-1}$ for all $x \in F$. Say $a = v^{p^e} H_{2n-2e}$ where $v \in H_{n-e}$; then

$$\delta(x)^{p^e} \equiv (v^x v^{-1})^{p^e} \bmod H_{2n-2e}$$

for each x, because M is abelian. Using Lemma 4.10 we infer that

$$\delta(x) \equiv v^x v^{-1} (\text{mod } H_{2n-3e}),$$

for all $x \in F$. But $n \geq k = 3e + 1$, so $2n - 3e \geq n + 1$ and we conclude that $\pi_{n+1} \circ \delta \sim 1$. The result follows by induction.

It is now easy to deduce Proposition 4.24. Put $H \rtimes F = G$, let p^e be the exact power of p that divides $|F|$, and write π for the natural epimorphism $G \to G/H_{3e+1}$. Then $\pi(G)$ is finite, so we may choose finitely many complements $F_1, \ldots, F_r$ to H in G so that for any complement B to H, we have $\pi(B) = \pi(F_i)$ for some $i \leq r$.

Now fix such a value of i, and suppose that $\pi(B) = \pi(F_i)$. Replacing F by F_i in Lemma 4.25, we see that B is conjugate to F_i if and only if the cocycle $\delta_B : F_i \to H$ is trivial. On the other hand, we have $\pi \circ \delta_B = 1$ because $\pi(B) = \pi(F_i)$, and it follows by Proposition 4.27 that δ_B is trivial. Thus B is conjugate to F_i.

Hence every complement to H in G is conjugate to one of $F_1, \ldots, F_r$, and the proof is complete.

4.5 The Lie algebra

The procedure of passing from the uniform pro-p group G to the $\mathbb{Z}_p$-module $(G, +)$, described in Section 4.3, involves 'forgetting' a lot of information about the structure of G, since all free $\mathbb{Z}_p$-modules of a given rank are isomorphic. More information can be saved by defining yet another operation. We keep the notation of Sections 4.2 and 4.3, so G denotes a uniform pro-p group of rank d, and $G_i = P_i(G)$ for each i.

Definition For $x, y \in G$ and $n \in \mathbb{N}$,

$$(x, y)_n = [x^{p^n}, y^{p^n}]^{p^{-2n}}.$$

This makes sense because $[x^{p^n}, y^{p^n}] \in [G_{n+1}, G_{n+1}] \leq G_{2n+2}$.

4.28 Lemma *If* $n > 1$, $x, y \in G$ *and* $u, v \in G_n$, *then*

$$(xu, yv)_n \equiv (x, y)_n \equiv (x, y)_{n-1} \pmod{G_{n+1}},$$

and for all $m > n$

$$(x, y)_m \equiv (x, y)_n \pmod{G_{n+2}}.$$

Proof Noting that $[G_{2n}, G_{n+1}] \leq G_{3n+1}$, and using Lemma 4.10, we see (as in the proof of Lemma 4.11) that

$$(xu, yv)_n \equiv (x, y)_n \pmod{G_{n+1}}.$$

Now if $a \in G_i$ and $b \in G_j$ then $[a^p, b] \equiv [a, b]^p \pmod{G_{2i+j}}$ and $[a, b^p] \equiv [a, b]^p \pmod{G_{i+2j}}$. Taking $a = x^{p^n}$ and $b = y^{p^{n-1}}$ this gives

$$[x^{p^n}, y^{p^n}] \equiv [x^{p^n}, y^{p^{n-1}}]^p \pmod{G_{3n+1}}.$$

Taking $a = x^{p^{n-1}}$ and $b = y^{p^{n-1}}$ gives

$$[x^{p^n}, y^{p^{n-1}}] \equiv [x^{p^{n-1}}, y^{p^{n-1}}]^p \pmod{G_{3n}}.$$

Therefore (in view of Lemma 4.10)

$$[x^{p^n}, y^{p^n}] \equiv [x^{p^{n-1}}, y^{p^{n-1}}]^{p^2} \pmod{G_{3n+1}}$$
$$= (x, y)_{n-1}^{p^{2n}}.$$

Extracting p^{2n}th roots and using Lemma 4.10 again we obtain

$$(x, y)_n \equiv (x, y)_{n-1} \pmod{G_{n+1}}.$$

The final claim follows by induction on $m - n$ (on replacing n by $n+1$).

Thus for given $x, y \in G$, $((x, y)_n)$ is a Cauchy sequence, and we can make the following definition:

4.29 Definition For $x, y \in G$,

$$(x, y) = \lim_{n \to \infty} (x, y)_n.$$

4.30 Theorem. *With the operation (,), the $\mathbb{Z}_p$-module $(G, +)$ becomes a Lie algebra over $\mathbb{Z}_p$.*

A direct proof of this theorem, in the same spirit as the proof of Proposition 4.13, is outlined in the exercises, below. Theorem 4.30 will be proved by a different route in Chapter 7. There, we define an injective mapping log from G into a certain associative $\mathbb{Q}_p$-algebra $\widehat{A}$, and establish that it has the following properties:

$$\log(x + y) = \log x + \log y$$
$$\log(\lambda x) = \lambda \log x$$
$$\log(x, y) = (\log x)(\log y) - (\log y)(\log x).$$

From this, it follows at once that the operation $(\ ,\ )$ is $\mathbb{Z}_p$-bilinear, anti-commutative, and satisfies the Jacobi identity, which is precisely the claim of Theorem 4.30.

A further consequence is worth mentioning at this point, though a full discussion must be postponed until Chapter 9. For x and $y \in G$, we have

$$\log(xy) = \Phi(\log x, \log y)$$

where $\Phi(U, V)$ is the *Campbell–Hausdorff formula*: this is a certain infinite sum of terms in the Lie algebra generated by U and V, which under suitable conditions converges to an element of this Lie algebra. This shows that xy can be recovered from the Lie algebra structure of $(G, +)$ alone, and hence that, unlike the $\mathbb{Z}_p$-module $(G, +)$, this Lie algebra captures all the information in the given pro-p group G.

If a subgroup or quotient of G is itself a uniform group, we would like to know that the additive and Lie structures induced from G are the 'right' ones; this is assured by

4.31 Proposition *Let H be a uniform closed subgroup of G, and let $N \lhd_c G$ be such that G/N is uniform. Then*

(i) *the inclusion map $H \to G$ is a monomorphism of Lie algebras $(H, +, (\ ,\ )) \to (G, +, (\ ,\ ))$; in particular, H is a subalgebra of the Lie algebra $(G, +, (\ ,\ ))$;*

(ii) *N is uniform;*

(iii) *N is an ideal in the $\mathbb{Z}_p$-Lie algebra $(G, +, (\ ,\ ))$; and the additive cosets of N in G are the same as the multiplicative cosets, so $(G/N, +, (\ ,\ )) = (G, +, (\ ,\ ))/(N, +, (\ ,\ ))$; moreover, the natural epimorphism $* : G \to G/N$ is an epimorphism of $\mathbb{Z}_p$-Lie algebras from $(G, +, (\ ,\ ))$ onto $(G/N, +, (\ ,\ ))$.*

Proof (i) follows directly from the definitions, as the topology on H is just the subspace topology induced from G.

If $x \in G$ and $x^{p^n} \in N$ then $x \in N$, since G/N is torsion-free by Theorem 4.5. As G^p consists of pth powers it follows that $G^p \cap N = N^p$, whence N/N^p is abelian, showing (if p is odd) that N is powerful, and hence uniform by Theorem 4.5; the same argument applies if $p = 2$, considering instead N/N^4. This establishes (ii).

Now let $a, b \in G$ and put $c_n = a +_n b$. Then $(c_n^*)^{p^n} = a^{*p^n} b^{*p^n}$, so in G/N we have $a^* +_n b^* = c_n^*$. It follows by continuity that $a^* + b^* = \lim_{n\to\infty} c_n^* = (\lim_{n\to\infty} c_n)^* = (a + b)^*$. A similar argument shows that

$*$ respects the bracket operation; and it is easy to see that $*$ respects the operation of $\mathbb{Z}_p$. Thus $*$ is a Lie algebra homomorphism as claimed.

Since N is the kernel of $*$ it follows that N is an ideal in $(G, +, (\ ,\ ))$. Finally, for $a, b \in G$ we have

$$a + N = b + N \Leftrightarrow a - b \in N \Leftrightarrow (a - b)^* = 0 \Leftrightarrow a^* = b^* \Leftrightarrow aN = bN,$$

showing that $(G, +)/(N, +) = G/N$. This concludes the proof of (iii). $\blacksquare$

In Section 7.2 we shall see that, conversely, suitable Lie subalgebras of $(G, +)$ are in fact subgroups of G.

4.6 Generators and relations

We show in this section that every pro-p group of finite rank has a finite presentation by 'generators and relations', in the sense appropriate to pro-p groups. (The results of this section will not be needed elsewhere in the book.)

We saw in Exercise 1.20 that for each finite set X there exists a 'free pro-p group on X', namely the pro-p completion of the (ordinary) free group on X. To simplify notation, we now denote this 'free pro-p group' by $F(X)$. For any subset R of $F(X)$, we write

$$\langle X; R \rangle = F(X)/\overline{\langle R^{F(X)} \rangle}$$

where $\langle R^{F(X)} \rangle$ denotes the normal closure of R in $F(X)$. We say that $\langle X; R \rangle$ is a *presentation* for a pro-p group G if G is isomorphic to $\langle X; R \rangle$. The presentation is *finite* if R as well as X is finite, and in this case G is said to be *finitely presented*.

Suppose now that G is a pro-p group and X is a finite topological generating set for G. The identity map on X induces an epimorphism $\pi : F(X) \rightarrow G$. For any subset R of $\ker \pi$, we say that 'the relations $R = 1$ hold in G'; and then π induces an epimorphism π^* from the group $\langle X; R \rangle$ onto G. If R satisfies the condition: $\overline{\langle R^{F(X)} \rangle} = \ker \pi$, then π^* is an isomorphism and $\langle X; R \rangle$ is a presentation for G.

4.32 Proposition *Let G be a uniform pro-p group of dimension d, and let $\{x_1, \ldots, x_d\}$ be a topological generating set for G. Then G has a presentation $\langle x_1, \ldots, x_d; R \rangle$ where*

$$R = \{ [x_i, x_j] x_1^{\lambda_1(i,j)} \ldots x_d^{\lambda_d(i,j)} \mid 1 \le i < j \le d \},$$

and, for each m, i and j, $\lambda_m(i,j) \in p\mathbb{Z}_p$ if p is odd, $\lambda_m(i,j) \in 4\mathbb{Z}_2$ if $p = 2$.

Proof Since G is powerful, $[x_j, x_i] \in \overline{G^p}$ if p is odd, $[x_j, x_i] \in \overline{G^4}$ if $p = 2$. It follows from Theorem 3.6 and Proposition 3.7 that $[x_j, x_i] = \prod_{m=1}^{d} x_m^{\lambda_m(i,j)}$ where each $\lambda_m(i,j)$ lies in $p\mathbb{Z}_p$ (if p is odd) or in $4\mathbb{Z}_2$ (if $p = 2$). Thus the relations $R = 1$ hold in G.

Let $H = \langle x_1, \dots, x_d; R \rangle$ and put $H_i = P_i(H)$, $G_i = P_i(G)$ for each i. Let $\pi^* : H \to G$ denote the natural epimorphism. Then $H_i \pi^* \leq G_i$ for each i. Now the relations $R = 1$ which hold in H imply that H is powerful. It follows by Theorem 3.6 that $|H_i/H_{i+1}| \leq |H/H_2| \leq p^d$ for each i, whence

$$|H/H_{n+1}| \leq p^{nd} = |G/G_{n+1}|$$

for each n. Since π^* is an epimorphism and $H_{n+1}\pi^* \leq G_{n+1}$, this shows that $\ker \pi^* \leq H_{n+1}$, for each n. But $\bigcap_{n=1}^{\infty} H_{n+1} = 1$, so π^* is injective. Thus G is isomorphic to H, as required.

To deal with pro-p groups of finite rank in general, we use the following elementary fact:

4.33 Lemma. *Let G be a pro-p group and K an open normal subgroup of G. If K is finitely presented then so is G.*

Proof Arguing by induction on the index $|G : K|$, we reduce to the case where $|G : K| = p$. Thus $G = K\langle y \rangle$ where $y^p \in K$. Suppose $\langle X; R \rangle$ is a finite presentation of K, coming from an epimorphism $\pi : F(X) \to K$. There exists $v \in F(X)$ such that $y^p = v\pi$, and for each $x \in X$ there exists $w_x \in F(X)$ such that $(x\pi)^y = w_x\pi$. Now take $Y = X \cup \{t\}$, where $t \notin X$, and define an epimorphism $\overline{\pi} : F(Y) \to G$ by $x\overline{\pi} = x\pi$ for $x \in X$ and $t\overline{\pi} = y$. Put

$$S = \{t^p v^{-1}\} \cup \{x^t w_x^{-1} \mid x \in X\} \subseteq F(Y),$$

let N be the closure in $F(Y)$ of $\langle (R \cup S)^{F(Y)} \rangle$, and put $M = \ker \overline{\pi}$. Clearly $N \leq M$. The relations $S = 1$ which hold in $F(Y)/N$ show that $F(X)N \lhd F(Y)$ and that $|F(Y) : F(X)N| \leq p$ (we are identifying $F(X)$ with its image in $F(Y)$, as we may by Exercise 1.20 (ii)). Since $F(Y)\overline{\pi} = G$ and $(F(X)N)\overline{\pi} = K$ it follows that $M \leq F(X)N$, whence $M = (M \cap F(X))N$. But

$$M \cap F(X) = \ker \pi = \overline{\langle R^{F(X)} \rangle} \leq N \, ;$$

consequently $M = N$. It follows that $\langle Y; R \cup S \rangle$ is a presentation for G.

In view of Corollary 4.3, the following is now immediate:

4.34 Theorem *Every pro-p group of finite rank is finitely presented.*

One may ask what is the minimal number of relations required to present a given pro-p group. For a finitely generated pro-p group G, define t(G) by

$$\mathrm{t}(G) = \inf\{|R| \mid G \text{ has a presentation } \langle X; R \rangle \text{ with } |X| = \mathrm{d}(G)\} .$$

Let us prove

4.35 Theorem. *Let G be a finitely generated powerful pro-p group. Put $d = \dim(G)$ and $r = \mathrm{d}(G) = \mathrm{rk}(G)$. Then*

$$\binom{r}{2} \le \mathrm{t}(G) \le \binom{r}{2} + r - d .$$

In particular if G is uniform then $\mathrm{t}(G) = \binom{d}{2}$.

Proof Consider the second inequality first. Write $G_i = P_i(G)$ for each i, and define d_i by $p^{d_i} = |G_i : G_{i+1}|$. Thus

$$r = d_1 \ge d_2 \ge \ldots \ge d_k = d$$

for some k, where G_k is uniform. It follows from Theorem 3.6 (see Exercise 3.12) that G has a generating set $\{x_1, \ldots, x_r\}$ such that, for each i, G_i is generated by $\{x_1^{p^{i-1}}, \ldots, x_{d_i}^{p^{i-1}}\}$.

Now whenever $d_i \ge m > d_{i+1}$, we have

$$x_m^{p^i} = x_1^{\mu_1(m)} \ldots x_{d_{i+1}}^{\mu_{d_{i+1}}(m)} = \mathbf{x}^{\mu(m)}, \quad \text{say} ,$$

with $\mu_n(m) \in p^i \mathbb{Z}_p$ for each n. Also, as in the proof of Proposition 4.32, whenever $1 \le i < j \le r$ there is a relation

$$[x_j, x_i] = x_1^{\lambda_1(i,j)} \ldots x_r^{\lambda_r(i,j)} = \mathbf{x}^{\lambda(i,j)} , \quad \text{say} ,$$

with $\lambda_n(i,j) \in p\mathbb{Z}_p$ (or $4\mathbb{Z}_2$ if $p = 2$), for each n. Let H be the group

$$\left\langle x_1, \ldots, x_r; [x_i, x_j]\mathbf{x}^{\lambda(i,j)} \; (1 \le i < j \le r), \right.$$

$$\left. x_m^{-p^i}\mathbf{x}^{\mu(m)} \; (d_i \ge m > d_{i+1}, \; 1 \le i < k) \right\rangle .$$

Writing $H_i = P_i(H)$, we see that H is powerful and that $|H/H_n| \le$

$|G/G_n|$ for each n. Arguing as before, we may conclude that $H \cong G$. As there are $\binom{r}{2}$ relators of the form $[x_i, x_j]\mathbf{x}^\lambda$, and $r - d$ relators of the form $x_m^{-p^i}\mathbf{x}^\mu$, it follows that $\mathrm{t}(G) \le \binom{r}{2} + r - d$.

For the other inequality, suppose we have a presentation $\langle X; R \rangle$ for G with $|X| = r$ and $|R| = t$. It is easy to see that then G/G_2 has the presentation $\langle X; R, x_1^p, \dots, x_r^p \rangle$ where $X = \{x_1, \dots, x_r\}$. But G/G_2 is an elementary abelian p-group of rank r, and so $\mathrm{t}(G/G_2) = r(r+1)/2$ (see Exercise 10). Hence

$$t + r \ge r(r+1)/2$$

and so $t \ge \binom{r}{2}$, as claimed.

An upper bound for $\mathrm{t}(G)$, when G is *any* pro-p group of finite rank, is given in Exercise 11. It is also possible to give a lower bound in that case, but this depends on more sophisticated methods (the 'Golod–Shafarevich inequality': see Interlude D).

It must be emphasised that all of these results refer to presentations of groups *within the category of pro-p groups*: thus if, for example, G is a finite p-group, then G has a presentation *as a p-group* on $\mathrm{d}(G)$ generators and $\mathrm{t}(G)$ relations, but the number of relations needed to define G as an abstract group may conceivably be greater than $\mathrm{t}(G)$. Whether this is actually the case, for any finite p-group, is at present unknown.

Notes

Both the name and the theory of uniformly powerful pro-p groups developed in this chapter are new. However, this class of groups is contained in the class of 'groupes p-saturables' defined in [L], Chapter III, 3.1.6. Lazard calls a group p-saturable if it possesses a filtration satisfying certain conditions, and has 'finite rank' (in a sense different from ours). When $p \ge 3$, the natural filtration by the lower p-series will do; when $p = 2$, this is not adequate, and one has to use the filtration induced by the *norm* on the group algebra $\mathbb{Z}_2[G]$, constructed in Exercise 7.10.

It is not clear to us (despite the claim made in the Introduction to the first edition, and in Interlude C) whether every p-saturable group is in fact a uniform pro-p group. It follows fairly directly from the definitions that if G is p-saturable with respect to an *integer-valued* filtration then G is uniform if $p \ge 3$, and G^2 is uniform if $p = 2$.

Given that uniform pro-p groups are p-saturable, the existence of both the 'multiplicative' and the 'additive' systems of co-ordinates is due to Lazard, as is the Lie algebra of §4.5 ([L] Chapter III, Ex. 2.1.10).

Theorem 4.23 is from Segal (1999). Most of the other results are new.

Exercises

1. Let a, b be elements of a group and suppose that $c = [b, a^{-1}]$ commutes with both a and b. Show that for $h \in \mathbb{N}$,

$$a^h b^h = (ab)^h c^{h(h-1)/2}.$$

Now let G be a pro-p group and put $G_i = P_i(G)$. Let $k, n \in \mathbb{N}$ and put $m = k + \min\{k, n - \epsilon\}$, where $\epsilon = 0$ if p is odd, $\epsilon = 1$ if $p = 2$. Show that the mapping $x \mapsto x^{p^n}$ induces a homomorphism from G_k into G_{k+n}/G_{k+m}.

In Exercises 2–7, G denotes a uniform pro-p group as in §4.5.

2. (i) Prove that $(x, y) = -(y, x)$.
(ii) Show that $(x, y) \in G_{2\epsilon}$ where $\epsilon = 1$ if $p \neq 2$, $\epsilon = 2$ if $p = 2$.
[*Hint*: when $p = 2$, see Exercise 2.4(ii).]

3. Let $\lambda \in \mathbb{Z}_p$ and suppose that $\lambda \equiv a \pmod{p^n}$ where $a \in \mathbb{Z}$. Show that $(\lambda x, y) \equiv (ax, y) \pmod{G_{n+1}}$.

4. Verify the following identities:

(i) $[x^{p^n}, z^{p^n}]^{p^{-n}} \cdot [y^{p^n}, z^{p^n}]^{p^{-n}} = ((x, z)_n +_n (y, z)_n)^{p^n}$.
(ii) $[x^{p^n} y^{p^n}, z^{p^n}]^{p^{-n}} = (x +_n y, z)_n^{p^n}$.
(iii) $[x^{p^n} y^{p^n}, z^{p^n}]^{p^{-n}} \equiv [x^{p^n}, z^{p^n}]^{p^{-n}} [y^{p^n}, z^{p^n}]^{p^{-n}} \pmod{G_{2n+3}}$.

Hence show that $(x, z)_n +_n (y, z)_n \equiv (x +_n y, z)_n \pmod{G_{n+3}}$, and deduce that $(x, z) + (y, z) = (x + y, z)$.
[*Hint for (iii)*: Put $a = [x^{p^n}, z^{p^n}]^{p^{-n}}$, $b = [y^{p^n}, z^{p^n}]^{p^{-n}}$ and note that $(ab)^{p^n} \equiv a^{p^n} b^{p^n} \pmod{G_{3n+3}}$, by Exercise 1.]

5. Deduce from Exercises 2–4 that $(\ ,\ )$ is bilinear with respect to the $\mathbb{Z}_p$-module structure on $(G, +)$.

6. Let U and V be normal subgroups of a group, such that $[U, V, V, V] = 1$. Show that for $a \in U$, $b \in V$ and $h \in \mathbb{N}$,

$$[a, b^h] = [a, b]^h [a, b, b^{h(h-1)/2}].$$

Now let $\ell, k, n \in \mathbb{N}$ and put $m = k + \min\{k, n - \epsilon\}$ where $\epsilon = 0$ if p is odd, $\epsilon = 1$ if $p = 2$. Show that if $a \in G_\ell$ and $b \in G_k$ then

$$[a, b^{p^n}] \equiv [a, b]^{p^n} \pmod{G_{k+\ell+m}}.$$

7. Prove the following:

(i) $((x, -y)_n, z)_n^{p^{4n}} \equiv [x^{p^n}, y^{-p^n}, z^{p^n}]^{p^n} \pmod{G_{5n+1}}$.
[*Hint:* Take $a = [x^{p^n}, y^{-p^n}]$ and $b = z^{p^n}$ in Exercise 6.]

(ii) $[x^{p^n}, y^{-p^n}, z^{p^n}] \equiv [x^{p^n}, y^{-p^n}, z^{p^n}]^{y^{p^n}} \pmod{G_{4n+4}}$.

(iii) $((x, y), z) + ((y, z), x) + ((z, x), y) \equiv -(A + B + C) \pmod{G_{n+2}}$,
where $A = ((x, -y)_n, z)_n$, $B = ((y, -z)_n, x)_n$, $C = ((z, -x)_n, y)_n$.

(iv) $(A +_n B)^{p^{3n}} \equiv A^{p^{3n}} B^{p^{3n}} \pmod{G_{4n+5}}$, provided $n \geq 2$.
[*Hint:* Use Exercise 1.]

(v) $A^{p^{3n}} B^{p^{3n}} \equiv C^{-p^{3n}} \pmod{G_{4n+1}}$.
[*Hint:* Use (i), (ii) and the Hall–Witt identity.]

(vi) Deduce that $((x, y), z) + ((y, z), x) + ((z, x), y) = 0$.

8. Let G be a pro-p group of finite rank, and let K be the *FC-centre* of G (defined by $K = \{x \in G \mid |G : C_G(x)| \text{ is finite}\}$). (i) Show that $K = C_G(H)$ for every open uniform subgroup of H of G. (ii) Deduce that G/K is isomorphic to a subgroup of $\mathrm{GL}_d(\mathbb{Z}_p)$ where $d = \dim(G)$. (iii) Deduce that if G is also torsion-free, then there is an exact sequence

$$1 \to \mathbb{Z}_p^e \to G \to \mathrm{GL}_d(\mathbb{Z}_p)$$

where $e \leq d$.

9. Let G be a uniform pro-p group and suppose that G has an abelian open normal subgroup. Show that $G/Z(G)$ is finite; deduce that in fact $G \cong \mathbb{Z}_p^d$ for some d.
[*Hint:* Use Exercise 8(i); note that for any group G, if $G/Z(G)$ is finite then $[G, G]$ is finite (Schur's theorem, see Hall (1969), §8).]

10. ([L] III, 3.1.8) Let G be a pro-p group of finite rank. Prove that

$$\dim(G) = \lim_{k \to \infty} \frac{\log_p |G : G^{p^k}|}{k}.$$

[*Hint:* examine the proof of Theorem 3.16.]

11. Let G be a pro-p group of rank r and dimension d. Show that $t(G) \leq \binom{r}{2} + r - d + r(r + 1)(2 + \lambda(r))$.
[*Hint:* G has a powerful normal subgroup H of index at most $r(2 + \lambda(r))$. Apply Theorem 4.35 to H, and use the proof of Lemma 4.33 to estimate the number of additional relations required to define G.]

12. (i) Prove that for each $d \geq 1$ there exists a group $H = H(d)$ with $d(H) = d$ having a subgroup $Z \leq Z(H)$ such that $H/Z \cong C_p^{(d)}$ and $Z \cong C_p^{(d(d+1)/2)}$.

[*Hint*: Given $H(d) = H$, construct $H(d+1)$ as follows. Put $A = H/Z$ and define $\alpha : H \times A \to H \times A$ by $(h, a)\alpha = (h, \bar{h}.a)$ where $\bar{h} = hZ$; then put $H(d+1) = (H \times A) \rtimes C_{p^2}$, where the generator of C_{p^2} acts like α on $H \times A$.]

(ii) Let F be the free group on $d \geq 2$ generators, put $F_2 = [F, F]F^p$ and $F_3 = [F_2, F]F_2^p$. Deduce from (i) that $F_2/F_3 \cong C_p^{(d(d+1)/2)}$.

(iii) Show that if $A = C_p^{(d)}$ then $t(A) = d(d+1)/2$.

[*Hint*: consider epimorphisms from F onto A, and use (ii).]

13. Let G be a powerful non-abelian pro-p group of rank 2 and dimension 2.

(i) Show that G is uniform, G contains a unique normal procyclic subgroup N such that G/N is procyclic, and that N has a complement in G.

(ii) Deduce that G has a presentation $\langle x, y; [x, y]x^{-p^e} \rangle$ for some uniquely determined positive integer e.

[*Hint for (ii)*: Suppose that x generates N and Nz generates G/N; then $x^z = x^\lambda$ for some $\lambda \in \mathbb{Z}_p$. Show that $\lambda = 1 + p^e\mu$ where $e \geq 1$ and $p \nmid \mu$; then show that $\lambda^\tau = 1 + p^e$ for some p-adic unit τ, and take $y = z^\tau$. (The existence of τ follows from Theorem 5.2, applied to the group $GL_1(\mathbb{Z}_p)$).]

The next three exercises are from Barnea and Shalev (1997); they prove a little more, by a different method.

14. Let G be a pro-p group of finite rank and put $G_n = \overline{G^{p^{n-1}}}$ for each $n \geq 1$. Let $H \leq_c G$.

(i) Show that for all sufficiently large n, both G_n and $H \cap G_n$ are uniform.

[*Hint*: look at the proof of Theorem 4.8.]

(ii) Suppose that in fact both G and H are uniform. Show that for some non-negative integer c,

$$H \cap G_n = (H \cap G_c)^{p^{n-c}} \text{ for all } n \geq c.$$

[*Hint*: Use 4.14(iii) and 4.17 to translate this into a statement about $\mathbb{Z}_p$-modules.]

(iii) Deduce that in the general case there exists $b \geq 0$ such that $H \cap G_n \leq H^{p^{n-b}}$ for all $n \geq b$.

15. Let G, G_n and H be as in Exercise 14. Show that

$$\lim_{n \to \infty} \frac{\log |HG_n : G_n|}{\log |G : G_n|} = \frac{\dim H}{\dim G}.$$

[*Hint:* Exercises 10 and 14.]

The limit on the left here is the *Hausdorff dimension* of H in G, relative to the metric on G given by

$$d(a, b) = \inf\{|G : G_n|^{-1} \mid ab^{-1} \in G_n\}.$$

16. Let G be a finitely generated pro-p group. Prove that G *has finite rank if and only if G contains no infinite closed subgroup of Hausdorff dimension zero.*

[*Hint:* For 'only if' use Exercise 15. 'If' depends on a deep theorem of Zel'manov (1992): *every finitely generated periodic pro-p group is finite.* It follows that if G has infinite rank then G contains an infinite procyclic subgroup H; use Theorem 3.16 to find the Hausdorff dimension of H.]

5

Automorphism groups

The main result to be established in this chapter is that the automorphism group of a pro-p group of finite rank is itself virtually a pro-p group of finite rank. An important special case is the automorphism group of $\mathbb{Z}_p^d$, namely the group $\mathrm{GL}_d(\mathbb{Z}_p)$ of all invertible $d \times d$ matrices over $\mathbb{Z}_p$, and we begin by discussing this group in some detail.

5.1 The group $\mathrm{GL}_d(\mathbb{Z}_p)$

We fix a positive integer d, and write $\Gamma = \mathrm{GL}_d(\mathbb{Z}_p)$. Then Γ is a Hausdorff topological group, with the p-adic topology (the subspace topology induced from the natural topology on the space $\mathrm{M}_d(\mathbb{Z}_p)$ of all $d \times d$ matrices over $\mathbb{Z}_p$). In fact Γ is both closed and open as a subspace of $\mathrm{M}_d(\mathbb{Z}_p)$: for if $a \in \mathrm{M}_d(\mathbb{Z}_p)$ then $a \in \Gamma$ if and only if $\det a \not\equiv 0 \pmod{p}$, so every matrix $b \equiv a \pmod{p}$ satisfies $b \in \Gamma \Leftrightarrow a \in \Gamma$; this shows that Γ is the union of at most p^{d^2} additive cosets of $p\mathrm{M}_d(\mathbb{Z}_p)$. Hence Γ is compact. A base for the neighbourhoods of 1 in Γ is given by the 'congruence subgroups'

$$\Gamma_i = \{\gamma \in \Gamma \mid \gamma \equiv 1_d \pmod{p^i}\},$$

for $i \geq 0$. Since $\Gamma/\Gamma_i \cong \mathrm{GL}_d(\mathbb{Z}/p^i\mathbb{Z})$ for $i \geq 1$, we have

$$|\Gamma : \Gamma_1| = (p^d - 1)(p^d - p)\ldots(p^d - p^{d-1})$$
$$|\Gamma_1 : \Gamma_i| = p^{d^2(i-1)} \quad \text{for } i \geq 1.$$

It follows that Γ is profinite and that Γ_1 is a pro-p group.

Once we have defined analytic groups, it will be clear that Γ is a compact p-adic analytic group; a fundamental property of such groups is that they contain an open powerful finitely generated pro-p subgroup,

87

and we now verify this directly for $\Gamma = GL_d(\mathbb{Z}_p)$. The key step is the following simple variation on Hensel's Lemma (we keep the notation just introduced):

5.1 Lemma. *If p is odd and $n \geq 2$, or $p = 2$ and $n \geq 3$, then every element of Γ_n is the pth power of an element of Γ_{n-1}.*

Proof The claim is that for any $a \in M_d(\mathbb{Z}_p)$ we can solve

$$1 + p^n a = (1 + p^{n-1}x)^p \qquad (*)$$

with $x \in M_d(\mathbb{Z}_p)$. The solution is by successive approximation. To begin with, $(1 + p^{n-1}a)^p \equiv 1 + p^n a \pmod{p^{n+1}}$ (provided n lies in the stated range). Put $x_1 = a$, and suppose inductively that we have found, for some $r \geq 1$, a matrix x_r, commuting with a, such that $(1 + p^{n-1}x_r)^p \equiv 1 + p^n a \pmod{p^{n+r}}$. Say

$$(1 + p^{n-1}x_r)^p = 1 + p^n a + p^{n+r}c \,.$$

Now put

$$z = (1 + p^{n-1}x_r)^{-(p-1)}c \,,$$

and let $x_{r+1} = x_r - p^r z$; note that x_r commutes with c, hence with z, and that x_{r+1} commutes with a. A direct calculation shows that

$$(1 + p^{n-1}x_{r+1})^p \equiv 1 + p^n a \pmod{p^{n+r+1}} \,.$$

Thus we obtain a convergent sequence (x_r) in $M_d(\mathbb{Z}_p)$, whose limit x satisfies $(*)$.

5.2 Theorem. *For each i let $\Gamma_i = \{\gamma \in GL_d(\mathbb{Z}_p) \mid \gamma \equiv 1_d \pmod{p^i}\}$. Put $G = \Gamma_1$ if p is odd, $G = \Gamma_2$ if $p = 2$. Then G is a uniform pro-p group and $\dim(G) = \mathrm{rk}(G) = \mathrm{d}(G) = d^2$. Also $P_i(G) = \Gamma_{i+\epsilon}$ for all i, where $\epsilon = 0$ if $p \neq 2$, $\epsilon = 1$ if $p = 2$.*

Proof We have $P_1(G) = G = \Gamma_{1+\epsilon}$ by definition. Suppose $r \geq 1$ and $P_r(G) = \Gamma_{r+\epsilon}$. Then a trivial calculation shows that $P_r(G)^p[P_r(G), G] \leq \Gamma_{r+1+\epsilon}$, and Lemma 5.1 shows that $\Gamma_{r+1+\epsilon} \leq \Gamma_{r+\epsilon}^p = P_r(G)^p$. Since $\Gamma_{r+1+\epsilon}$ is a closed subgroup of G it follows that $P_{r+1}(G) = \Gamma_{r+1+\epsilon}$. Thus by induction we have $P_i(G) = \Gamma_{i+\epsilon}$ for all i, and on the way we have shown that $P_{i+1}(G) = P_i(G)^p$ for all i. Taking $i = 1$, we see that G is powerful (when $p = 2$, note that $[\Gamma_2, \Gamma_2] \leq \Gamma_4 \leq \Gamma_2^4$); and since $P_2(G) = \Gamma_{2+\epsilon}$ is open in G, Theorem 1.14 shows that G is finitely generated. Since $|\Gamma_i : \Gamma_{i+1}| = p^{d^2}$ is constant for all $i \geq 1$, G is uniform. Finally, since $G/\Phi(G) = \Gamma_{1+\epsilon}/\Gamma_{2+\epsilon}$ is elementary abelian of order p^{d^2}, it needs exactly d^2 generators, whence $\dim(G) = \mathrm{rk}(G) = \mathrm{d}(G) = d^2$.

Note that we have established that $\mathrm{GL}_d(\mathbb{Z}_p)$ has finite rank, without any serious matrix calculations, by appealing to the theory of powerful groups from Chapter 3. Once we have shown that every pro-p group of finite rank has a faithful linear representation over $\mathbb{Z}_p$, this will provide yet another characterisation for the pro-p groups of finite rank.

5.2 The automorphism group of a profinite group

Now we move on to consider automorphism groups in general. For a profinite group G, $\mathrm{Aut}(G)$ denotes the group of all topological automorphisms of G (recall that if G is finitely generated and pro-p, this means *all* automorphisms, by Corollary 1.21). $\mathrm{Aut}(G)$ has a natural topology, the 'congruence topology': a base for the neighbourhoods of 1 is given by the subgroups

$$\Gamma(N) = \{\gamma \in \mathrm{Aut}(G) \mid [G, \gamma] \subseteq N\}$$

as N runs over the open normal subgroups of G (generalising the case of $\mathrm{GL}_d(\mathbb{Z}_p) = \mathrm{Aut}(\mathbb{Z}_p^d)$, discussed above). This makes $\mathrm{Aut}(G)$ into a Hausdorff topological group (see Exercise 1). Note that for $\gamma \in \mathrm{Aut}(G)$, we have $\gamma \in \Gamma(N)$ if and only if γ fixes N and induces the trivial automorphism on G/N.

In general, $\mathrm{Aut}(G)$ will not itself be a profinite group (see Exercise 3). However, we have

5.3 Theorem. *If G is a finitely generated profinite group then* $\mathrm{Aut}(G)$ *is a profinite group.*

Proof Let us write $\Gamma = \mathrm{Aut}(G)$. Proposition 1.6 shows that every open normal subgroup of N of G contains an open topologically characteristic subgroup, N_0 say. Then $\Gamma(N)$ contains $\Gamma(N_0)$. Since $\Gamma(N_0)$ is the kernel of the induced action of Γ on the finite group G/N_0, the index $|\Gamma : \Gamma(N_0)|$ is finite. Thus in Γ there is a base for the neighbourhoods of 1 consisting of subgroups of finite index, so to show that Γ is profinite it will suffice to verify that it is *complete* (see Exercise 1.1). Denote the set of all open topologically characteristic subgroups in G by $\mathcal{C}$, and let $(\gamma_N)_{N \in \mathcal{C}}$ be a Cauchy net in Γ, with respect to the neighbourhood base at 1

$\{\Gamma(N) \mid N \in \mathcal{C}\}$: thus for each $N \in \mathcal{C}$ there exists $M(N) \in \mathcal{C}$ such that $\gamma_S^{-1}\gamma_T \in \Gamma(N)$ whenever $S, T \in \mathcal{C}$ and $S \leq M(N), T \leq M(N)$. We have to show that the net $(\gamma_N)_{N \in \mathcal{C}}$ converges in Γ.

For $g \in G$ and N, S, T as above we have

$$g^{\gamma_S} \equiv (g^{\gamma_S})^{\gamma_S^{-1}\gamma_T} = g^{\gamma_T} \quad (\text{mod } N).$$

Thus the family $(g^{\gamma_N})_{N \in \mathcal{C}}$ is a Cauchy net in G, which therefore converges to an element of G which we shall denote g^{γ}. This defines a map $\gamma : G \to G$; to complete the proof, we will show that $\gamma \in \Gamma$ and that the net (γ_N) converges to γ.

For each $N \in \mathcal{C}$ there exists $N_1 \in \mathcal{C}$ such that $g^{\gamma_S} \equiv g^{\gamma}$ (mod N) for all $S \in \mathcal{C}$ with $S \leq N_1$. Here N_1 depends on g, but if $g, h \in G$ we can find $S \in \mathcal{C}$ such that $g^{\gamma_S} \equiv g^{\gamma}$ (mod N), $h^{\gamma_S} \equiv h^{\gamma}$ (mod N) and $(gh^{-1})^{\gamma_S} \equiv (gh^{-1})^{\gamma}$ (mod N). Since γ_S induces an automorphism on G/N, it follows that γ induces an automorphism on G/N. As $\mathcal{C}$ is a base for the neighbourhoods of 1 in G, this implies that γ is an automorphism of G, and also that γ is continuous. Thus $\gamma \in \Gamma$.

Now let $N \in \mathcal{C}$, and let X be a set of coset representatives for G/N (so X is finite). As above, we can find $N_1 \in \mathcal{C}$ such that whenever $S \in \mathcal{C}$ and $S \leq N_1$, $t^{\gamma_S} \equiv t^{\gamma}$ (mod N) for each $t \in X$. Then for $g = xt \in G$, with $x \in N$ and $t \in X$, we have

$$g^{\gamma_S\gamma^{-1}} = x^{\gamma_S\gamma^{-1}} t^{\gamma_S\gamma^{-1}} \equiv xt = g \quad (\text{mod } N),$$

showing that $\gamma_S\gamma^{-1} \in \Gamma(N)$. Thus the net (γ_N) converges to γ as required.

The following elementary lemma will be needed in the next section:

5.4 Lemma. *Let G be a profinite group and H an open normal subgroup of G with centre Z. Let Ξ be a subgroup of $\mathrm{Aut}(G)$ which acts trivially on H and induces the trivial action on G/H. Then there is an injective continuous homomorphism $\theta : \Xi \to Z^{(m)}$ where $m = \mathrm{d}(G/H)$. If Ξ is compact then θ is a topological isomorphism of Ξ onto a closed subgroup of $Z^{(m)}$.*

Proof Fix a generating set $\{x_1 H, \ldots, x_m H\}$ for G/H, and for $\gamma \in \Xi$ define

$$\gamma\theta = ([x_1, \gamma], \ldots, [x_m, \gamma]) \in G^{(m)}.$$

Then $\gamma\theta$ clearly determines the action of γ on G, so the map θ is injective. For any $x \in G$ and $h \in H$,

$$[x, \gamma]^h = x^{-h}(x^h)^\gamma = [x^h, \gamma] = [[h, x^{-1}]x, \gamma] = [x, \gamma]$$

since $h^\gamma = h$ and $[[h, x^{-1}], \gamma] \in [H, \gamma] = 1$. Hence $[G, \gamma] \subseteq Z$, and so $\gamma\theta \in Z^{(m)}$. If $\alpha, \beta \in \Xi$ and $x \in G$ then

$$[x, \alpha\beta] = [x, \beta][x, \alpha]^\beta = [x, \alpha][x, \beta],$$

since Z is abelian and β acts trivially on Z. Thus θ is a homomorphism of Ξ into $Z^{(m)}$.

To show that θ is continuous, it now suffices to show that for any neighbourhood U of 1 in $Z^{(m)}$ there exists $N \lhd_o G$ such that $(\Xi \cap \Gamma(N))\theta \subseteq U$. Now there exists $N \lhd_o G$ such that $(N \cap Z)^{(m)} \subseteq U$: this subgroup N clearly has the required property.

If Ξ is compact then so is $\Xi\theta$, and the final claim follows from the standard property of compact Hausdorff spaces.

5.3 Automorphism groups of pro-p groups

5.5 Proposition. *Let G be a finitely generated pro-p group. Then $\Gamma(\Phi(G))$ is a pro-p group.*

Proof Write $G_n = P_n(G)$ for each n. The family (G_n) is a base for the neighbourhoods of 1 in G, and consists of characteristic subgroups; also $G_2 = \Phi(G)$ (see §1.2). It follows that the subgroups $\Gamma(G_n)$, $n \geq 2$, are normal in $\Gamma(G_2) = \Gamma(\Phi(G))$, and form a base for the neighbourhoods of 1 in $\Gamma(G_2)$. Hence, by Theorem 5.3, it will suffice to show that, for each $n \geq 2$, $\Gamma(G_2)/\Gamma(G_n)$ is a p-group. Now $\Gamma(G_2)/\Gamma(G_n)$ acts faithfully on the finite p-group G/G_n, and induces the trivial action on $G/G_2 = (G/G_n)/\Phi(G/G_n)$: this gives the result (see Exercise 4).

A profinite group G is said to have a property $\mathcal{P}$ *virtually* if G has an open normal subgroup H such that H has $\mathcal{P}$.

5.6 Theorem. *Let G be a finitely generated profinite group. If G is virtually a pro-p group then $\mathrm{Aut}(G)$ is also virtually a pro-p group.*

Proof By Proposition 1.6, G has a topologically characteristic open pro-p subgroup H, and H is finitely generated (by Proposition 1.7). Then $\Phi(H)$ is open and topologically characteristic in G. Let $\Delta = \Gamma(\Phi(H))$

be the kernel of the action of $\mathrm{Aut}(G)$ on $G/\Phi(H)$. Then $\Delta \lhd_o \mathrm{Aut}(G)$, and we claim that Δ is a pro-p group.

Let $\pi : \Delta \to \mathrm{Aut}(H)$ be the restriction map, and put $\Xi = \ker \pi$. It is easy to see that π is continuous, since for any $N \lhd_o H$ we have $(\Delta \cap \Gamma(N))\pi \subseteq \Gamma_H(N)$, where $\Gamma_H(N) = \{\gamma \in \mathrm{Aut}(H) \mid [H, \gamma] \subseteq N\}$. Hence Ξ is a closed normal subgroup of Δ, so Ξ is compact and Lemma 5.4 shows that Ξ is topologically isomorphic to a closed subgroup of $Z(H)^{(m)}$, where $m = \mathrm{d}(G/H)$ is finite. Thus Ξ is a pro-p group.

Since Δ is compact and π is continuous, Δ/Ξ is topologically isomorphic to the closed subgroup $\Delta\pi$ of $\mathrm{Aut}(H)$. But $\Delta\pi \le \Gamma_H(\Phi(H))$, so $\Delta\pi$ is a pro-p group by Proposition 5.5. Putting the results together we see (by Proposition 1.11) that Δ is a pro-p group as claimed.

In a similar way, we can now establish the final result:

5.7 Theorem. *Let G be a profinite group. If G is virtually a pro-p group of finite rank, then so is* $\mathrm{Aut}(G)$.

Proof Certainly G is finitely generated. We can choose H as in the preceding proof so that, in addition, H is a uniform pro-p group, of dimension d, say (see §4.1). We keep the notation of the preceding proof. Then Ξ is topologically isomorphic to a closed subgroup of $Z(H)^{(m)}$; as this is now a pro-p group of finite rank, it follows that Ξ has finite rank.

By Corollary 4.18, there is an injective homomorphism $\mu : \mathrm{Aut}(H) \to \mathrm{GL}_d(\mathbb{Z}_p)$, and Lemma 4.14(iii) shows that μ is continuous. Thus the composition $\pi\mu : \Delta \to \mathrm{GL}_d(\mathbb{Z}_p)$ is continuous and maps Δ onto a closed subgroup of $\mathrm{GL}_d(\mathbb{Z}_p)$. Theorem 5.2 now shows that $\Delta\pi\mu$ has finite rank. As in the preceding proof, we see that Δ/Ξ is topologically isomorphic to $\Delta\pi\mu$.

Putting the results together shows that Δ has finite rank. We already know that Δ is a pro-p group and that $\Delta \lhd_o \mathrm{Aut}(G)$, so the proof is complete.

5.4 Finite extensions

Theorem 5.7 has a nice application to the following question: given a pro-p group G of finite rank and a finite group F, how many different extensions can there be of G by F, or of F by G ? We begin with

5.8 Theorem *There are only finitely many isomorphism types of extensions of a pro-p group G of finite rank by a given finite group F.*

By an *extension* of G by F we mean a group E containing G as a normal subgroup such that $E/G \cong F$. As we shall see, all such extensions can be embedded into a sort of 'universal container'; this is constructed as follows (henceforth, G denotes a pro-p group of finite rank):

5.9 Lemma *Let $K = G \rtimes \mathrm{Aut}(G)$ and put $W = K \wr F$. Then for every extension E of G by F there exists an injective homomorphism $\theta_E : E \to W$ such that $(\theta_E)_{|G} = \Delta_{|G}$ where $\Delta : K \to W$ denotes the 'diagonal' embedding of K into the base group of W.*

Recall that the *wreath product* $K \wr F$ is the semidirect product $K^F \rtimes F$, where F acts on the *base group* K^F by permuting the factors; identifying K with one of these factors, we think of the base group as being the direct product $\prod_{f \in F} K^f$. For $x \in K$ we then have

$$\Delta(x) = \prod_{f \in F} x^f.$$

Proof Let E be an extension of G by F; thus we have an epimorphism $\pi : E \to F$ with kernel G. Choose a transversal T to the cosets of G in E, with $1 \in T$, and for $t \in T$ denote by $t^\natural$ the automorphism induced on G by conjugation with t. We define a map $\psi : E \to W$ by putting

$$\psi(x) = \prod_{t \in T} (\overline{tx}^{-1} xt)^{\pi(t)} \cdot \pi(x),$$

where $\overline{tx} \in T$ denotes the representative of the coset Gtx. A direct calculation shows that ψ is a homomorphism.

Now if $x \in G$ then $\overline{tx} = t$ for each $t \in T$, so

$$\psi(x) = \prod_{t \in T} (x^{t^\natural})^{\pi(t)}$$
$$= \tau^{-1} \Delta(x) \tau,$$

where $\tau = \prod_{t \in T} (t^\natural)^{\pi(t)}$, an element of the base group of W. Thus the map $\theta_E : E \to W$ given by

$$\theta_E(x) = \tau \psi(x) \tau^{-1}$$

is a homomorphism which restricts to Δ on G. Finally, if $\theta_E(x) = 1$ then $\psi(x) = 1$, so $x \in \ker \pi = G$, and so $x = 1$ since Δ is injective. Thus θ_E is injective as claimed.

We can now complete the

Proof of Theorem 5.8 We know from Theorem 5.7 that $\operatorname{Aut}(G)$ is profinite, and virtually a pro-p group of finite rank. The same therefore can be said of K, and hence of W (by Exercise 3.1). Put $G^* = \Delta(G) \leq W$ and let N be the normaliser of G^* in W. Then G^* is closed in W, being compact, hence N is closed, and it follows that N/G^* is again virtually a pro-p group of finite rank, and profinite.

Now Theorem 4.23 shows that the finite subgroups of N/G^* lie in finitely many conjugacy classes; let C_i/G^* $(i = 1, \ldots, s)$ be representatives for these classes. Suppose E is an extension of G by F. Then

$$G^* = \Delta(G) = \theta_E(G) \lhd \theta_E(E),$$

so $\theta_E(E)/G^*$ is a finite subgroup of N/G^*, and therefore conjugate to one of the groups C_i/G^*. But then $\theta_E(E)$ is conjugate to C_i, showing that $E \cong \theta_E(E) \cong C_i$. The theorem follows.

5.10 Corollary *Let m be a positive integer. Then there are only finitely many isomorphism types of group E containing G as a subgroup of index at most m.*

Proof In G there are only finitely many open subgroups of index at most $m!$ (by Proposition 1.6); call them $G_1, \ldots, G_s$. Up to isomorphism, there are only finitely many groups of order at most $m!$; call them $F_1, \ldots, F_t$. Now suppose that $|E : G| \leq m$. Then for some $i \leq s$ we have $G_i \lhd E$, and then $E/G_i \cong F_j$ for some $j \leq t$. The corollary is now obvious.

We can also consider extensions the other way round:

5.11 Theorem *There are only finitely many isomorphism types of profinite extensions of a finite group F by a given pro-p group G of finite rank.*

This depends on

5.12 Lemma *Let E be a profinite group of finite rank r and let F be a finite normal subgroup of E such that E/F is a torsion-free pro-p group. Say $|F| = mp^e$ with $p \nmid m$. Then E has an open subgroup K such that $K \cap F = 1$ and $|E : K| \mid mp^{f(r,e)}$, for a certain function f.*

Proof Let H be a Sylow pro-p subgroup of E (see Exercise 1.12). Then $FH = E$ and $|E : H| = m$. Theorem 3.10 shows that H has a powerful open subgroup Q of index at most $p^{r+r\lambda(r)}$. Now put $K = Q^{p^e}$. Then

$|Q : K| \mid p^{er}$, and every element of K is the p^eth power of an element of Q, by Theorem 3.6. As $Q/(Q \cap F)$ is torsion-free and $(Q \cap F)^{p^e} = 1$ it follows that $K \cap F = 1$. This gives the result, with $f(r,e) = r + r\lambda(r) + er$.

Suppose now that we have a profinite group E containing the given finite group F as a normal subgroup, with E/F isomorphic to the given group G. We may choose in G a uniform open subgroup G_1, by Corollary 4.3. Let E_1 be the inverse image of G_1 in E. By Lemma 5.12, E_1 has an open subgroup K, with $K \cap F = 1$, such that $|E_1 : K| \leq mp^{f(r,e)}$, where m and e depend only on F, and r is at most the sum of the ranks of G and F. Then $|E : K| \leq |G : G_1| mp^{f(r,e)}$. On the other hand, K is isomorphic to KF/F, a subgroup of index at most $mp^{f(r,e)}/|F|$ in $E_1/F \cong G_1$. Hence there are only finitely many possibilities for the isomorphism type of K (Proposition 1.6), and it follows by Corollary 5.10 that there are only finitely many possibilities for the isomorphism type of E. This completes the proof of Theorem 5.11.

Notes

The material of §§1–3 is presumably well known (apart from the 'uniformly powerful' terminology). Theorem 5.7 may be new. §4 is from Segal (1999).

Exercises

1. Let G be a profinite group. Show that there is a unique topology on $\mathrm{Aut}(G)$ making $\mathrm{Aut}(G)$ into a topological group and having the family $(\Gamma(N))_{N \lhd_o G}$ as a base for the neighbourhoods of 1, and that this topology is Hausdorff (it is the *existence* rather then the *uniqueness* which is in question here).

2. Let G be a profinite group. Show that the map

$$G \times \mathrm{Aut}(G) \to G; \quad (g, \gamma) \mapsto g^\gamma$$

is continuous. Show also that if G is finitely generated, then the congruence topology on $\mathrm{Aut}(G)$ is the coarsest topology (making $\mathrm{Aut}(G)$ into a topological group) for which this statement is true.

3. Give an example of a profinite group G such that $\mathrm{Aut}(G)$ is not profinite.

4. Show that if G is a finite p-group then $\Gamma(\Phi(G))$ is a p-group.
[*Hint:* See §0.5. Alternatively, argue as follows. Write $G_n = \gamma_n(G)$. (i) Show that if $\gamma \in \Gamma(\Phi(G))$ then $\gamma^{p^m} \in \Gamma(G_2)$ for some m. (ii) By applying Lemma 5.4 to the action of G on G_n/G_{n+2}, show that $[G_n, G_2] \leq G_{n+2}$ for each n. (iii) By induction on n, show that if $\gamma \in \Gamma(G_2)$ then $[G_n, \gamma] \leq G_{n+1}$ for each n. (iv) Suppose $G_{c+1} = 1$. Using induction on c, and Lemma 5.4, show that $\Gamma(G_2)$ is a p-group.]

5. Suppose that the profinite group G has finite rank r and is virtually pro-p. Show that $\mathrm{Aut}(G)$ is virtually (pro-p of rank $\leq 2r^2$).

6. Let $G = \Gamma_{1+\epsilon}$ be as in Theorem 5.2. Show that G is topologically generated by the set

$$\left\{ 1 + p^{1+\epsilon} e_{ij} \mid i, j = 1, \dots, d \right\}$$

where e_{ij} is the matrix with (i, j)-entry 1 and all other entries 0.

7. Let G be a pro-p group of finite rank and let F be a finite p-group. Deduce that there are only finitely many isomorphism types of pro-p groups which are extensions of F by G from the fact that G is finitely presented (Theorem 4.34). Use Theorem 4.35 to give an upper bound for the number of such isomorphism types in terms of $\mathrm{rk}(G)$ and $|F|$.
[*Hint:* Suppose $F \lhd H$ and $H/F \cong G$; write down a presentation for H as a pro-p group, in terms of a presentation for G, the multiplication table of F and an action of H on F.]

Interlude A

'Fascicule de résultats': pro-p groups of finite rank

1. *Let G be a pro-p group. Each of the following conditions is necessary and sufficient for G to have finite rank:*

(a) *G is finitely generated and virtually powerful* (Theorem 3.13);

(b) *G is finitely generated and virtually uniform* (Corollary 4.3);

(c) *$|G : \overline{G^{p^k}}|$ or $|G : G^{p^k}|$ grows at most polynomially with k* (Theorem 3.16);

(c*) *($p \neq 2$) G is finitely generated, and $|G : \overline{G^{p^k}}| < p^{p^k + k - 1}$ for some k* (Corollary 11.19);

(d) *G has polynomial subgroup growth* (Theorem 3.19);

(d*) *G is finitely generated, and $\sigma_k(G) < p^{ck^2}$ for all large k, where $c < \frac{1}{8}$* (Theorem 11.7);

(e) *G is the product of finitely many procyclic subgroups* (Theorem 3.17);

(e*) *G is finitely (or countably) generated as a $\mathbb{Z}_p$-powered group* (Theorem 3.17);

(f) *for some finite k, the k-generator subgroups of G have positive lower density* (Exercise 3.3);

(g) *G is finitely generated, and for some n, G does not involve $C_p \wr C_{p^n}$* (Exercise 3.4);

(h) *G is finitely generated, and no infinite closed subgroup of G has Hausdorff dimension zero* (Exercise 4.16);

(i) *G is finitely generated, and for some n (or for infinitely many n), $D_n(G) = D_{n+1}(G)$* (Theorem 11.4, Proposition 11.3);

(j) *G is finitely generated, and for some n (or for infinitely many n), there exists h with $p^h > n$ such that $\gamma_n(G)$ consists of p^hth powers in G (or such that $\gamma_n(G) \leq \overline{G^{p^h}}\gamma_{n+1}(G)$)* (Corollary 11.17);

(k) *G is finitely generated, and the graded Lie algebra $\bigoplus D_n(G)/D_{n+1}(G)$ is nilpotent* (Exercise 11.7);

(l) *G is finitely generated, and the 'Golod–Shafarevich sequence' $c_n(G)$ grows at most polynomially (or for some $n > 1$, $c_n(G) < p(n)$)* (Exercise 7.5, Proposition 12.17);

(m) *the 'subgroup-counting zeta-function' $\zeta_{G,p}(s)$ is a rational function of p^{-s}* (Interlude C);

(n) *G is isomorphic to a closed subgroup of $\mathrm{GL}_d(\mathbb{Z}_p)$ for some d* (Theorem 7.19, Theorem 5.2);

(o) *G is a p-adic analytic group* (Corollary 8.34).

2. *Let G be a pro-p group of finite rank. Then* $\dim(G)$ *is equal to each of the following numbers:*

(a) $d(H)$ *where H is any uniform open subgroup of G* (Definition 4.7);

(b) $\dim(N) + \dim(G/N)$ *where N is any closed normal subgroup of G* (Theorem 4.8);

(c) *the rank of the free $\mathbb{Z}_p$-module $(H, +)$ where H is any uniform open subgroup of G* (Theorem 4.17);

(d) $\lim_{k \to \infty} k^{-1} \log_p |G : G^{p^k}|$ (Exercise 4.10);

(e) *the order of the pole at 1 of the function* $\mathrm{gocha}(G; T)$ (Corollary 12.19);

(f) $\dim_{\mathbb{Q}_p}(L)$ *where L is the Lie algebra $\mathcal{L}(G)$* (Theorem 9.11);

(g) *the dimension of any chart belonging to any p-adic analytic group structure on G* (Theorem 8.36).

Part II
Analytic groups

Part II

6

Normed algebras

In this utilitarian chapter we introduce some simple analytic concepts, and establish some of their basic features. The proofs in Sections 1–4 are exercises in analysis of the most elementary kind, and will mostly only be sketched; however we present a number of arguments in detail, as the analysis is non-Archimedean and perhaps somewhat unfamiliar. Section 5 is devoted to a single more substantial result, the Campbell–Hausdorff formula.

While Sections 2–4 concentrate on p-adic analysis, Section 6 extends some of the results to 'pro-p rings' of more general type; these will be needed in Chapter 13.

6.1 Normed rings

Rings are assumed to have an identity element distinct from 0; the identity of a ring R is denoted 1_R.

6.1 Definition A *norm* on a ring R is a function $\|\cdot\| : R \to \mathbb{R}$ such that for all $a, b \in R$

(N1) $\|a\| \geq 0$; $\|a\| = 0$ if and only if $a = 0$;

(N2) $\|1_R\| = 1$ and $\|ab\| \leq \|a\| \|b\|$; and

(N3) $\|a \pm b\| \leq \max\{\|a\|, \|b\|\}$.

If these hold then $(R, \|\cdot\|)$ is said to be a *normed ring*.

Norms of this kind, satisfying the 'ultrametric inequality' (N3), are called *non-Archimedean*. These are the only ones considered in this chapter. It is easy to see that (N3) implies the following inequalities, which will often be used without special mention:

$$\|a_1 + \cdots + a_n\| \leq \max_{1 \leq i \leq n} \|a_i\| \quad \text{if } n \text{ is finite;} \tag{1}$$

$$\|a \pm b\| = \max\{\|a\|, \|b\|\} \quad \text{unless} \quad \|a\| = \|b\| \tag{2}$$

The distance function $(a, b) \mapsto \|a - b\|$ is a metric on a normed ring $(R, \|\cdot\|)$; all topological terms applied to R will refer to the topology defined by this metric. A sequence (a_n) of elements in R is a *Cauchy sequence* if for each $\varepsilon > 0$ there exists an integer N_ε such that $\|a_n - a_m\| < \varepsilon$ whenever $n > N_\varepsilon$ and $m > N_\varepsilon$.

6.2 Definition (i) The normed ring $(R, \|\cdot\|)$ is *complete* if every Cauchy sequence in R converges to an element of R.

(ii) A normed ring $(\widehat{R}, \|\cdot\|)$ is called a *completion of* R if (a) R is a dense subring of $\widehat{R}$ and the norm on $\widehat{R}$ extends the norm on R and (b) $\widehat{R}$ is complete.

6.3 Proposition *Let* $(R, \|\cdot\|)$ *be a normed ring. Then there exists a* *completion* $(\widehat{R}, \|\cdot\|)$ *of* $(R, \|\cdot\|)$, *which is unique up to isomorphism – that is, if* $(\widehat{R}', \|\cdot\|)$ *is a second completion, then there exists a norm-preserving isomorphism* $\phi \colon \widehat{R} \to \widehat{R}'$ *which restricts to the identity on* R.

Proof The existence of a completion follows by a standard procedure which we briefly recall. We can define a ring structure on the set C of Cauchy sequences in R by defining addition and multiplication componentwise. Define a *null sequence* to be a sequence (x_i) such that $\lim_{i \to \infty} \|x_i\| = 0$. The null sequences form an ideal N in C. Put $\widehat{R} = C/N$, and note that R embeds into $\widehat{R}$ via the diagonal map $a \mapsto (a)$. We identify R with its image in $\widehat{R}$ and define a norm $\|\cdot\|$ on $\widehat{R}$ as follows: if (a_i) is a Cauchy sequence representing an element r in $\widehat{R}$, then $\|r\| = \lim_{i \to \infty} \|a_i\|$. This norm is well defined and its restriction to (the copy of) R in $\widehat{R}$ gives the original norm on R. It is a familiar exercise to show that R is dense in $\widehat{R}$ and that $(\widehat{R}, \|\cdot\|)$ is complete and unique up to isomorphism.

Proposition 6.3 allows us to talk about *the* completion of a normed ring $(R, \|\cdot\|)$.

6.4 Example The function $|\cdot| \colon \mathbb{Q} \to \mathbb{R}$ defined by

$$|0| = 0; \quad |a| = p^{-k} \quad \text{if} \quad a = p^k m/n \quad \text{with} \quad k, m, n \in \mathbb{Z} \quad \text{and} \quad p \nmid mn$$

is a norm on $\mathbb{Q}$. The completion of $(\mathbb{Q}, |\cdot|)$ is the p-adic field $(\mathbb{Q}_p, |\cdot|)$.

(As throughout the book, the symbol $|\cdot|$ denotes the p-adic absolute value on $\mathbb{Q}_p$.) The Theorem of Ostrowski (see Schikhof (1984), Theorem 10.1) states that, up to equivalence, the p-adic norms are the only non-trivial norms defined on $\mathbb{Q}$. (The ordinary absolute value on $\mathbb{Q}$ is *not* a norm in our sense as it fails (N3).)

The following easily verified lemma generalises the definition of the p-adic norm on $\mathbb{Z}$. Most of the normed rings discussed in the later chapters arise in this manner.

6.5 Lemma *Let R be a ring and*

$$R = R_0 \supseteq R_1 \supseteq \cdots \supseteq R_i \supseteq \cdots$$

a chain of ideals such that

- $\bigcap_{i \in \mathbb{N}} R_i = 0;$
- *for all $i, j \in \mathbb{N}$, $R_i R_j \subseteq R_{i+j}$.*

Fix a real number $c > 1$, and define $\|\cdot\| : R \to \mathbb{R}$ by

$$\|0\| = 0; \qquad \|a\| = c^{-k} \text{ if } a \in R_k \setminus R_{k+1}.$$

Then $(R, \|\cdot\|)$ is a normed ring.

In the situation of Lemma 6.5, the norm extends in a natural way to the inverse limit $\varprojlim (R/R_i)$, which is then isomorphic to the completion $(\widehat{R}, \|\cdot\|)$ of $(R, \|\cdot\|)$ (see Exercise 2).

We shall be particularly interested in norms defined on $\mathbb{Q}_p$-algebras A, compatible with the action of $\mathbb{Q}_p$ on A:

6.6 Definition Let A be a $\mathbb{Q}_p$-algebra. Then $(A, \|\cdot\|)$ is a *normed $\mathbb{Q}_p$-algebra* if $\|\cdot\|$ is a norm on the ring A and the following holds:

(N4) $\|\lambda a\| = |\lambda| \cdot \|a\|$ for all $a \in A$ and $\lambda \in \mathbb{Q}_p$.

We leave it to the reader to check the following important example:

6.7 Example The ring $\mathrm{M}_n(\mathbb{Q}_p)$ of $n \times n$ matrices over $\mathbb{Q}_p$ is a normed $\mathbb{Q}_p$-algebra with norm given by $\|(a_{ij})\| = \max\{|a_{ij}| \mid i, j = 1, \dots, n\}$.

6.2 Sequences and series

Throughout this section, we assume that $(R, \|\cdot\|)$ is a *complete normed ring*. Because our norm satisfies the ultrametric inequality (N3), elementary analysis in such a ring is particularly simple, as the following propositions indicate. However, caution should be exercised as non-Archimedean analysis has curious features, some of which are illustrated in the exercises.

In order to deal with multiple series it is convenient to introduce the following rather general notion of convergence.

6.8 Definition Let T be a countably infinite set and let $n \mapsto a_n$ be a map of T into R. Let $a, s \in R$.

(i) The family $(a_n)_{n \in T}$ *converges to* a, written

$$\lim_{n \in T} a_n = a,$$

if for each $\varepsilon > 0$ there exists a finite subset T' of T such that $\|a - a_n\| < \varepsilon$ for all $n \in T \setminus T'$.

(ii) The series $\sum_{n \in T} a_n$ *converges with sum* s, written

$$\sum_{n \in T} a_n = s,$$

if for each $\varepsilon > 0$ there exists a finite subset T' of T such that for all finite sets T'' for which $T' \subseteq T'' \subseteq T$ we have $\left\| s - \sum_{n \in T''} a_n \right\| < \varepsilon$.

It is easy to see that the limits and infinite sums defined in this way commute with the operations of taking finite sums and multiplication by constants; we shall use these facts without special mention. Note that the condition in (ii) is rather a strong one; part (iii) of the next proposition indicates that is analogous to *absolute* convergence in the real case.

6.9 Proposition *Let T be a countably infinite set and let $n \mapsto a_n$ be a map from T into R. Let $i \mapsto n(i)$ be a bijection from $\mathbb{N}$ onto T.*

(i) $\lim_{n \in T} a_n = a$ *if and only if* $\lim_{i \to \infty} a_{n(i)} = a$.

(ii) *The series* $\sum_{n \in T} a_n$ *converges in R if and only if* $\lim_{n \in T} a_n = 0$.

(iii) $\sum_{n \in T} a_n = s$ *if and only if* $\sum_{i=0}^{\infty} a_{n(i)} = s$.

(iv) *If* $\sum_{n \in T} a_n = s$ *then* $\|s\| \leq \sup\{\|a_n\| \mid n \in T\}$.

(v) *If* $\sum_{n \in T} a_n = s$ *and for some* $m \in T$, $\|a_m\| > \|a_n\|$ *for all* $n \in T \setminus \{m\}$, *then* $\|s\| = \|a_m\|$.

Proof We leave (i) as an exercise.

(ii) Suppose that $\sum_{n \in T} a_n = s$. Let $\varepsilon > 0$. There exists T', a finite subset of T, such that for any finite set T'' with $T' \subseteq T'' \subseteq T$, we have $\left\| s - \sum_{n \in T''} a_n \right\| < \varepsilon$. Let $m \in T \setminus T'$ and put $T'' = T' \cup \{m\}$. Then

$$\|a_m\| \leq \max \left\{ \left\| s - \sum_{n \in T''} a_n \right\|, \left\| s - \sum_{n \in T'} a_n \right\| \right\} < \varepsilon.$$

Thus $\lim_{n \in T} a_n = 0$.

Before proving the converse, we give the proof of

(iii) We show that $\sum_{i=0}^{\infty} a_{n(i)} = s$ implies $\sum_{n \in T} a_n = s$; the converse is clear from Definition 6.8.

Let $\varepsilon > 0$; then there exists $N_1 \in \mathbb{N}$ such that for all $M > N_1$, $\left\| s - \sum_{i=0}^{M} a_{n(i)} \right\| < \varepsilon$. We have just shown that $\lim_{n \in T} a_n = 0$; hence there exists $N_2 \in \mathbb{N}$ such that for all $i > N_2$, $\|a_{n(i)}\| < \varepsilon$. Put $N = \max\{N_1, N_2\}$ and $T' = \{n(i) \mid 0 \leq i \leq N\}$. Then for any finite set T'' with $T' \subseteq T'' \subseteq T$, the inequality (1) gives

$$\left\| s - \sum_{n \in T''} a_n \right\| \leq \max \left\{ \left\| s - \sum_{i=0}^{N} a_{n(i)} \right\|, \|a_n\| \mid n \in T'' \setminus T' \right\} < \varepsilon.$$

Thus $\sum_{n \in T} a_n$ converges to s in the sense of Definition 6.8.

(ii), *converse*: Assume that $\lim_{n \in T} a_n = 0$, and put $s_k = \sum_{i=0}^{k} a_{n(i)}$ for each $k \in \mathbb{N}$. Let $\varepsilon > 0$; then there exists a finite subset T' of T such that for all $n \in T \setminus T'$, $\|a_n\| < \varepsilon$. Put $N = \max\{i \mid n(i) \in T'\}$. Then for all $k > j \geq N$ we have

$$\|s_k - s_j\| \leq \max \left\{ \|a_{n(i)}\| \mid j < i < k \right\} < \varepsilon.$$

So (s_k) is a Cauchy sequence. Since R is complete, there exists $s \in R$ such that

$$s = \lim_{k \to \infty} s_k = \sum_{i=0}^{\infty} a_{n(i)},$$

so the series $\sum_{n \in T} a_n$ converges by (iii).

(iv) Put $\sigma = \sup\{\|a_n\| \mid n \in T\}$. It is clearly enough to consider the case where $\sigma > 0$. Then convergence of the series implies that there exists a finite subset T' of T such that $\left\| s - \sum_{n \in T'} a_n \right\| < \sigma$, and we have

$$\|s\| \leq \max \left\{ \left\| s - \sum_{n \in T'} a_n \right\|, \|a_n\| \mid n \in T' \right\} \leq \sigma$$

as required.

(v) Now we have $\sigma = \|a_m\| > \|a_n\|$ for all $n \neq m$. It follows from (ii) that $\|a_n\| \leq \frac{1}{2}\sigma$ for all but finitely many $n \in T$, and hence that $\sup\{\|a_n\| \mid n \in T \setminus \{m\}\} < \sigma$. Hence by (iv) we have

$$\|s - a_m\| = \left\| \sum_{n \in T \setminus \{a_m\}} a_n \right\| \leq \sup\{\|a_n\| \mid n \in T \setminus \{m\}\} < \|a_m\|,$$

which implies $\|s\| = \|a_m\|$, by (2).

6.10 Proposition *Let T be the disjoint union of a countable family $\{T_\lambda \mid \lambda \in \Lambda\}$ of countable sets T_λ. Suppose that $\sum_{n \in T} a_n$ is a convergent series in R, with sum s. Then each of the series $\sum_{n \in T_\lambda} a_n$ converges in R, with sum s_λ, say, and $\sum_{\lambda \in \Lambda} s_\lambda = s$.*

Proof Part (ii) of Proposition 6.9 shows that $\lim_{n \in T} a_n = 0$; this clearly implies that $\lim_{n \in T_\lambda} a_n = 0$ whenever T_λ is infinite. Using Proposition 6.9(ii) in the reverse direction we deduce that for each $\lambda \in \Lambda$ there exists $s_\lambda \in R$ such that $\sum_{n \in T_\lambda} a_n = s_\lambda$.

Now let $\varepsilon > 0$. We must show that there exists a finite set $\Lambda' \subseteq \Lambda$ such that for each finite set Λ'' with $\Lambda' \subseteq \Lambda'' \subseteq \Lambda$ we have $\|s - \sum_{\lambda \in \Lambda''} s_\lambda\| < \varepsilon$. By hypothesis, there exists a finite set $T' \subseteq T$ such that for each finite set T'' with $T' \subseteq T'' \subseteq T$ we have $\|s - \sum_{n \in T''} a_n\| < \varepsilon$. As in the proof of Proposition 6.9(ii) above, this implies that $\|a_n\| < \varepsilon$ whenever $n \in T \setminus T'$. Now put $\Lambda' = \{\lambda \in \Lambda \mid T_\lambda \cap T' \neq \varnothing\}$, a finite subset of Λ. Then for each $\lambda \in \Lambda$ we have

$$\left\| s_\lambda - \sum_{n \in T_\lambda \cap T'} a_n \right\| = \left\| \sum_{n \in T_\lambda \setminus T'} a_n \right\| < \varepsilon.$$

It follows that if Λ'' is any finite subset of Λ with $\Lambda' \subseteq \Lambda''$ then

$$\left\| s - \sum_{\lambda \in \Lambda''} s_\lambda \right\| = \left\| \left(s - \sum_{n \in T'} a_n \right) - \sum_{\lambda \in \Lambda''} \left(s_\lambda - \sum_{n \in T_\lambda \cap T'} a_n \right) \right\| < \varepsilon$$

as required.

6.11 Corollary (Double series) *Let S_1 and S_2 be countable sets. Suppose that for each $(m, n) \in S_1 \times S_2$, a_{mn} is an element of R, and that $\lim_{(m,n) \in S_1 \times S_2} a_{mn} = 0$. Then the double series $\sum_{m \in S_1}(\sum_{n \in S_2} a_{mn})$ and $\sum_{n \in S_2}(\sum_{m \in S_1} a_{mn})$ both converge and their common sum is $\sum_{(m,n) \in S_1 \times S_2} a_{mn}$.*

Proof That the series $\sum_{(m,n) \in S_1 \times S_2} a_{mn}$ converges is guaranteed by Proposition 6.9(ii). To show that the first of the double series converges to the same value, write $T = S_1 \times S_2 = \bigcup_{m \in S_1} T_m$ where $T_m = \{m\} \times S_2$, and apply Proposition 6.10. A similar argument applies to the second double series.

6.12 Corollary ('Cauchy multiplication of series'). *Suppose that $(T, *)$ is a countable set with a binary operation $*$, and that $\sum_{n \in T} a_n$ and $\sum_{n \in T} b_n$ are convergent series in R. Then, for each $n \in T$, the series*

$$\sum_{\substack{(r,s) \in T \times T \\ r*s=n}} a_r b_s$$

converges with sum c_n, say, and the series $\sum_{n \in T} c_n$ converges with

$$\sum_{n \in T} c_n = (\sum_{n \in T} a_n)(\sum_{n \in T} b_n).$$

Proof The hypotheses imply that $\lim_{n \in T} a_n = \lim_{n \in T} b_n = 0$, from which it follows that $\lim_{(r,s) \in T \times T} a_r b_s = 0$. Therefore $\sum_{(r,s) \in T \times T} a_r b_s$ converges. We now apply Proposition 6.10 to conclude that for each n the series for c_n converges and that $\sum_{n \in T} c_n = \sum_{(r,s) \in T \times T} a_r b_s$. Finally, Corollary 6.11 shows that

$$\sum_{(r,s) \in T \times T} a_r b_s = \sum_{r \in T}(\sum_{s \in T} a_r b_s) = \sum_{r \in T} a_r(\sum_{s \in T} b_s) = (\sum_{r \in T} a_r)(\sum_{s \in T} b_s).$$

The result follows.

As a final application of Proposition 6.9, we mention the following important result on the uniqueness of power series:

6.13 Proposition *Let A be a complete normed $\mathbb{Q}_p$-algebra and let a_n ($n \in \mathbb{N}$) be elements of A. Suppose that there exists a neighbourhood V of 0 in $\mathbb{Q}_p$ such that*

$$\sum_{n \in \mathbb{N}} \lambda^n a_n = 0 \quad \text{for all } \lambda \in V.$$

Then $a_n = 0$ for all $n \in \mathbb{N}$.

Proof Suppose that not all a_n are zero, and let m be minimal such that $a_m \neq 0$. Let $\lambda_0 \in V$ be non-zero and set $r = |\lambda_0|$. Since the series $\sum_{n \in \mathbb{N}} \lambda_0^n a_n$ converges, there exists $C > 0$ such that $r^n \|a_n\| =$

$\|\lambda_0^n a_n\| < C$ for all $n \in \mathbb{N}$. Choose $\lambda \in V$ such that $0 < |\lambda| < r \min\{C^{-1} r^m \|a_m\|, 1\}$. Now let $n > m$; then

$$\|\lambda^n a_n\| \le r^{n-m-1} \|a_n\| \cdot |\lambda|^{m+1}$$

$$< \left(\frac{C}{r^{m+1}}\right) |\lambda| \cdot |\lambda|^m < \|a_m\| \cdot |\lambda|^m = \|\lambda^m a_m\|.$$

Since $a_n = 0$ for $n < m$, it follows by Proposition 6.9(v) that

$$\left\| \sum_{n \in \mathbb{N}} \lambda^n a_n \right\| = \|\lambda^m a_m\| \ne 0,$$

which contradicts the hypothesis. The result follows.

6.3 Strictly analytic functions

Throughout this section, $(A, \|\cdot\|)$ will denote a *complete normed $\mathbb{Q}_p$-algebra*.

We begin by defining *formal power series* in non-commuting variables. We write

$$W = W(X_1, \dots, X_n) = W(\mathbf{X})$$

to denote the free monoid generated by a set of symbols $X_1, \dots, X_n$: the elements of W are the *words*

$$w(\mathbf{X}) = X_{i_1} X_{i_2} \dots X_{i_m},$$

where $i_1, \dots, i_m \in \{1, \dots, n\}$ and $m \ge 1$, together with the *empty word* ($m = 0$) which we denote by 1; here $m = \deg w$ is the *degree* of $w = w(\mathbf{X})$. The words are multiplied by concatenation.

6.14 Definition The *ring of formal power series in the (non-commuting) variables* $X_1, \dots, X_n$, denoted

$$\mathbb{Q}_p \langle\langle X_1, \dots, X_n \rangle\rangle \quad \text{or} \quad \mathbb{Q}_p \langle\langle \mathbf{X} \rangle\rangle,$$

is the set of all formal sums

$$F(\mathbf{X}) = \sum_{w \in W} a_w w \quad (a_w \in \mathbb{Q}_p \text{ for all } w),$$

made into a $\mathbb{Q}_p$-algebra with componentwise addition and scalar multiplication, and with multiplication given by

$$\sum_{w \in W} a_w w \sum_{w \in W} b_w w = \sum_{w \in W} c_w w \quad \text{where} \quad c_w = \sum_{\substack{u,v \in W \\ uv=w}} a_u b_v.$$

It is readily verified that $\mathbb{Q}_p \langle\langle \mathbf{X} \rangle\rangle$ with this definition is indeed a $\mathbb{Q}_p$-algebra.

6.15 Definition Let $\mathbf{x} = (x_1, \dots, x_n) \in A^n$. The formal power series $F(\mathbf{X}) = \sum_{w \in W} a_w w$ can be *evaluated* at $\mathbf{x}$ if the series $\sum_{w \in W} a_w w(\mathbf{x})$, obtained by substituting x_i for X_i ($i = 1, \dots, n$) in each word $w = w(\mathbf{X})$, converges in A; in this case we denote its sum by $F(\mathbf{x})$. The set of all such power series $F(\mathbf{X})$ is denoted $E_\mathbf{x}$.

(We shall sometimes loosely say '$F(\mathbf{x})$ exists' to mean '$F(\mathbf{X})$ can be evaluated at $\mathbf{x}$'.) Note that, according to Proposition 6.9(ii), $F(\mathbf{X}) \in E_\mathbf{x}$ if and only if $\lim_{w \in W} a_w w(\mathbf{x}) = 0$. The following lemma is easily proved:

6.16 Lemma *Let* $\mathbf{x} = (x_1, \dots, x_n) \in A^n$.
(i) *The subset* $E_\mathbf{x}$ *is a subalgebra of* $\mathbb{Q}_p \langle\langle \mathbf{X} \rangle\rangle$.
(ii) *The mapping* $F(\mathbf{X}) \mapsto F(\mathbf{x})$ *of* $E_\mathbf{x}$ *into* A *is a* $\mathbb{Q}_p$*-algebra homomorphism*.

We are now ready for the main definition of this section. Here, the set A^n is given the product topology; for $w \in W$ and $\mathbf{x} = (x_1, \dots, x_n) \in A^n$ we use the shorthand notation

$$w(\|\mathbf{x}\|) = w(\|x_1\|, \dots, \|x_n\|).$$

6.17 Definition Let $f : D \to A$ be a mapping, where D is a non-empty open subset of A^n. Then f is *strictly analytic* on D if there exists $F(\mathbf{X}) = \sum_{w \in W} a_w w \in \mathbb{Q}_p \langle\langle \mathbf{X} \rangle\rangle$ such that, for each $\mathbf{x} = (x_1, \dots, x_n) \in D$,
(i) $\lim_{w \in W} |a_w| w(\|\mathbf{x}\|) = 0$, and
(ii) $f(\mathbf{x}) = F(\mathbf{x})$.
In this case we say that F *represents* f.

Condition (i) is a kind of 'absolute' convergence. It implies that $F(\mathbf{X})$ can be evaluated at $\mathbf{x}$, since $\|a_w w(\mathbf{x})\| \le |a_w| w(\|\mathbf{x}\|)$.

6.18 Lemma *Suppose that* f *is strictly analytic on a non-empty open set* $D \subseteq A^n$ *and that* f *is represented by* $F(\mathbf{X}) = \sum_{w \in W} a_w w \in \mathbb{Q}_p \langle\langle \mathbf{X} \rangle\rangle$. *Then there exists* $k \in \mathbb{N}$ *such that* $p^{k \deg w} a_w \in \mathbb{Z}_p$ *for all* $w \in W \setminus \{1\}$.

Proof Equivalently, we must show that $|a_w| \le p^{k \deg w}$. Since D is non-empty and open there exists $\mathbf{x} \in D$ with $x_i \ne 0$ for $i = 1, \dots, n$; thus

$\min_i \|x_i\| = p^r$ for some integer r. Then for each $w \in W$ we have

$$|a_w| w(\|\mathbf{x}\|) \geq |a_w| p^{r \deg w}.$$

Condition (i) of Definition 6.17 now implies that $|a_w| p^{r \deg w}$ is bounded above for all $w \in W$, say by p^s where $s \in \mathbb{N}$. The result follows with $k = \max\{s - r, 0\}$.

6.19 Proposition *Let D be a non-empty open subset of A^n. If f is a strictly analytic function on D then f is continuous on D.*

Proof Let $F(\mathbf{X}) = \sum_{w \in W} a_w w \in \mathbb{Q}_p \langle\langle \mathbf{X} \rangle\rangle$ represent f and let $\mathbf{x} \in D$. Since D is open we can choose a positive integer r such that $\|x_i\| > p^{-r}$ whenever $x_i \neq 0$ and such that the set

$$D' = \{\mathbf{y} \in A^n \mid \|x_i - y_i\| \leq p^{-r} \text{ for each } i\}$$

is an open subset of D containing $\mathbf{x}$. Moreover, D' contains an element $\mathbf{z}$ with the properties

$$\|z_i\| = \|x_i\| \quad \text{if } x_i \neq 0$$
$$\|z_i\| = p^{-r} \quad \text{if } x_i = 0$$

and, by (N3), $\|y_i\| \leq \|z_i\|$ for each i whenever $\mathbf{y} \in D'$.

We now prove that f is continuous at $\mathbf{x}$. Let $\varepsilon > 0$. There exists a finite subset $W' \subseteq W$ such that $|a_w| w(\|\mathbf{z}\|) < \varepsilon$ for all $w \in W \setminus W'$, and hence $|a_w| w(\|\mathbf{y}\|) < \varepsilon$ for all $\mathbf{y} \in D'$ and all $w \in W \setminus W'$. It follows that

$$\|a_w(w(\mathbf{x}) - w(\mathbf{y}))\| < \varepsilon \tag{3}$$

for each $w \in W \setminus W'$ and all $\mathbf{y} \in D'$.

On the other hand, when $w = X_{i_1} \ldots X_{i_m}$ and $\mathbf{y} \in D'$ we have

$$\|w(\mathbf{x}) - w(\mathbf{y})\| = \left\| \begin{array}{c} (x_{i_1} - y_{i_1})x_{i_2} \ldots x_{i_m} + y_{i_1}(x_{i_2} - y_{i_2})x_{i_3} \ldots x_{i_m} + \cdots \\ \cdots + y_{i_1} \ldots y_{i_{m-1}}(x_{i_m} - y_{i_m}) \end{array} \right\|$$
$$\leq \max_i \|z_i\|^{m-1} \max_i \|x_i - y_i\|.$$

Hence there exists $\delta > 0$ (with $\delta < p^{-r}$) such that, if $\|x_i - y_i\| < \delta$ for each i, then $|a_w| \cdot \|w(\mathbf{x}) - w(\mathbf{y})\| < \varepsilon$ whenever $w \in W'$ (note that $\deg w = m$ takes only finitely many values as w ranges over the finite set W'). Thus (3) holds for all such $\mathbf{y}$ when $w \in W'$. We now apply Proposition 6.9(iv) to infer that

$$\|f(\mathbf{x}) - f(\mathbf{y})\| \leq \sup_{w \in W} \|a_w(w(\mathbf{x}) - w(\mathbf{y}))\| < \varepsilon$$

for all $\mathbf{y}$ satisfying $\|x_i - y_i\| < \delta$ for $1 \leq i \leq n$. Thus f is continuous at $\mathbf{x}$ as required.

Our next task is to define the *exponential* and *logarithm* functions. First we need

6.20 Lemma *For each positive integer n, $v(n!) \leq (n - 1)/(p - 1)$.*

Proof Recall that $|i| = p^{-v(i)}$ for each i. Now

$$v(n!) = \sum_{i=1}^{n} v(i).$$

The number of integers i between 1 and n for which $v(i) \geq j$ is equal to the integer part of n/p^j. Thus

$$v(n!) \leq \sum_{j=1}^{k} \frac{n}{p^j} = \frac{n(1 - p^{-k})}{p - 1} \leq \frac{n - 1}{p - 1}$$

where $p^k \leq n < p^{k+1}$.

For the precise value of $v(n!)$ see Exercise 5.

6.21 Definition Two formal power series in $\mathbb{Q}_p \langle\langle X \rangle\rangle$ (one variable) are defined by

$$\mathcal{E}(X) = \sum_{n=0}^{\infty} \frac{1}{n!} X^n$$

$$\mathcal{L}(X) = \sum_{n=1}^{\infty} \frac{(-1)^{n+1}}{n} X^n.$$

Writing

$$A_0 = \begin{cases} \{x \in A \mid \|x\| \leq p^{-1}\} & \text{if } p \neq 2 \\ \{x \in A \mid \|x\| \leq 2^{-2}\} & \text{if } p = 2, \end{cases} \tag{4}$$

we now have

6.22 Proposition *There exist strictly analytic functions*

$$\exp : A_0 \to 1 + A_0, \qquad \log : 1 + A_0 \to A_0$$

such that for all $x \in A_0$

$$\exp(x) = \mathcal{E}(x), \qquad \log(1 + x) = \mathcal{L}(x).$$

Proof Let $x \in A_0$; then $\|x\| = p^{-r}$ for some r satisfying $r > 1/(p-1)$. Now it follows from Lemma 6.20 that $|n| \geq |n!| \geq p^{-(n-1)/(p-1)}$, so for each $n \geq 1$ we have

$$\left\| \frac{(-1)^{n+1} x^n}{n} \right\| \leq \left\| \frac{x^n}{n!} \right\| \leq p^{-nr + (n-1)/(p-1)}.$$

The condition on r implies that $\lim_{n \to \infty} p^{-nr + (n-1)/(p-1)} = 0$. It follows by Proposition 6.9(ii) that the two series $\mathcal{E}(x)$ and $\mathcal{L}(x)$ both converge. Moreover, the inequality above implies that $(-1)^{n+1} x^n / n$ and $x^n / n!$ both lie in A_0 if $n \geq 1$. Then Proposition 6.9(iv) shows that

$$\mathcal{L}(x) = \sum_{n=1}^{\infty} \frac{(-1)^{n+1} x^n}{n} \in A_0$$

and

$$\mathcal{E}(x) = 1 + \sum_{n=1}^{\infty} \frac{x^n}{n!} \in 1 + A_0.$$

The result follows.

Remark For later reference we note that for $x \in A_0$ the above inequality implies

$$\left\| \frac{(-1)^{n+1} x^n}{n} \right\| \leq \left\| \frac{x^n}{n!} \right\| \leq \|x\|. \tag{5}$$

We shall show that these functions satisfy the usual identities for log and exp. For this we need to discuss the *composition* of formal power series.

6.23 Definition Let

$$G(\mathbf{Y}) = \sum_{v \in W(\mathbf{Y})} b_v w \in \mathbb{Q}_p \langle\langle \mathbf{Y} \rangle\rangle,$$

$$F_i(\mathbf{X}) = \sum_{w \in W(\mathbf{X})} a_{iw} w \in \mathbb{Q}_p \langle\langle \mathbf{X} \rangle\rangle \qquad i = 1, \ldots, m,$$

where $\mathbf{X} = (X_1, \ldots, X_n)$ and $\mathbf{Y} = (Y_1, \ldots, Y_m)$. Assume that for $i =$

$1, \ldots, m$, the 'constant term' a_{i1} is equal to 0. For $v(\mathbf{Y}) = Y_{i_1} \ldots Y_{i_d} \in W(\mathbf{Y})$, define the coefficients $c_{vw} \in \mathbb{Q}_p$ by

$$v(F_1(\mathbf{X}), \ldots, F_m(\mathbf{X})) = F_{i_1}(\mathbf{X}) \ldots F_{i_d}(\mathbf{X}) = \sum_{w \in W(\mathbf{X})} c_{vw} w(\mathbf{X}).$$

The *composite* of G and $\mathbf{F} = (F_1, \ldots, F_m)$ is defined to be the formal power series

$$(G \circ \mathbf{F})(\mathbf{X}) = \sum_{w \in W(\mathbf{X})} \left(\sum_{v \in W(\mathbf{Y})} b_v c_{vw} \right) w(\mathbf{X}).$$

Note that this is a well-defined formal power series since $c_{vw} = 0$ whenever $\deg v > \deg w$.

The next theorem establishes that under certain conditions the operations of composition and evaluation of formal power series commute.

6.24 Theorem *Let $(A, \|\cdot\|)$ be a complete normed $\mathbb{Q}_p$-algebra, and suppose that $F_1(\mathbf{X}), \ldots, F_m(\mathbf{X})$ and $G(\mathbf{Y})$ are formal power series satisfying the conditions of Definition 6.23. Suppose that $F_1(\mathbf{X}), \ldots, F_m(\mathbf{X})$ can all be evaluated at some point $\mathbf{x} \in A^n$. For each i put $\tau_i = \sup\{\|a_{iw} w(\mathbf{x})\| \mid w \in W(\mathbf{X})\}$, and suppose that*

$$\lim_{v \in W(\mathbf{Y})} |b_v| v(\tau_1, \ldots, \tau_m) = 0.$$

Then $G(F_1(\mathbf{x}), \ldots, F_m(\mathbf{x}))$ and $(G \circ \mathbf{F})(\mathbf{x})$ both exist and are equal.

Proof The term $c_{vw} w(\mathbf{X})$ given in Definition 6.23 is the sum of a finite number of terms of the form

$$a_{i_1 w_1} \ldots a_{i_d w_d} w_1(\mathbf{X}) \ldots w_d(\mathbf{X})$$

where $v(\mathbf{Y}) = Y_{i_1} \ldots Y_{i_d}$, each $w_i(\mathbf{X}) \in W(\mathbf{X})$ and $w_1(\mathbf{X}) \ldots w_d(\mathbf{X}) = w(\mathbf{X})$. Thus

$$\|c_{vw} w(\mathbf{x})\| \leq \max_{w_1 \ldots w_d = w} \|a_{i_1 w_1} w_1(\mathbf{x})\| \ldots \|a_{i_d w_d} w_d(\mathbf{x})\|$$

$$\leq \tau_{i_1} \ldots \tau_{i_d} = v(\tau_1, \ldots, \tau_m)$$

for each v and w.

We claim that $\lim_{(v,w) \in W(\mathbf{Y}) \times W(\mathbf{X})} b_v c_{vw} w(\mathbf{x}) = 0$. Let $\varepsilon > 0$. The hypothesis on b_v and the inequality above imply that there exists some finite subset $V' \subseteq W(\mathbf{Y})$ such that, for all (v, w) with $v \notin V'$, $\|b_v c_{vw} w(\mathbf{x})\| < \varepsilon$. On the other hand $\lim_{w \in W(\mathbf{X})} a_{iw} w(\mathbf{x}) = 0$

for $i = 1, \ldots, m$ and so, for each fixed $v \in W(\mathbf{Y})$, we have $\lim_{w \in W(\mathbf{X})} c_{vw} w(\mathbf{x}) = 0$: this follows from the first inequality above if we note that $\|a_{iw} w(\mathbf{x})\|$ is bounded for all i and w. Thus for each $v \in V'$ only finitely many $w \in W(\mathbf{X})$ satisfy $\|b_v c_{vw} w(\mathbf{x})\| \geq \varepsilon$; as V' is finite this inequality holds for only finitely many pairs $(v, w) \in W(\mathbf{Y}) \times W(\mathbf{X})$, and our claim is established.

Corollary 6.11 now shows that

$$\sum_{v \in W(\mathbf{Y})} \left(\sum_{w \in W(\mathbf{X})} b_v c_{vw} w(\mathbf{x}) \right) = \sum_{w \in W(\mathbf{X})} \left(\sum_{v \in W(\mathbf{Y})} b_v c_{vw} w(\mathbf{x}) \right). \quad (6)$$

By Lemma 6.16, evaluation is an algebra homomorphism, so

$$v(F_1(\mathbf{x}), \ldots, F_m(\mathbf{x})) = \sum_{w \in W(\mathbf{X})} c_{vw} w(\mathbf{x});$$

therefore

$$G(F_1(\mathbf{x}), \ldots, F_m(\mathbf{x})) = \sum_{v \in W(\mathbf{Y})} b_v \left(\sum_{w \in W(\mathbf{X})} c_{vw} w(\mathbf{x}) \right),$$

which is equal to the left-hand side of (6). On the other hand,

$$(G \circ \mathbf{F})(\mathbf{x}) = \sum_{w \in W(\mathbf{X})} \left(\sum_{v \in W(\mathbf{Y})} b_v c_{vw} \right) w(\mathbf{x}),$$

which is equal to the right-hand side of (6). The result follows.

6.25 Corollary *Let $x \in A_0$. Then*
(i) $\log(\exp(x)) = x$;
(ii) $\exp(\log(1 + x)) = 1 + x$;
(iii) $\log((1 + x)^n) = n \log(1 + x)$ *for each $n \in \mathbb{Z}$;*
(iv) $\exp(nx) = (\exp(x))^n$ *for each $n \in \mathbb{Z}$.*

Proof We sketch the proof of (i); the others follow in a similar fashion. Let $F(X) = \mathcal{E}(X) - 1$. A standard identity (see Exercise 7) asserts that

$$(\mathcal{L} \circ F)(X) = X.$$

Put $\tau = \sup_{1 \leq n \in \mathbb{N}} \|x^n/n!\|$. We see from (5) above that $\tau \leq \|x\|$, from which it follows that $\lim_{n \to \infty} |(-1)^{n+1}(1/n)| \tau^n = 0$. We may thus invoke Theorem 6.24 to deduce that

$$\log(\exp(x)) = \mathcal{L}(F(x)) = (\mathcal{L} \circ F)(x) = x.$$

Exercises 7 and 8 indicate that there are some subtleties to watch out for when composing strictly analytic functions.

To conclude this section, we introduce the *Campbell–Hausdorff series*; in the following chapters this will provide the essential link between an analytic pro-p group and its associated Lie algebra.

6.26 Definition Let

$$P(X, Y) = \mathcal{E}(X)\mathcal{E}(Y) - 1 \in \mathbb{Q}_p \langle\langle X, Y \rangle\rangle$$
$$C(X, Y) = \mathcal{E}(-X)\mathcal{E}(-Y)\mathcal{E}(X)\mathcal{E}(Y) - 1 \in \mathbb{Q}_p \langle\langle X, Y \rangle\rangle .$$

The *Campbell–Hausdorff series* $\Phi(X, Y)$ is defined by

$$\Phi(X, Y) = (\mathcal{L} \circ P)(X, Y);$$

the *commutator Campbell–Hausdorff series* $\Psi(X, Y)$ is defined by

$$\Psi(X, Y) = (\mathcal{L} \circ C)(X, Y)$$

6.27 Proposition. *Let* $x, y \in A_0$. *Then both* Φ *and* Ψ *can be evaluated at* (x, y), *and*

$$\Phi(x, y) = \log(\exp(x) \cdot \exp(y))$$
$$\Psi(x, y) = \log(\exp(-x) \cdot \exp(-y) \cdot \exp(x) \cdot \exp(y)).$$

Proof We prove the claim regarding Φ; the other claim is proved similarly. Let $\tau = \sup\{\|(x^n/n!) \cdot (y^m/m!)\| \mid (n, m) \in \mathbb{N}^2 \setminus (0, 0)\}$. Then $\tau \leq \|x\| \cdot \|y\|$ by (5), and it follows that

$$\lim_{n \to \infty} |(-1)^{n+1}/n| \tau^n = 0.$$

Theorem 6.24 now shows that

$$\Phi(x, y) = (\mathcal{L} \circ P)(x, y) = \mathcal{L}(P(x, y)) = \log(\exp(x) \cdot \exp(y)).$$

The most remarkable feature of the Campbell–Hausdorff series is that it can be expressed as an infinite sum of *Lie elements*. The algebra $\mathbb{Q}_p \langle\langle X, Y \rangle\rangle$, like any associative algebra, can be made into a Lie algebra by defining the *Lie bracket* operation

$$(U_1, U_2) = U_1 U_2 - U_2 U_1.$$

For $r > 2$ we put

$$(U_1, U_2, \ldots, U_r) = ((U_1, \ldots, U_{r-1}), U_r),$$

and for a vector $\mathbf{e} = (e_1, \ldots, e_n)$ of positive integers we write

$$\langle e \rangle = e_1 + \cdots + e_n$$
$$(X, Y)_{\mathbf{e}} = (X, \underbrace{Y, \ldots, Y}_{e_1}, \underbrace{X, \ldots, X}_{e_2}, \ldots).$$

6.28 Theorem (Campbell–Hausdorff formula) *Let* $\Phi(X, Y) = \sum_{n \in \mathbb{N}} u_n(X, Y)$ *where* $u_n(X, Y)$ *is the sum of terms of degree* n. *Then*

$$u_0(X, Y) = 0, \quad u_1(X, Y) = X + Y, \quad u_2(X, Y) = \frac{1}{2}(XY - YX);$$

and for each $n \geq 3$

$$u_n(X, Y) = \sum_{\langle \mathbf{e} \rangle = n-1} q_{\mathbf{e}}(X, Y)_{\mathbf{e}},$$

where each $q_{\mathbf{e}}$ *is a rational constant satisfying*

$$p^{n-1} q_{\mathbf{e}} \in p\mathbb{Z}_p \quad \text{if } p \neq 2$$
$$2^{2n-2} q_{\mathbf{e}} \in 4\mathbb{Z}_2 \quad \text{if } p = 2.$$

Moreover, writing $\epsilon = 1$ *if* $p \neq 2$, $\epsilon = 2$ *if* $p = 2$, *we have*

$$\lim_{\langle \mathbf{e} \rangle \to \infty} \left| p^{\epsilon \langle \mathbf{e} \rangle} q_{\mathbf{e}} \right| = 0.$$

We postpone the proof until §6.5. A similar result holds for the series $\Psi(X, Y)$. However, we shall only require knowledge of the first few terms; these are easily calculated from the definition, giving

$$\Psi(X, Y) = XY - YX + (\text{terms of degree} \geq 3).$$

The following result will be important in Chapter 9, where we use the Campbell–Hausdorff series to define a group law on suitable Lie algebras:

6.29 Proposition *The Campbell–Hausdorff series satisfies the following identity:*

$$\Phi \circ (H_1, \Phi \circ \mathbf{H}_{23}) = \Phi \circ (\Phi \circ \mathbf{H}_{12}, H_3)$$

where $H_i(X_1, X_2, X_3) = X_i$ *and* $\mathbf{H}_{ij}(X_1, X_2, X_3) = (X_i, X_j)$.

Proof This is an expression of the associative law in the algebra $\mathbb{Q}_p \langle\langle X, Y, Z \rangle\rangle$, which implies that

$$\mathcal{E}(X)(\mathcal{E}(Y)\mathcal{E}(Z)) = (\mathcal{E}(X)\mathcal{E}(Y))\mathcal{E}(Z).$$

Recall that $\Phi = \mathcal{L} \circ P$ where $P(Y, Z) = \mathcal{E}(Y)\mathcal{E}(Z) - 1$. Using the standard identity $(\mathcal{E} \circ \mathcal{L})(T) = 1 + T$ (see Exercise 6), and the fact that composition of power series is associative, we have

$$\mathcal{E}(Y)\mathcal{E}(Z) = 1 + P(Y, Z) = ((\mathcal{E} \circ \mathcal{L}) \circ P))(Y, Z) = (\mathcal{E} \circ \Phi)(Y, Z)$$

$$\Rightarrow \mathcal{E}(X)(\mathcal{E}(Y)\mathcal{E}(Z)) - 1 = P \circ (H_1, \Phi \circ \mathbf{H}_{23})(X, Y, Z).$$

Similarly,

$$(\mathcal{E}(X)\mathcal{E}(Y))\mathcal{E}(Z) - 1 = P \circ (\Phi \circ \mathbf{H}_{12}, H_3)(X, Y, Z).$$

Thus $P \circ (H_1, \Phi \circ \mathbf{H}_{23}) = P \circ (\Phi \circ \mathbf{H}_{12}, H_3)$, and the result follows after composing on the left with $\mathcal{L}$, by the associativity of composition.

6.4 Commuting indeterminates

Many of the results of the preceding section have analogues in the context of *commuting* variables; these we now briefly sketch.

The set of formal power series over $\mathbb{Q}_p$ in n *commuting* indeterminates $X_1, \ldots, X_n$ is denoted $\mathbb{Q}_p[[\mathbf{X}]]$. The elements of $\mathbb{Q}_p[[\mathbf{X}]]$ will be written as formal sums $F(\mathbf{X}) = \sum_{\mathbf{m} \in \mathbb{N}^n} b_{\mathbf{m}} \mathbf{X}^{\mathbf{m}}$ with each $b_{\mathbf{m}} \in \mathbb{Q}_p$, where $\mathbf{X}^{\mathbf{m}} = X_1^{m_1} \ldots X_n^{m_n}$. The set $\mathbb{Q}_p[[\mathbf{X}]]$ is a $\mathbb{Q}_p$-algebra under pointwise addition and scalar multiplication, and the product is given by

$$\sum_{\mathbf{m} \in \mathbb{N}^n} a_{\mathbf{m}} \mathbf{X}^{\mathbf{m}} \sum_{\mathbf{m} \in \mathbb{N}^n} b_{\mathbf{m}} \mathbf{X}^{\mathbf{m}} = \sum_{\mathbf{m} \in \mathbb{N}^n} c_{\mathbf{m}} \mathbf{X}^{\mathbf{m}}$$

where

$$c_{\mathbf{m}} = \sum_{\mathbf{s}+\mathbf{t}=\mathbf{m}} a_{\mathbf{s}} b_{\mathbf{t}}.$$

There is an obvious 'forgetful' mapping

$$^{-} : \mathbb{Q}_p \langle\langle \mathbf{X} \rangle\rangle \to \mathbb{Q}_p[[\mathbf{X}]];$$

writing $W_{\mathbf{m}}(\mathbf{X})$ to denote the (finite) set of words on $\mathbf{X}$ in which X_i occurs exactly m_i times for $i = 1, \ldots, n$, we define

$$\overline{\sum_{w \in W(\mathbf{X})} a_w w(\mathbf{X})} = \sum_{\mathbf{m} \in \mathbb{N}^n} \left(\sum_{w \in W(\mathbf{m})} a_w \right) \mathbf{X}^{\mathbf{m}}.$$

It is straightforward to verify

6.30 Lemma *The map* $\bar{\ }$: $\mathbb{Q}_p \langle\langle \mathbf{X} \rangle\rangle \to \mathbb{Q}_p[[X]]$ *is an epimorphism of* $\mathbb{Q}_p$-*algebras; its kernel is the ideal generated by* $\{X_i X_j - X_j X_i \mid i,j = 1,\dots,n\}$.

6.31 Definition Let

$$G(\mathbf{Y}) = \sum_{\mathbf{s} \in \mathbb{N}^m} b_{\mathbf{s}} \mathbf{Y}^{\mathbf{s}} \in \mathbb{Q}_p[[\mathbf{Y}]],$$

$$F_i(\mathbf{X}) = \sum_{\mathbf{t} \in \mathbb{N}^m} a_{i\mathbf{t}} \mathbf{X}^{\mathbf{t}} \in \mathbb{Q}_p[[\mathbf{X}]] \qquad i = 1,\dots,m,$$

where $\mathbf{X} = (X_1,\dots,X_n)$ and $\mathbf{Y} = (Y_1,\dots,Y_m)$. Assume that for $i = 1,\dots,m$, the 'constant term' a_{i0} is equal to 0. For $\mathbf{s} \in \mathbb{N}^m$, define the coefficients $c_{\mathbf{st}} \in \mathbb{Q}_p$ by

$$F_1(\mathbf{X})^{s_1} \dots F_m(\mathbf{X})^{s_m} = \sum_{\mathbf{t} \in \mathbb{N}^n} c_{\mathbf{st}} \mathbf{X}^{\mathbf{t}}.$$

The *composite* of G and $\mathbf{F} = (F_1,\dots,F_m)$ is defined to be the formal power series

$$(G \circ \mathbf{F})(\mathbf{X}) = \sum_{\mathbf{t} \in \mathbb{N}^n} \left(\sum_{\mathbf{s} \in \mathbb{N}^m} b_{\mathbf{s}} c_{\mathbf{st}} \right) \mathbf{X}^{\mathbf{t}}.$$

Comparing this with Definition 6.23, it is a routine exercise to verify

6.32 Lemma *Suppose that* $F_1(\mathbf{X}),\dots,F_m(\mathbf{X}) \in \mathbb{Q}_p \langle\langle \mathbf{X} \rangle\rangle$ *and* $G(\mathbf{Y}) \in \mathbb{Q}_p \langle\langle \mathbf{Y} \rangle\rangle$ *are formal power series satisfying the conditions of Definition 6.23. Then* $\overline{\mathbf{F}} = (\overline{F}_1,\dots,\overline{F}_m)$ *satisfies the conditions of Definition 6.31 and*

$$\overline{G \circ \mathbf{F}} = \overline{G} \circ \overline{\mathbf{F}}.$$

From now on, $(A, \|\cdot\|)$ will denote a complete normed $\mathbb{Q}_p$-algebra, as in the preceding section.

6.33 Definition. The power series $F(\mathbf{X}) = \sum_{\mathbf{m} \in \mathbb{N}^n} b_{\mathbf{m}} \mathbf{X}^{\mathbf{m}} \in \mathbb{Q}_p[[\mathbf{X}]]$ can be *evaluated* at $\mathbf{x} = (x_1,\dots,x_n) \in A^n$ if the series $\sum_{\mathbf{m} \in \mathbb{N}^n} b_{\mathbf{m}} x_1^{m_1} \dots x_n^{m_n}$ converges in A; in this case we denote its sum by $F(\mathbf{x})$.

6.34 Proposition *Let* $F(\mathbf{X}) \in \mathbb{Q}_p \langle\langle \mathbf{X} \rangle\rangle$ *and suppose that* $F(\mathbf{x})$ *exists where* $\mathbf{x} = (x_1,\dots,x_n) \in A^n$. *If* $x_i x_j = x_j x_i$ *for each* $i,j = 1,\dots,n$ *then* $\overline{F(\mathbf{X})} \in \mathbb{Q}_p[[\mathbf{X}]]$ *can be evaluated at* $\mathbf{x}$ *and* $\overline{F}(\mathbf{x}) = F(\mathbf{x})$.

Proof Say $F(\mathbf{X}) = \sum_{w \in W} a_w w$; then $\overline{F(\mathbf{X})} = \sum_{\mathbf{m} \in \mathbb{N}^n} b_{\mathbf{m}} \mathbf{X}^{\mathbf{m}}$ where $b_{\mathbf{m}} = \sum_{w \in W_{\mathbf{m}}} a_w$. Now W is the union of the countable family $\{W_{\mathbf{m}} \mid \mathbf{m} \in \mathbb{N}^n\}$ of disjoint sets $W_{\mathbf{m}}$, each of which is finite. The hypothesis asserts that the series $\sum_{w \in W} a_w w(\mathbf{x})$ converges in A with sum $F(\mathbf{x})$. Put $s_{\mathbf{m}} = \sum_{w \in W_{\mathbf{m}}} a_w w(\mathbf{x})$ for each $\mathbf{m} \in \mathbb{N}^n$. Since $x_i x_j = x_j x_i$ for each $i, j = 1, \dots, n$, we have

$$s_{\mathbf{m}} = \left(\sum_{w \in W_{\mathbf{m}}} a_w \right) x_1^{m_1} \dots x_n^{m_n} = b_{\mathbf{m}} x_1^{m_1} \dots x_n^{m_n}.$$

It follows by Proposition 6.10 that $\sum_{\mathbf{m} \in \mathbb{N}^n} b_{\mathbf{m}} x_1^{m_1} \dots x_n^{m_n}$ converges to the sum $F(\mathbf{x})$; this is the claim of the proposition.

We can now deduce the following analogue to Theorem 6.24:

6.35 Theorem *Let* $(A, \|\cdot\|)$ *be a complete normed* $\mathbb{Q}_p$-*algebra, and let* $F_1(\mathbf{X}), \dots, F_m(\mathbf{X})$ *and* $G(\mathbf{Y})$ *be formal power series satisfying the conditions of Definition 6.31. Suppose that* $F_1(\mathbf{X}), \dots, F_m(\mathbf{X})$ *can all be evaluated at some point* $\mathbf{x} = (x_1, \dots, x_m) \in A^n$, *where* $x_i x_j = x_j x_i$ *for each pair* (i, j). *For each* i *put* $\tau_i = \sup\{\|a_{i\mathbf{t}} x_1^{t_1} \dots x_n^{t_n}\| \mid \mathbf{t} \in \mathbb{N}^n\}$, *and suppose that*

$$\lim_{\mathbf{s} \in \mathbb{N}^m} |b_{\mathbf{s}}| \tau_1^{s_1} \dots \tau_m^{s_m} = 0.$$

Then $G(F_1(\mathbf{x}), \dots, F_m(\mathbf{x}))$ *and* $(G \circ \mathbf{F})(\mathbf{x})$ *both exist and are equal.*

Proof It is clear that we can choose $K_1(\mathbf{X}), \dots, K_m(\mathbf{X}) \in \mathbb{Q}_p \langle\langle \mathbf{X} \rangle\rangle$ and $L(\mathbf{Y}) \in \mathbb{Q}_p \langle\langle \mathbf{Y} \rangle\rangle$ satisfying the hypothesis of Theorem 6.24 with respect to the point $\mathbf{x} \in A^n$ and with the property that $\overline{K_i}(\mathbf{X}) = F_i(\mathbf{X})$ for each $i = 1, \dots, m$ and $\overline{L}(\mathbf{Y}) = G(\mathbf{Y})$. From Theorem 6.24 we may conclude that $L(K_1(\mathbf{x}), \dots, K_m(\mathbf{x}))$ and $(L \circ \mathbf{K})(\mathbf{x})$ both exist and are equal. The result now follows by 6.32 and 6.34.

A typical application is

6.36 Corollary *Let* $x, y \in A_0$ *and suppose that* $xy = yx$. *Then*

$$\exp(x - y) = \exp(x)(\exp(y))^{-1}.$$

Here we take $F(X_1, X_2) = X_1 + X_2$ and $G(Y) = \text{Exp}(Y) = \overline{\mathcal{E}(Y)}$, the exponential series. Then $(G \circ F)(X_1, X_2) = G(X_1)G(X_2)$ by a standard identity. Let $\mathbf{x} = (x_1, x_2) \in A_0 \times A_0$. Then the hypotheses of Theorem

6.35 are satisfied, since now $\tau_i = \|x_i\|$ for $i = 1, 2$, so we may infer that $\exp(x_1 + x_2) = \exp(x_1)\exp(x_2)$. The result follows by Corollary 6.25(iv), putting $x_1 = x$ and $x_2 = -y$.

We conclude this section with a result that will be important in Chapter 9:

6.37 Theorem ('Inverse Function Theorem') *Let* $U_1(\mathbf{X}), \ldots, U_d(\mathbf{X}) \in \mathbb{Q}_p[[X_1, \ldots, X_d]]$, *and let* $m \in \mathbb{N}$. *Assume that for each* i,
   (a) U_i *has constant term* 0;
   (b) *all coefficients of* U_i *lie in* $p^m \mathbb{Z}_p$; *and*
   (c) *the linear part of* $U_i(\mathbf{X})$ *is* $\sum_{j=1}^d b_{ij} X_j$, *where the matrix* $(p^{-m}b_{ij})$ *lies in* $\mathrm{GL}_d(\mathbb{Z}_p)$.
Then there exist power series $V_1(\mathbf{X}), \ldots, V_d(\mathbf{X}) \in \mathbb{Q}_p[[X_1, \ldots, X_d]]$, *with zero constant terms, such that*

$$(V_i \circ \mathbf{U})(\mathbf{X}) = (U_i \circ \mathbf{V})(\mathbf{X}) = X_i \quad (i = 1, \ldots, d).$$

For each $\mathbf{y} \in p^{m+1}\mathbb{Z}_p^d$, *we have* $\mathbf{V}(\mathbf{y}) \in p\mathbb{Z}_p^d$ *and* $U_i(\mathbf{V}(\mathbf{y})) = y_i$ ($i = 1, \ldots, d$).

Proof Consider first the special case where $m = 0$ and (b_{ij}) is the identity matrix. Thus

$$U_i(\mathbf{X}) = X_i + \sum_{j \geq 2} U_i^{[j]}(\mathbf{X}),$$

where in general $W^{[j]}$ will stand for the homogeneous part of degree j in a power series W. We want to find power series $V_1, \ldots, V_d$ such that $(\mathbf{U} \circ \mathbf{V})(\mathbf{X}) = \mathbf{X}$. Put

$$V_i^{[0]} = 0, \qquad V_i^{[1]} = X_i$$

for $i = 1, \ldots, d$. Then $(\mathbf{U} \circ \mathbf{V})(\mathbf{X}) = \mathbf{X}$ is equivalent to

$$\sum_{j \geq 2} V_i^{[j]}(\mathbf{X}) = -\sum_{k \geq 2} U_i^{[k]}\left(\mathbf{X} + \sum_{j \geq 2} V^{[j]}(\mathbf{X})\right). \tag{7}$$

Write $W_i(\mathbf{X})$ for the right-hand side of (7). Then for $t \geq 2$, $W_i^{[t+1]}(\mathbf{X})$ is the same as the homogeneous part of degree $t + 1$ in the polynomial

$$-\sum_{k=2}^{t+1} U_i^{[k]}\left(\mathbf{X} + \sum_{j=2}^{t} V^{[j]}(\mathbf{X})\right). \tag{8}$$

Also, $W_i^{[2]}(\mathbf{X}) = -U_i^{[2]}(\mathbf{X})$. So we can construct the $V_i^{[t]}$ recursively by putting $V_i^{[2]} = -U_i^{[2]}$ and, for $t \geq 2$, $V_i^{[t+1]} = W_i^{[t+1]}$, since the polynomial (8) depends only on $V_1^{[j]}, \ldots, V_d^{[j]}$ for $j \leq t$.

Note that, by construction, $\mathbf{V}$ has coefficients in $\mathbb{Z}_p$, $V_i^{[0]} = 0$ and $V_i^{[1]}(\mathbf{X}) = X_i$ for each i. Hence we can repeat the construction with $\mathbf{V}$ in place of $\mathbf{U}$, to obtain power series $S_1, \ldots, S_d$, say, such that $(\mathbf{V} \circ \mathbf{S})(\mathbf{X}) = \mathbf{X}$. Then (Exercise 6.9)

$$\mathbf{U}(\mathbf{X}) = (\mathbf{U} \circ (\mathbf{V} \circ \mathbf{S}))(\mathbf{X}) = ((\mathbf{U} \circ \mathbf{V}) \circ \mathbf{S})(\mathbf{X}) = \mathbf{S}(\mathbf{X}),$$

so $(\mathbf{V} \circ \mathbf{U})(\mathbf{X}) = \mathbf{X}$, as required.

The general case is reduced to the case just considered by a change of variables. To begin with, let us assume still that $m = 0$, but allow $(b_{ij}) = B$ to be an arbitrary matrix in $\mathrm{GL}_d(\mathbb{Z}_p)$. Put $C = (c_{ij}) = B^{-1}$. For any matrix $A = (a_{ij}) \in \mathrm{GL}_d(\mathbb{Z}_p)$, we denote by $\mathbf{A}(\mathbf{X})$ the d-tuple of formal power series

$$\left(\sum_j a_{1j} X_j, \ldots, \sum_j a_{dj} X_j \right).$$

Now put

$$\mathbf{U}^*(\mathbf{X}) = (\mathbf{U} \circ \mathbf{C})(\mathbf{X}).$$

Then $U_i^{*[0]} = 0$ and $U_i^{*[1]}(\mathbf{X}) = X_i$ for each i. So by the special case above, there exist $V_i^*(\mathbf{X}) \in \mathbb{Z}_p[[\mathbf{X}]]$, $1 \leq i \leq d$, with zero constant terms, such that

$$(\mathbf{U}^* \circ \mathbf{V}^*)(\mathbf{X}) = (\mathbf{V}^* \circ \mathbf{U}^*)(\mathbf{X}) = \mathbf{X}.$$

Now let

$$\mathbf{V}(X) = (\mathbf{C} \circ \mathbf{V}^*)(\mathbf{X}).$$

Then

$$(\mathbf{V} \circ \mathbf{U})(\mathbf{X}) = (\mathbf{C} \circ \mathbf{V}^* \circ \mathbf{U})(\mathbf{X})$$
$$= (\mathbf{C} \circ \mathbf{V}^* \circ \mathbf{U}^* \circ \mathbf{B})(\mathbf{X}) = \mathbf{X}$$

since clearly $\mathbf{U}^* \circ \mathbf{B} = \mathbf{U}$ and $(\mathbf{C} \circ \mathbf{B})(\mathbf{X}) = \mathbf{X}$. Similarly,

$$(\mathbf{U} \circ \mathbf{V})(\mathbf{X}) = (\mathbf{U}^* \circ \mathbf{B} \circ \mathbf{C} \circ \mathbf{V}^*)(\mathbf{X}) = \mathbf{X}.$$

Now consider the general case, with $m \geq 0$. Put

$$\overline{\mathbf{U}}(\mathbf{X}) = p^{-m}\mathbf{U}(\mathbf{X}).$$

By the previous case, there exist $\overline{V_1}, \ldots, \overline{V_d} \in \mathbb{Z}_p[[\mathbf{X}]]$ such that $(\overline{\mathbf{U}} \circ \overline{\mathbf{V}})(\mathbf{X}) = (\overline{\mathbf{V}} \circ \overline{\mathbf{U}})(\mathbf{X}) = \mathbf{X}$. Put

$$\mathbf{V}(\mathbf{X}) = \overline{\mathbf{V}}(p^{-m}\mathbf{X}).$$

One checks immediately that then $(\mathbf{U} \circ \mathbf{V})(\mathbf{X}) = (\mathbf{V} \circ \mathbf{U})(\mathbf{X}) = \mathbf{X}$.

Finally, let $\mathbf{y} \in p^{m+1}\mathbb{Z}_p^d$. Then $\mathbf{V}(\mathbf{y}) = \overline{\mathbf{V}}(p^{-m}\mathbf{y})$, which converges to an element of $p\mathbb{Z}_p^d$ since $p^{-m}\mathbf{y} \in p\mathbb{Z}_p^d$ and all coefficients of $\overline{\mathbf{V}}$ lie in $\mathbb{Z}_p$. It is clear that the hypotheses of Theorem 6.35 are fulfilled if we take $G = U_i$ and $\mathbf{F} = \mathbf{V}$, so we may infer that

$$U_i(\mathbf{V}(\mathbf{y})) = (U_i \circ \mathbf{V})(\mathbf{y}) = y_i$$

for each i.

6.5 The Campbell–Hausdorff formula

The Campbell–Hausdorff series $\Phi(X, Y)$ was defined in §6.4 by

$$\Phi(X, Y) = (\mathcal{L} \circ P)(X, Y),$$

where

$$P(X, Y) = \mathcal{E}(X)\mathcal{E}(Y) - 1 \in \mathbb{Q}_p \langle\langle X, Y \rangle\rangle$$

and $\mathcal{E}(X)$, $\mathcal{L}(X)$ are the formal power series given in Definition 6.21; they represent the functions $\exp(x)$ and $\log(1 + x)$ respectively. According to Propositions 6.22 and 6.27, if A is a complete normed $\mathbb{Q}_p$-algebra and $x, y \in A_0$ then

$$\Phi(x, y) = \log(\exp(x) \cdot \exp(y)) \in A_0.$$

Recall that $x \in A_0$ if and only if $\|x\| \leq p^{-\epsilon}$ where

$$\epsilon = 1 \quad \text{if} \ \ p \neq 2, \quad \epsilon = 2 \quad \text{if} \ \ p = 2,$$

notation that we keep throughout this section. In view of Corollary 6.25 this implies

$$\exp \Phi(x, y) = \exp(x) \cdot \exp(y). \tag{9}$$

Writing $W = W(X, Y)$, we have

$$\Phi(X, Y) = \sum_{w \in W} a_w w,$$

where $a_w \in \mathbb{Q}$ for each w (because both the series $\mathcal{E}$ and $\mathcal{L}$ have rational coefficients). For each $n \in \mathbb{N}$ we define a polynomial u_n (in the non-commuting variables X and Y) by

$$u_n(X, Y) = \sum_{\deg w = n} a_w w;$$

thus $\Phi(X, Y) = \sum_{n \in \mathbb{N}} u_n(X, Y)$. It takes a short calculation to verify that

$$u_0(X, Y) = 0, \quad u_1(X, Y) = X + Y, \quad u_2(X, Y) = \frac{1}{2}(X, Y),$$

where $(X, Y) = XY - YX$.

We now restate

6.28 Theorem *For each $n \geq 3$ we have*

$$u_n(X, Y) = \sum_{\langle \mathbf{e} \rangle = n-1} q_{\mathbf{e}}(X, Y)_{\mathbf{e}},$$

where each $q_{\mathbf{e}} \in \mathbb{Q}$ satisfies

$$p^{\epsilon(n-1)} q_{\mathbf{e}} \in p^{\epsilon} \mathbb{Z}_p.$$

Moreover,

$$\lim_{\langle \mathbf{e} \rangle \to \infty} \left| p^{\epsilon \langle \mathbf{e} \rangle} q_{\mathbf{e}} \right| = 0.$$

The notation here is as in §6.4: $\mathbf{e}$ runs over vectors $\mathbf{e} = (e_1, \ldots, e_s)$ of positive integers (with $s \geq 1$), $\langle \mathbf{e} \rangle = e_1 + \cdots + e_s$ and

$$(X, Y)_{\mathbf{e}} = (X, \underbrace{Y, \ldots, Y}_{e_1}, \underbrace{X, \ldots, X}_{e_2}, \ldots).$$

Remark It is in fact the case that

$$p^{\lceil \frac{n-1}{p-1} \rceil} q_{\mathbf{e}} \in \mathbb{Z}_p$$

for all $n \geq 2$ (this is clearly stronger than the assertion made above). For the proof, which depends on the theory of free Lie algebras, see Exercise 11.

Before embarking on the proof of the theorem, we mention an important consequence:

6.38 Corollary *Let $L \cong \mathbb{Z}_p^d$ be a Lie algebra over $\mathbb{Z}_p$, and suppose that $(L, L) \subseteq p^{\epsilon} L$. Let $x, y \in L$, and for $n \in \mathbb{N}$ define the element $u_n(x, y) \in \mathbb{Q}_p L$ by $u_n(x, y) = \sum_{\langle \mathbf{e} \rangle = n-1} q_{\mathbf{e}}(x, y)_{\mathbf{e}}$. Then*

(i) $u_n(x, y) \in L$ for all $n \in \mathbb{N}$;

(ii) *the series*

$$\widetilde{\Phi}(x, y) = \sum_{n \in \mathbb{N}} u_n(x, y)$$

converges in L;

(iii) $\widetilde{\Phi}(x, y) - (x + y) \in pL$, *and*

(iv) *if* $p = 2$ *then* $\widetilde{\Phi}(x, y) - (x + y) - \frac{1}{2}(x, y) \in 4L$.

Proof The hypothesis implies that if $\langle \mathbf{e} \rangle = n - 1$ then $(x, y)_\mathbf{e} \in p^{\epsilon(n-1)}L$. Since $p^{\epsilon(n-1)}q_\mathbf{e} \in \mathbb{Z}_p$ by the theorem, it follows that

$$p^{\epsilon(n-1)}q_\mathbf{e}(x, y)_\mathbf{e} \in p^{\epsilon(n-1)}L.$$

Therefore $q_\mathbf{e}(x, y)_\mathbf{e} \in L$, giving (i).

To prove (ii), it suffices to show that the partial sums of the given series form a Cauchy sequence in L (here we think of L as an additive pro-p group). Let $N \in \mathbb{N}$. The second part of the theorem shows that for all sufficiently large $n \in \mathbb{N}$ we have $p^{\epsilon(n-1)}q_\mathbf{e} \in p^N \mathbb{Z}_p$ when $\langle \mathbf{e} \rangle = n - 1$. As above, this implies that

$$p^{\epsilon(n-1)}q_\mathbf{e}(x, y)_\mathbf{e} \in p^{N+\epsilon(n-1)}L,$$

which shows in turn that $u_n(x, y) \in p^N L$ for all sufficiently large n. The result follows.

The argument above shows that if $n \geq 3$ then in fact $u_n(x, y) \in p^\epsilon L$. Parts (iii) and (iv) follow, since $p^\epsilon L$ is closed in L (and $\frac{1}{2} \in \mathbb{Z}_p$ when p is odd).

We now start on the proof of the theorem, and begin with some general observations. Let $(A, \|\cdot\|)$ be a complete normed $\mathbb{Q}_p$-algebra. A $\mathbb{Q}_p$-linear map $\phi : A \to A$ is called *bounded* if there exists a real number $r > 0$ such that

$$\|a\phi\| \leq r\|a\| \quad \text{for all } a \in A.$$

Let B be the set of all bounded $\mathbb{Q}_p$-linear maps $A \to A$ and define the norm $\|\cdot\|_0$ on B by

$$\|\phi\|_0 = \sup_{a \in A} \frac{\|a\phi\|}{\|a\|}.$$

We leave it as an exercise to prove that $(B, \|\cdot\|_0)$ is a complete normed $\mathbb{Q}_p$-algebra. We shall be especially interested in the following class of elements of B. For each $a \in A$, we define *right multiplication, left multiplication* and *adjoint* by

$$u\mathbf{r}_a = ua, \quad u\mathbf{l}_a = au, \quad u\mathrm{ad}(a) = ua - au$$

for all $u \in A$. Note that $\mathbf{r}_a$, $\mathbf{l}_a$ and $\mathrm{ad}(a)$ all lie in B with norms bounded by $\|a\|$. In particular, it follows that if $a \in A_0$ then each of them lies in B_0. The maps $\mathbf{r} : A \to B$, $\mathbf{l} : A \to B$ and $\mathrm{ad} : A \to B$ are continuous, and clearly $\mathbb{Q}_p$-linear. Combining this with the following identities

$$\mathbf{l}_{ab} = \mathbf{l}_b \mathbf{l}_a, \quad \mathbf{r}_{ab} = \mathbf{r}_a \mathbf{r}_b,$$

we may deduce that, provided $a \in A_0$,

$$\mathbf{l}_{\exp(a)} = \exp(\mathbf{l}_a), \quad \mathbf{r}_{\exp(a)} = \exp(\mathbf{r}_a).$$

For a related result concerning the map ad, see Exercise 12. What we need now is the following key result:

6.39 Lemma *For all $a, b \in A$ and $n \in \mathbb{N}$,*

$$u_n(\mathrm{ad}(a), \mathrm{ad}(b)) = \mathrm{ad}(u_n(a, b)).$$

Proof Let $a, b \in A_0$. Then $\mathbf{l}_a, \mathbf{l}_b \in B_0$ and we have

$$\begin{aligned}
\exp(\mathbf{l}_b)\exp(\mathbf{l}_a) = \mathbf{l}_{\exp(b)}\mathbf{l}_{\exp(a)} &= \mathbf{l}_{\exp(a)\exp(b)}\\
&= \mathbf{l}_{\exp \Phi(a,b)} \quad \text{by (9)}\\
&= \exp(\mathbf{l}_{\Phi(a,b)}).
\end{aligned}$$

Similarly, $\exp(\mathbf{r}_a)\exp(\mathbf{r}_b) = \exp(\mathbf{r}_{\Phi(a,b)})$. Since left multiplication commutes with right multiplication and $\mathrm{ad}(a) = \mathbf{r}_a - \mathbf{l}_a$ we may apply (9) again with Corollary 6.36 to deduce that

$$\begin{aligned}
\exp(\Phi(\mathrm{ad}(a), \mathrm{ad}(b))) &= \exp(\mathrm{ad}(a))\exp(\mathrm{ad}(b))\\
&= \exp(\mathbf{r}_a)\exp(\mathbf{l}_a)^{-1}\exp(\mathbf{r}_b)\exp(\mathbf{l}_b)^{-1}\\
&= \exp(\mathbf{r}_a)\exp(\mathbf{r}_b)(\exp(\mathbf{l}_b)\exp(\mathbf{l}_a))^{-1}\\
&= \exp(\mathbf{r}_{\Phi(a,b)})\exp(\mathbf{l}_{\Phi(a,b)})^{-1}\\
&= \exp(\mathrm{ad}(\Phi(a,b))).
\end{aligned}$$

It follows that $\Phi(\mathrm{ad}(a), \mathrm{ad}(b)) = \mathrm{ad}(\Phi(a,b))$, so we have

$$\sum_{n \in \mathbb{N}} u_n(\mathrm{ad}(a), \mathrm{ad}(b)) = \mathrm{ad}(\sum_{n \in \mathbb{N}} u_n(a, b)); \tag{10}$$

this holds for all $a, b \in A_0$.

Now let a and b be arbitrary elements from A with norms bounded by p^r. Then for all $\lambda \in \mathbb{Q}_p$ with $|\lambda| \leq p^{-r-2}$ we have $\lambda a, \lambda b \in A_0$. Since the polynomials $u_n(X, Y)$ are homogeneous of degree n we conclude that for all $a, b \in A$ and all such $\lambda \in \mathbb{Q}_p$,

$$\sum_{n \in \mathbb{N}} \lambda^n u_n(\mathrm{ad}(a), \mathrm{ad}(b)) = \sum_{n \in \mathbb{N}} u_n(\mathrm{ad}(\lambda a), \mathrm{ad}(\lambda b))$$

$$= \mathrm{ad}(\sum_{n \in \mathbb{N}} u_n(\lambda a, \lambda b)) \qquad \text{by (10)}$$

$$= \sum_{n \in \mathbb{N}} \mathrm{ad}(u_n(\lambda a, \lambda b)))$$

$$= \sum_{n \in \mathbb{N}} \lambda^n \mathrm{ad}(u_n(a, b))),$$

because ad is continuous and linear. The lemma follows by Proposition 6.13.

Let us write

$$c_w = c(i_1, j_1, \ldots, i_k, j_k) \quad \text{where } w = X^{i_1} Y^{j_1} \ldots X^{i_k} Y^{j_k}$$

(here $i_1 \geq 0$ and all other exponents are ≥ 1). The following lemma lies at the heart of the proof:

6.40 Lemma *Let* $n \geq 2$. *Then* $n u_n(X, Y)$ *is equal to*

$$\sum_{i_1 = 1} c(\mathbf{f})(X, Y)_{(j_1, i_2, \ldots, i_k, j_k)} - \sum_{\substack{(i_1, j_1) = (0,1) \\ i_2 \geq 1}} c(\mathbf{f})(X, Y)_{(j_1, i_2 - 1, j_2, \ldots, i_k, j_k)},$$

the sum being over all $2k$-tuples $\mathbf{f} = (i_1, j_1, \ldots, i_k, j_k)$ *with* $k \geq 1$ *and* $\langle \mathbf{f} \rangle = n$.

Proof We start by constructing a certain complete normed $\mathbb{Q}_p$-algebra A, as follows. Let $\mathbb{Q}_p \langle X, Y \rangle$ be the $\mathbb{Q}_p$-subalgebra of $\mathbb{Q}_p \langle\langle X, Y \rangle\rangle$ generated by X and Y, i.e. the non-commuting polynomial ring in X and Y. We construct a $\mathbb{Q}_p$-algebra P by adjoining to $\mathbb{Q}_p \langle X, Y \rangle$ an indeterminate t, satisfying the multiplication rules

$$Xt = (t-1)X, \qquad Yt = (t-1)Y;$$

thus P is a so-called *skew polynomial ring over* $\mathbb{Q}_p \langle X, Y \rangle$. The elements of P can be uniquely written as finite sums of the form $\sum_w f_w(t) w$, where each $w \in W(X, Y)$ and each $f_w(t) \in \mathbb{Q}_p[t]$. If we define $\|\cdot\| : P \to \mathbb{R}$ by

$$\left\| \sum a_{wi} t^i w \right\| = \max_{w,i} |a_{wi}|,$$

$(P, \|\cdot\|)$ becomes a normed $\mathbb{Q}_p$-algebra, as the reader may verify. Finally, let $(A, \|\cdot\|)$ be the completion of $(P, \|\cdot\|)$.

Note that A contains the non-commuting polynomial ring $\mathbb{Q}_p \langle X, Y \rangle$ as a subalgebra, and that if $w \in W$ has degree $n \geq 1$ then

$$wt = (t - n)w,$$

by the definition of multiplication in P. It follows that if $v(X, Y) \in \mathbb{Z} \langle X, Y \rangle$ is any homogeneous polynomial of degree $n \geq 1$ then

$$\mathrm{tad}(v(X, Y)) = n v(X, Y). \tag{11}$$

Put $\xi = \mathrm{ad}(X)$, $\eta = \mathrm{ad}(Y)$. Let $w = X^{i_1} Y^{j_1} \ldots X^{i_k} Y^{j_k}$. Suppose that $i_1 \geq 1$. Then

$$
\begin{aligned}
tw(\xi, \eta) &= \mathrm{tad}(X)\xi^{i_1 - 1}\eta^{j_1} \ldots \xi^{i_k}\eta^{j_k} \\
&= X\xi^{i_1 - 1}\eta^{j_1} \ldots \xi^{i_k}\eta^{j_k} \\
&= \begin{cases} 0 & \text{if } i_1 > 1 \\ (X, Y)_{(j_1, i_2, \ldots, i_k, j_k)} & \text{if } i_1 = 1. \end{cases}
\end{aligned}
$$

Similarly, if $i_1 = 0$ then

$$
\begin{aligned}
tw(\xi, \eta) &= \mathrm{tad}(Y)\eta^{j_1 - 1}\xi^{i_2} \ldots \xi^{i_k}\eta^{j_k} \\
&= \begin{cases} 0 & \text{if } j_1 > 1 \\ (Y, X)_{(i_2, \ldots, i_k, j_k)} & \text{if } j_1 = 1. \end{cases}
\end{aligned}
$$

this is equal to 0 if $i_2 = 0$ and equal to

$$-(X, Y)_{(1, i_2 - 1, j_2, \ldots, i_k, j_k)}$$

if $j_1 = 1$ and $i_2 \geq 1$.

Now applying (11) we obtain

$$
\begin{aligned}
n u_n(X, Y) &= \mathrm{tad}(u_n(X, Y)) \\
&\quad - t u_n(\xi, \eta) \qquad \text{by Lemma 6.39} \\
&= \sum_{\deg(w) = n} c_w t w(\xi, \eta) \\
&= \sum_{i_1 = 1} c(\mathbf{f})(X, Y)_{(j_1, i_2, \ldots, i_k, j_k)} \\
&\quad - \sum_{\substack{(i_1, j_1) = (0, 1) \\ i_2 \geq 1}} c(\mathbf{f})(X, Y)_{(j_1, i_2 - 1, j_2, \ldots, i_k, j_k)},
\end{aligned}
$$

in view of the preceding paragraph.

The lemma shows that $u_n(X, Y)$ can be written in the form $\sum q_e(X, Y)_e$ where each q_e is a sum of terms $\pm n^{-1}c(\mathbf{f})$ with $\langle \mathbf{f} \rangle - 1 = \langle e \rangle = n - 1$. The following therefore completes the proof of Theorem 6.28:

6.41 Lemma *Let* $\mathbf{f} = (i_1, j_1, \ldots, i_k, j_k)$, *and suppose that* $\langle \mathbf{f} \rangle = n \geq 3$ *and that* $i_1 = 1$ *or* $j_1 = 1$. *Then*

$$p^{\epsilon(n-1)}n^{-1}c(\mathbf{f}) \in p^{h(n)}\mathbb{Z}_p$$

where $h(n) \geq \epsilon$ *for all* n *and* $h(n) \to \infty$ *as* $n \to \infty$.

Proof The terms in the power series $P(X, Y) = \mathcal{E}(X)\mathcal{E}(Y) - 1$ are each of the simple form

$$\frac{1}{i!j!}X^i Y^j,$$

with $i + j \geq 1$. It follows that each term of degree $n \geq 1$ in the series $\Phi(X, Y) = (\mathcal{L} \circ P)(X, Y)$ has the form

$$\frac{(-1)^{k+1}}{k} \cdot \frac{1}{i_1!j_1!\ldots i_k!j_k!}X^{i_1}Y^{j_1}\ldots X^{i_k}Y^{j_k}$$

where $1 \leq k \leq n$, $i_1 + j_1 + \cdots + i_k + j_k = n$, and $i_\ell + j_\ell \geq 1$ for each ℓ. Some of these terms will coalesce, since some of the exponents i_ℓ and j_ℓ may be zero; however, if we put

$$f(n) = n \cdot \mathrm{lcm}\left\{ k \prod_{\ell=1}^{k} i_\ell!j_\ell! \,\Big|\, \sum_{\ell=1}^{k} i_\ell + j_\ell = n, \quad i_\ell + j_\ell \geq 1 \text{ for each } \ell \right\},$$

it is clear that $f(n) \cdot n^{-1}c(\mathbf{f}) \in \mathbb{Z}$ for each $\mathbf{f}$ with $\langle \mathbf{f} \rangle = n$. To show that $h(n) \geq \epsilon$ it would therefore suffice to verify that

$$v(f(n)) \leq \epsilon(n - 2) \tag{12}$$

when $n \geq 3$. Unfortunately this is false (though the lemma is true) for $n = p = 3$; we refer to Exercise 10 for the cases $n = 3, 4$, and now prove (12) for $n \geq 5$.

Assume that $s_\ell = i_\ell + j_\ell \geq 1$ for each ℓ. Then $i_\ell!j_\ell!$ divides $s_\ell!$, so Lemma 6.20 gives

$$v\left(k \prod_{\ell=1}^{k} i_\ell!j_\ell! \right) \leq \frac{k-1}{p-1} + \sum_{\ell=1}^{k} \frac{s_\ell - 1}{p-1} = \frac{n-1}{p-1} \tag{13}$$

where $\sum_{\ell=1}^{k} s_\ell = n$. On the other hand, observe that for any integer $e \geq 2$ we have $2(p-1)e + 1 \leq p^e$, unless $p = 2$ in which case we require $e \geq 3$ (to see this, expand $(1 + (p-1))^e$); taking $e = v(n)$ gives

$$v(n) \leq \max \left\{ \epsilon, \frac{n-1}{2(p-1)} \right\}. \tag{14}$$

One verifies trivially that

$$\epsilon + \frac{n-1}{p-1} \leq \epsilon(n-2)$$

$$\frac{3(n-1)}{2(p-1)} \leq \epsilon(n-2)$$

when $n \geq 5$; hence (12) follows from (13) and (14).

To show that $h(n)$ tends to infinity with n is easier: it suffices to show that

$$v(f(n)) - \epsilon(n-1) \to -\infty;$$

but from (13) we have

$$v(f(n)) - \epsilon(n-1) \leq v(n) + \frac{n-1}{p-1} - \epsilon(n-1)$$

$$\leq \log_p n - \frac{\epsilon}{2}(n-1),$$

and the result follows since $\epsilon > 0$.

6.6 Power series over pro-p rings

The reader is advised to skip this section until she or he is about to read Chapter 13 where we introduce the concept of analytic groups over rings more general than $\mathbb{Q}_p$. For this we shall need suitable generalisations of some of the results established earlier in this chapter, where the role of the p-adic integers is taken by a ring of more general type, and it is the purpose of this section to supply these.

Throughout, all rings are assumed to be *commutative, Noetherian* and *with identity*. A ring R is *local* if R has a unique maximal ideal $\mathfrak{m} \neq 0$; one refers to 'the local ring $(R, \mathfrak{m})$'. The quotient $R/\mathfrak{m}$ is called the *residue field* of R.

If $(R, \mathfrak{m})$ is a local ring, the powers of the maximal ideal $\mathfrak{m}^n$ ($n \in \mathbb{N}$) define a filtration on R called the $\mathfrak{m}$-*adic* filtration. This filtration in turn defines a topology on R called the $\mathfrak{m}$-*adic topology*, in which the system $(\mathfrak{m}^n)_{n \in \mathbb{N}}$ forms a base for the neighbourhoods of 0.

Definition A local ring $(R, \mathfrak{m})$ is a *pro-p ring* if

- the residue field $R/\mathfrak{m}$ is finite, of characteristic p, and
- R is complete with respect to the $\mathfrak{m}$-adic topology.

A pro-p ring which is an integral domain will be called a *pro-p domain*. It follows from Krull's Intersection Theorem (Atiyah and Macdonald (1969), Corollary 10.18) that the powers of the maximal ideal in a pro-p domain intersect in 0, so the topology is Hausdorff. Familiar examples of pro-p rings are the (complete) *discrete valuation rings* with finite residue field: a pro-p ring is a discrete valuation ring – we shall say *DVR* – if and only if its maximal ideal is principal.

As was the case with pro-p groups, there is an alternative definition in terms of inverse limits: see Exercise 13.

The fundamental structure theorem of I.S Cohen shows how every pro-p domain can be built up starting from either $\mathbb{Z}_p$ or $\mathbb{F}_p$:

6.42 Theorem *Let* $(R, \mathfrak{m})$ *be a pro-p domain. Put* $R_0 = \mathbb{Z}_p$ *if* $\mathrm{char}(R) = 0$, $R_0 = \mathbb{F}_p$ *if* $\mathrm{char}(R) = p$. *Then* R *contains a subring* $R_1 = R_0[[T_1, \ldots, T_r]]$ *such that* R *is finitely generated as an* R_1-*module.* R_1 *is a pro-p domain with maximal ideal* $\mathfrak{m}_1 = (p, T_1, \ldots, T_r)$.

This is Theorem 6.3 in Dieudonné (1967). It is also implicit in Matsumura (1986), §29 or Bourbaki (1983), Chapter IX, §2. Since R is an integral extension of R_1 it has the same Krull dimension $\mathrm{Dim}(R)$; if this dimension is 1 then R_1 is a principal ideal domain, so we may infer

6.43 Corollary *Let* R *be a pro-p domain of Krull dimension 1. Then* R *is a finitely generated free module over a subring of the form* $\mathbb{Z}_p$ *or* $\mathbb{F}_p[[T]]$.

Since the powers of $\mathfrak{m}$ intersect in zero, we can use Lemma 6.5 to make $(R, \mathfrak{m})$ into a normed ring: fixing a positive real number $c < 1$, we define the norm $\|\cdot\| : R \to \mathbb{R}$ by $\|a\| = c^k$ if $a \in \mathfrak{m}^k \backslash \mathfrak{m}^{k+1}$, $\|0\| = 0$.

If $\mathrm{Dim}(R) = 1$ then R is a DVR; in this case, the maximal ideal $\mathfrak{m}$ is principal, and we choose a generator π, so $\mathfrak{m} = \pi R$. Thus $\|\pi\| = c$; we extend the norm to the field of fractions K by

$$\|x\| = \|\pi^n x\| \, c^{-n}$$

where n is so large that $\pi^n x \in R$. This makes K into a complete valued field, and the arguments of earlier sections work just as well with K in place of $\mathbb{Q}_p$.

If R is not a DVR serious difficulties arise: although it may be possible to extend the norm from R to K, we no longer obtain a locally compact field, and elementary analysis over K breaks down (see Exercise 14). In this case, therefore, we restrict consideration to power series with coefficients in R. To save repetition, we fix the notation

$$\Lambda = \begin{cases} K \text{ if } R \text{ is a DVR} \\ \\ R \text{ if } R \text{ is not a DVR.} \end{cases}$$

Definition A power series $F(\mathbf{X}) = \sum_{\mathbf{m} \in \mathbb{N}^n} a_{\mathbf{m}} X_1^{m_1} \dots X_n^{m_n} \in \Lambda[[\mathbf{X}]]$ can be *evaluated* at $\mathbf{x} = (x_1, \dots, x_n) \in \Lambda^n$ if the series $\sum_{\mathbf{m} \in \mathbb{N}^n} a_{\mathbf{m}} x_1^{m_1} \dots x_n^{m_n}$ converges in Λ. In this case, its sum is denoted $F(\mathbf{x})$.

(Whenever we use the notation $F(\mathbf{x})$, this is meant to imply that F can be evaluated at $\mathbf{x}$.)

We shall write $\mathfrak{m}^{(n)}$ to denote the set of n-tuples of elements of $\mathfrak{m}$, to distinguish it from the nth power of the maximal ideal $\mathfrak{m}^n$ (we continue to write Λ^n for n-tuples from Λ).

6.44 Lemma *If $F(\mathbf{X}) \in R[[\mathbf{X}]]$ then $F(\mathbf{X})$ can be evaluated at $\mathbf{x}$ for every $\mathbf{x} \in \mathfrak{m}^{(n)}$. If the constant term of $F(\mathbf{X})$ is in $\mathfrak{m}$ then $F(\mathbf{x}) \in \mathfrak{m}$.*

The following generalises Lemma 6.18:

6.45 Lemma *Let $F(\mathbf{X}) = \sum_{\alpha \in \mathbb{N}^n} a_\alpha X_1^{\alpha_1} \dots X_n^{\alpha_n} \in \Lambda[[\mathbf{X}]]$ and suppose that F can be evaluated on $\left(\mathfrak{m}^N\right)^{(n)}$. Then there exists $k \in \mathbb{N}$ such that*

$$a_\alpha \mathfrak{m}^{k\langle\alpha\rangle} \subseteq R \text{ for all } \alpha \neq \mathbf{0}.$$

Proof If R is not a DVR then each a_α is in R and we can take $k = 0$. Suppose that R is a DVR. Since F converges at $(\pi^N, \dots, \pi^N)$ there exists D such that $\left\| a_\alpha \pi^{N\langle\alpha\rangle} \right\| \leq 1$ whenever $\langle\alpha\rangle \geq D$. Let E be large enough so that $\pi^E a_\alpha \in R$ for each α with $\langle\alpha\rangle < D$. Then $k = \max\{D, E\}$ has the required property.

The next three results generalise earlier results, with similar proofs; see also Bourbaki (1989a), Chapter III §4.

6.46 Lemma *Let S_1 and S_2 be countably infinite sets. Suppose that for each $(m,n) \in S_1 \times S_2$, $a_{mn} \in \Lambda$ and that $\lim\limits_{(m,n)\in S_1 \times S_2} a_{mn} = 0$. Then the double series $\sum_{m\in S_1} \left(\sum_{n\in S_2} a_{mn} \right)$ and $\sum_{n\in S_2} \left(\sum_{m\in S_1} a_{mn} \right)$ both converge and their common sum is $\sum_{(m,n)\in S_1 \times S_2} a_{mn}$.*

6.47 Theorem *For $i = 1, \ldots, m$ let $F_i(\mathbf{X}) = \sum_{\mathbf{t}\in \mathbb{N}^n} a_{i\mathbf{t}} X_1^{t_1} \ldots X_n^{t_n} \in \Lambda\,[[\mathbf{X}]]$ be power series with zero constant terms. Let $G(\mathbf{Y}) = \sum_{\mathbf{s}\in \mathbb{N}^m} b_\mathbf{s} Y_1^{s_1} \ldots Y_m^{s_m} \in \Lambda\,[[\mathbf{Y}]]$. Suppose that $F_1, \ldots, F_m$ can be evaluated at $\mathbf{x} = (x_1, \ldots, x_n) \in \Lambda^n$, and for each i put*

$$\tau_i = \sup_{\mathbf{t}\in \mathbb{N}^n} \left\| a_{i\mathbf{t}} x_1^{t_1} \ldots x_n^{t_n} \right\|.$$

If $\lim_{\mathbf{s}\in \mathbb{N}^m} \|b_\mathbf{s}\| \, \tau_1^{s_1} \ldots \tau_m^{s_m} = 0$ then $(G \circ \mathbf{F})(\mathbf{x}) = G(F_1(\mathbf{x}), \ldots, F_n(\mathbf{x}))$.

6.48 Corollary *Let $F_i(\mathbf{X}) \in R[[\mathbf{X}]]$ be power series with zero constant terms and let $G(\mathbf{Y}) \in R[[\mathbf{Y}]]$. If $\mathbf{x} = (x_1, \ldots, x_n) \in \mathfrak{m}^n$ then $(G \circ \mathbf{F})(\mathbf{x}) = G(F_1(\mathbf{x}), \ldots, F_n(\mathbf{x}))$.*

The final result is fundamental to much of the theory we develop in Chapter 13. It requires a further hypothesis concerning the norm on R, which we state in terms of the *associated graded ring*

$$\mathrm{gr}(R) = \bigoplus_{n=0}^{\infty} \mathfrak{m}^n/\mathfrak{m}^{n+1},$$

namely that $\mathrm{gr}(R)$ be an *integral domain*. This is equivalent to the hypothesis

$$\|xy\| = \|x\|\,\|y\| \quad \text{for all } x, y \in R.$$

It holds if R is a DVR; it also holds if R is a power series ring over $\mathbb{F}_p$ or $\mathbb{Z}_p$, and more generally whenever R is a *regular* local ring, in which case $\mathrm{gr}(R)$ is isomorphic to a polynomial ring over the residue field (Atiyah and Macdonald (1969), Theorem 11.22).

6.49 Proposition *Assume that $\mathrm{gr}(R)$ is an integral domain. Suppose that $F(X_1, \ldots, X_d) \in \Lambda[[X_1, \ldots, X_d]]$ converges on some neighbourhood V of $\mathbf{0}$ in R^d and that*

$$F(\mathbf{x}) = 0 \quad \text{for all } \mathbf{x} \in V.$$

Then $F(X_1, \ldots, X_d) = 0$.

Proof Arguing by induction on d, we reduce to the case of a single variable. Suppose, then, that $F(X) = \sum_{n=0}^{\infty} a_n X^n$ and that not all the a_n are zero; let m be such that $a_m \neq 0$ but $a_n = 0$ for $n < m$. Since $F(0) = 0$ we have $m \geq 1$; and by Lemma 6.45 there exists $k \in \mathbb{N}$ such that $a_n \mathfrak{m}^{kn} \subseteq R$ for all $n \geq m$. Fix $y \in \mathfrak{m}^k \setminus \mathfrak{m}^{k+1}$, and suppose that $a_m y^m \in \mathfrak{m}^s \setminus \mathfrak{m}^{s+1}$. Now let t be so large that

$$\mathfrak{m}^t \subseteq V, \quad t > s.$$

Let $x \in \mathfrak{m}^t \setminus \mathfrak{m}^{t+1}$. Then $xy \in V$, so $F(xy) = 0$ and we have

$$a_m(xy)^m = -\sum_{n=m+1}^{\infty} a_n(xy)^n \in \mathfrak{m}^{t(m+1)}.$$

Hence

$$c^{t(m+1)} \geq \|a_m(xy)^m\| = \|a_m y^m\| \cdot \|x\|^m = c^s c^{tm},$$

a contradiction since $0 < c < 1$ and $t > s$.

The following corollary will be quoted in Chapter 13 as the 'uniqueness of power series':

6.50 Corollary *Assume that* $\mathrm{gr}(R)$ *is an integral domain. If two power series both converge on some neighbourhood of* $\mathbf{0}$ *in* R^d *and define the same function there then they are identical.*

Notes

For more on 'ultrametric analysis' one may consult Schikhoff (1984), Serre (1965) II, Chapters 1, 2. For the Campbell–Hausdorff formula see Serre (1965), II Chapter 5, Bourbaki (1989b), Chapter II §8, or [L], Chapter IV §3. The formula (without the p-adic estimates for the coefficients) is also proved in Jacobson (1962), Chapter 5, §5.

Exercises

1. Let $(R, \|\cdot\|)$ be a normed ring. Show that for each $a \in R$ and $\delta > 0$ the 'closed ball' $B_\delta(a) = \{x \in R \mid \|x - a\| \le \delta\}$ is both closed and open.

2. Let R be ring with a descending chain of ideals (R_i) as in Lemma 6.5. Let (R_i') be another such chain of ideals, cofinal with (R_i); that is, for each i there exists $j(i)$ such that $R_{j(i)}' \subseteq R_i$ and $R_{j(i)} \subseteq R_i'$.

(i) Show that the two metrics on R, defined using (R_i) and $c > 1$, and using (R_i') and $c' > 1$, give rise to the same concept of Cauchy sequence in R, and hence to the same completion $\hat{R}$.

(ii) Put $\tilde{R} = \varprojlim(R/R_i)$. Show that if $(r_i + R_i) \in \tilde{R}$ then (r_i) is a Cauchy sequence in R, and that if (s_i) is a Cauchy sequence in R then there is a subsequence $(s_{n(i)})$ such that $(s_{n(i)} + R_i) \in \tilde{R}$. Deduce that there is an isomorphism $\tilde{R} \to \hat{R}$ that sends $(r_i + R_i)$ to $\lim r_i$.

(iii) Identifying $\tilde{R}$ with $\hat{R}$ by this isomorphism, show that for each k, the closure $\overline{R_k}$ of R_k in $\hat{R}$ satisfies

$$\overline{R_k} = \{x \in \hat{R} \mid \|x\| \le c^{-k}\} = \varprojlim\left((R_k + R_i)/R_i\right)$$

(this shows that the norm on $\hat{R}$ can be defined by the chain of ideals $(\overline{R_k})$). Deduce that $\overline{R_k} \cap R = R_k$ and $\overline{R_k} + R = \hat{R}$, and hence that $\overline{R_k}/\overline{R_{k+1}} \cong R_k/R_{k+1}$.

(iv) Let (J_α) be a system of ideals of R directed w.r.t. reverse inclusion (i.e. for each α and β there exists γ such that $J_\alpha \cap J_\beta \supseteq J_\gamma$). Show that if (J_α) is cofinal with (R_i) then $\varprojlim(R/J_\alpha) \cong \tilde{R}$.

3. (i) Show that every linear map $\mathbb{Q}_p^n \to \mathbb{Q}_p^m$ is continuous (both spaces having the product topology). Now let V be a finite-dimensional vector space over $\mathbb{Q}_p$; deduce that V has a unique topology – the *p-adictopology* – given by identifying V with $\mathbb{Q}_p^n$ by choosing an arbitrary basis, and taking the product topology on $\mathbb{Q}_p^n$. Show that every vector subspace of V is closed in the p-adic topology.

(ii) Let A be a normed $\mathbb{Q}_p$-algebra and V a finite-dimensional vector subspace of A. Show that the inclusion map $V \to A$ is continuous with respect to the p-adic topology on V and the norm topology on A. Does the latter topology necessarily induce the former topology on V?

4. Let $(R, \|\cdot\|)$ be a complete normed ring.

(i) Let T be a countable set, let $i \mapsto n(i)$ be a bijection from $\mathbb{N}$ onto

T, and let $n \mapsto a_n$ be a map from T into R. Prove that $\lim_{n \in T} a_n = a$ if and only if $\lim_{i \to \infty} a_{n(i)} = a$.

(ii) Let $i \mapsto b_i$ and $i \mapsto c_i$ be maps from $\mathbb{N}$ into R, and σ a permutation of $\mathbb{N}$. Prove that if $\lim_{i \to \infty} b_i = b$ and $\sum_{i=0}^{\infty} c_i = c$ then $\lim_{i \to \infty} b_{\sigma(i)} = b$ and $\sum_{i=0}^{\infty} c_{\sigma(i)} = c$.

5. (i) Let $1 \le n \in \mathbb{N}$. Define $s(n) = \sum_{j=0}^{s} b_j$, where $n = b_0 + b_1 p + \cdots + b_s p^s$ with $0 \le b_i \le p - 1$. Prove that

$$v(n!) = (n - s(n))/(p - 1).$$

(ii) Show that $s(a + b) \le s(a) + s(b)$.

(iii) Deduce that if $n = a_1 + \cdots + a_k$ then $v(n!) \ge \sum_{i=1}^{k} v(a_i!)$. Hence show that $\prod_{i=1}^{k} a_i!$ divides $n!$.

(iv) Prove the last result another way. [*Hint:* partition a set of size n into subsets of size $a_1, \ldots, a_k$.]

6. For any power series $A(X) = \sum_{n=0}^{\infty} a_n X^n \in \mathbb{Q}_p \langle\langle X \rangle\rangle$ in one variable X we define the *derivative*

$$A'(X) = \sum_{n=1}^{\infty} n a_n X^{n-1}.$$

Put $F(X) = \mathcal{E}(X) - 1$.

(i) Prove that $F'(X) = \mathcal{E}(X)$ and $\mathcal{L}'(Y) = (1 + Y)^{-1}$.

(ii) Prove that for any power series $A(X)$ and $B(X)$ and any scalars a, b we have:

$$(aA + bB)' = aA' + bB'$$
$$(AB)' = A'B + AB'$$
$$(A \circ B)' = (A' \circ B) \cdot B' \quad \text{provided } B(0) = 0.$$

(iii) By considering $(\mathcal{L} \circ F)'$ and $(F \circ \mathcal{L})'$ prove that $(\mathcal{L} \circ F)(X) = X$ and $(\mathcal{E} \circ \mathcal{L})(Y) = 1 + Y$.

7. Show that in $\mathbb{Q}_2$, $\log(-1) = 0 = \log(1)$. Thus $\exp(\log(-1)) = 1$.

8. Define strictly analytic functions $f : \mathbb{Z}_p \to p\mathbb{Z}_p$ and $g : p\mathbb{Z}_p \to p\mathbb{Z}_p$ by $f(x) = x^p - x$ and $g(x) = \sum_{n=0}^{\infty} x^n = (1 - x)^{-1}$. Prove that $g \circ f : \mathbb{Z}_p \to p\mathbb{Z}_p$ is not a strictly analytic function on $\mathbb{Z}_p$.

9. Prove that composition of formal power series is an associative operation.

10. Let f be the arithmetical function defined in the proof of Lemma 6.41. (i) Verify that $f(3) = 36$ and $f(4) = 48$.

(ii) Show that $u_3(X, Y) = \frac{1}{12}(X, Y, Y) - \frac{1}{12}(X, Y, X)$.

(iii) Hence complete the proof of Lemma 6.41. [*For (ii)*, use the recipe provided by Lemma 6.40 and the proof of Lemma 6.41.]

11. Let $R = \mathbb{Z}\langle X, Y \rangle$ be the polynomial ring in two non-commuting variables over $\mathbb{Z}$, and let L be the Lie subring of R generated by X and Y. According to Theorem 1 of Bourbaki (1989b), Chapter II §3, L is a direct summand of R as $\mathbb{Z}$-module. Now let $n \geq 2$ and put $h = \left\lceil \frac{n-1}{p-1} \right\rceil$.

(i) Show that for some $m_1 \in \mathbb{N}$ not divisible by p and some $m_2 \in \mathbb{N} \setminus \{0\}$,

$$m_1 p^h u_n(X, Y) \in R, \quad m_2 u_n(X, Y) \in L;$$

deduce that $m_1 p^h u_n(X, Y) \in L$. [*Hint:* use Lemma 6.40 and the proof of Lemma 6.41.]

(ii) Prove that L is spanned as a $\mathbb{Z}$-module by X, Y and elements of the form $(X, Y)_{\mathbf{e}}$. [*Hint:* Show that if a and b are elements of a Lie ring then $((a, b)_{\mathbf{e}}, (a, b)_{\mathbf{f}})$ is equal to a $\mathbb{Z}$-linear combination of terms $(a, b)_{\mathbf{g}}$; argue by induction on $\langle \mathbf{f} \rangle$, using the Jacobi identity.]

(iii) Using (i) and (ii) show that $p^h q_{\mathbf{e}} \in \mathbb{Z}_p$ when $\langle \mathbf{e} \rangle = n - 1$.

12. Let $(A, \|\cdot\|)$ be a complete normed $\mathbb{Q}_p$-algebra, and let $a \in A_0$. Prove that

$$x \exp(\mathrm{ad}(a)) = (\exp a)^{-1} x (\exp a)$$

for all $x \in A$; here the notation is as in §6.5, so $\exp(\mathrm{ad}(a))$ is a (bounded) linear map on A. [*Hint:* Look at the beginning of the proof of Lemma 6.39.]

13. Let R be a commutative Noetherian ring. Prove that R is a pro-p ring (for some prime p) if and only if it is the inverse limit of an inverse system of finite local rings in which the maps are all surjective.

14. Let R be a pro-p ring such that $\mathrm{gr}(R)$ is an integral domain.

(i) Show that the norm on R can be extended to the field of fractions K, making K into a normed ring, with $\|xy\| = \|x\| \|y\|$ for all $x, y \in K$.

(ii) Suppose that R is not a DVR, and let $\{x_1, \dots, x_d\}$ be a system of parameters for R: that is, $d = \mathrm{Dim}(R) \geq 2$ and the ideal

$\mathfrak{q} = (x_1, \ldots, x_d)$ is $\mathfrak{m}$-primary (see Atiyah and Macdonald (1969), Theorem 11.14). Prove that $x_1^n \notin x_2 R$ for every $n \in \mathbb{N}$. Deduce that $\mathfrak{m}$ is not an open subset of K, in the topology on K defined by the norm. (Thus arguments relying on the assumption that a sequence tending to 0 must eventually lie inside $\mathfrak{m}$ are not available.)

(iii) Suppose that $R = R_0 [[T_1, \ldots, T_d]]$, where R_0 is a finite field and $d \geq 2$ or $R_0 = \mathbb{Z}_p$ and $d \geq 1$. Prove that K is not locally compact (in the topology defined by the norm).

[*Hint*: for (ii): if $\mathfrak{m}$ is open then $x_1^n x_2^{-1} \in \mathfrak{m}$ for sufficiently large n. For (iii): put $x = T_1, y = T_2$ if $d \geq 2$, $x = p$, $y = T_1$ otherwise. Suppose that K is locally compact; then for some $k \geq 1$ the sequence $\left(x^{k+n} y^{-n}\right)_{n=1}^{\infty}$ must contain a Cauchy subsequence. Show that $\left\| x^{k+n} y^{-n} - x^{k+m} y^{-m} \right\| \geq c^k$ for all n and m to derive a contradiction.]

7

The group algebra

In Chapter 4, we showed how to endow the underlying set of a uniform pro-p group G with an additive structure, and indicated that G could thereby be made into a Lie algebra over $\mathbb{Z}_p$. One of the purposes of the present chapter is to construct another Lie algebra, naturally isomorphic to that one. The new Lie algebra will be a subalgebra of the commutation Lie algebra on a certain *associative* algebra $\widehat{A}$, which we get by completing the group algebra $\mathbb{Q}_p[G]$ with respect to a suitable norm. Most of the work in the first section goes into setting up this norm. The Lie algebra L, and a bijective logarithm mapping $\log : G \to L$, are defined in §7.2, and this is used in §7.3, with Ado's Theorem, to show that G has a faithful linear representation over $\mathbb{Z}_p$. In the final section, we establish some ring-theoretic properties of the group algebras $\mathbb{Z}_p[G]$ and $\mathbb{F}_p[G]$ and of their completions.

7.1 The norm

Throughout this section, G denotes a finitely generated pro-p group. If $M \le N$ are open normal subgroups of G, the natural map $G/M \to G/N$ induces an epimorphism between the group algebras $\mathbb{Z}_p[G/M] \to \mathbb{Z}_p[G/N]$. Using these maps we may construct an inverse system of $\mathbb{Z}_p$-algebras: its inverse limit is denoted

$$\mathbb{Z}_p[[G]] = \varprojlim_{N \triangleleft_o G} \left(\mathbb{Z}_p[G/N] \right),$$

and is called the *completed group algebra* of G. Our purpose in this section is to show (a) how $\mathbb{Z}_p[[G]]$ arises as the completion of the group algebra $\mathbb{Z}_p[G]$ with respect to a certain *norm*, and (b) that if G is *uniform*, then the norm can be extended to a norm on the larger group algebra $\mathbb{Q}_p[G]$.

138

We write $R = \mathbb{Z}_p[G]$, $G_k = P_k(G)$ and

$$I_k = (G_k - 1)R = \ker(R \to \mathbb{Z}_p[G/G_k]),$$

the kernel of the natural epimorphism. Since the family (G_k) is cofinal with the family of all open normal subgroups of G (see Chapter 1), we may identify $\mathbb{Z}_p[[G]]$ with

$$\varprojlim_{k \in \mathbb{N}} (\mathbb{Z}/p^k\mathbb{Z})[G/G_k] \cong \varprojlim (R/(I_k + p^k R))$$

(see Exercise 1).

We begin by introducing a chain of ideals, cofinal with the chain $(I_k + p^k R)$ but better suited for defining a norm: it consists of the powers of the following ideal:

$$J = I_1 + pR.$$

Note that $I_1 = (G - 1)R = \sum_{x \in G}(x - 1)R$ is the augmentation ideal of R, and that J is the kernel of the natural epimorphism $R \to \mathbb{F}_p$ that sends each group element to 1.

The fact that the two chains of ideals (J^k) and $(I_k + p^k R)$ are indeed cofinal follows from

7.1 Lemma *Let* $k \geq 1$. *Then*

(i) $J^k \supseteq I_k + p^k R$;

(ii) *for each* $j \geq 1$, $I_k + p^j R \supseteq J^{m(k,j)}$ *where* $m(k,j) = j|G/G_k|$.

Proof We prove (ii) first. Put $n = |G/G_k|$. Now G/G_k is a finite p group, and it acts by right multiplication on the $\mathbb{F}_p$-vector space $\mathbb{F}_p[G/G_k]$, which has dimension equal to n. It follows that $(x_1 - 1)\cdots(x_n - 1) = 0$ in $\mathbb{F}_p[G/G_k]$ for any $x_1, ..., x_n \in G/G_k$ (see §0.4). The kernel of the natural map of R onto $\mathbb{F}_p[G/G_k]$ is just $I_k + pR$; hence $(g_1 - 1)\cdots(g_n - 1) \in I_k + pR$ for all $g_1, ..., g_n \in G$, and it follows that $J^n \subseteq I_k + pR$. This in turn implies that $J^{nj} \subseteq I_k + p^j R$ for each $j \geq 1$.

To prove (i) we argue by induction on k. When $k = 1$ it is true by definition; let $k > 1$ and suppose that $J^{k-1} \supseteq I_{k-1} + p^{k-1}R$. Since $p^k \in J^k$, what remains to show is that $I_k \subseteq J^k$. Since I_k is generated as an ideal by $G_k - 1$, this is equivalent to showing that G_k acts trivially by right multiplication on the quotient R/J^k. Now Corollary 1.20 shows that G_k is generated by elements of the form

$$x^p, \quad [x, y] \quad (x \in G_{k-1}, \quad y \in G).$$

Put $u = x - 1, v = y - 1$. Then

$$x^p - 1 = (u+1)^p - 1 = u^p + puw \tag{1}$$

for some $w \in R$, by the binomial theorem (since $p \mid \binom{p}{i}$ for $1 \le i \le p-1$); and

$$[x,y] - 1 = x^{-1}y^{-1}(xy - yx) = x^{-1}y^{-1}(uv - vu). \tag{2}$$

The inductive hypothesis gives $u \in J^{k-1}$. Since $p \in J$ and $v \in I_1 \subseteq J$, and $p \ge 2$, it follows that

$$x^p \equiv 1 \pmod{J^k}, \quad [x,y] \equiv 1 \pmod{J^k}.$$

Hence G_k acts trivially on R/J^k as required.

7.2 Corollary

$$\bigcap_{\ell=1}^{\infty} J^\ell = 0.$$

Proof Let $c = \sum_{i=1}^{n} \lambda_i x_i \in R$, where $x_1, ..., x_n$ are distinct elements of G and $\lambda_1, ..., \lambda_n \in \mathbb{Z}_p$ are non-zero. Let k be so large that (a) $x_i x_j^{-1} \notin G_k$ for all $i \ne j$, and (b) $\lambda_i \not\equiv 0 \pmod{p^k}$ for some i. Write

$$\psi_k : R \to (\mathbb{Z}_p/p^k\mathbb{Z}_p)[G/G_k]$$

for the natural map. Then $x_1\psi_k, ..., x_n\psi_k$ are distinct elements of G/G_k, and the coefficient of $x_i\psi_k$ in $c\psi_k$ is $\lambda_i + p^k\mathbb{Z}_p \ne 0$. So $c\psi_k \ne 0$. Since $\ker \psi_k = I_k + p^k R$ this implies that $c \notin I_k + p^k R$, and hence by Lemma 7.1 that $c \notin J^m$ for some m. The corollary follows.

Since $J^i J^j = J^{i+j}$ for all i and j, we may now invoke Lemma 6.5 and make the following definition:

7.3 Definition The norm $\| \cdot \|$ on $\mathbb{Z}_p[G]$ is defined by

$$\|c\| = p^{-k} \quad \text{if } c \in J^k \setminus J^{k+1}$$
$$\|0\| = 0.$$

It follows from Lemma 7.1 that the topology on R given by the norm induces on G the original topology of G (Exercise 2).

Let us write $\widehat{R}$ to denote the completion of $(R, \| \cdot \|)$. In view of Exercise 6.2, $\widehat{R}$ may be identified with $\varprojlim_k R/J^k$. Thus from Lemma 7.1 and the preceding discussion we see that

$$\widehat{R} \cong \varprojlim_k (\mathbb{Z}/p^k\mathbb{Z})[G/G_k] \cong \mathbb{Z}_p[[G]].$$

In §7.4 we shall use this isomorphism to examine some of the ring-theoretic properties of $\mathbb{Z}_p[[G]]$.

Having achieved our first goal (a), we move on to (b): if G is uniform, then the norm we have just defined on $\mathbb{Z}_p[G]$ can be extended to a norm on the group algebra $\mathbb{Q}_p[G]$.

For the remainder of this section, we fix the following notation:

$\{a_1, ..., a_d\}$ is a topological generating set for G where $d = d(G)$

$$b_i = a_i - 1 \quad \text{for} \quad i = 1, ..., d.$$

For $\alpha = (\alpha_1, ..., \alpha_d) \in \mathbb{N}^d$ and any d-tuple $\mathbf{v} = (v_1, ..., v_d)$, we write

$$\langle \alpha \rangle = \alpha_1 + \cdots + \alpha_d, \quad \mathbf{v}^\alpha = v_1^{\alpha_1}...v_d^{\alpha_d}.$$

The main results can now be stated:

7.4 Theorem (i) *If G is powerful then each element of $\mathbb{Z}_p[G]$ is equal to the sum of a convergent series*

$$\sum_{\alpha \in \mathbb{N}^d} \lambda_\alpha \mathbf{b}^\alpha \tag{3}$$

with $\lambda_\alpha \in \mathbb{Z}_p$ for each α.

(ii) *If G is uniform then the series (3) is uniquely determined by its sum.*

7.5 Theorem *Assume that G is uniform. If $c = \sum_{\alpha \in \mathbb{N}^d} \lambda_\alpha \mathbf{b}^\alpha \in \mathbb{Z}_p[G]$, where $\lambda_\alpha \in \mathbb{Z}_p$ for each α, then*

$$\|c\| = \sup_{\alpha \in \mathbb{N}^d} p^{-\langle \alpha \rangle} |\lambda_\alpha|.$$

If c is as in Theorem 7.5 and $\lambda \in \mathbb{Z}_p$, then the coefficient of $\mathbf{b}^\alpha$ in the series for λc is just $\lambda \lambda_\alpha$; so we may infer

7.6 Corollary *If G is uniform then*

$$\|\lambda c\| = |\lambda| \, \|c\|$$

for all $\lambda \in \mathbb{Z}_p$ and $c \in \mathbb{Z}_p[G]$.

Now if $a \in \mathbb{Q}_p[G]$ we can choose a non-zero $\lambda \in \mathbb{Z}_p$ such that $\lambda a \in \mathbb{Z}_p[G]$, and Corollary 7.6 ensures that $|\lambda|^{-1} \|\lambda a\|$ is independent of the choice of λ; so the following definition is unambiguous:

$$\|a\| = |\lambda|^{-1} \|\lambda a\|,$$

where $0 \neq \lambda \in \mathbb{Z}_p$ satisfies $\lambda a \in \mathbb{Z}_p[G]$.

The final result now follows easily:

7.7 Theorem *Let G be a uniform pro-p group. Then $(\mathbb{Q}_p[G], \|\cdot\|) = A$ is a normed $\mathbb{Q}_p$-algebra. The norm on A induces the original topology on G, and each element $g \in G$ satisfies $\|g - 1\| \leq p^{-1}$.*

The routine verification of conditions N1–N4 is left to the reader. The second statement follows from Exercise 2, and the final statement is clear, since $g - 1 \in J$ for all $g \in G$.

Note that if $p = 2$ we have

$$\|g^2 - 1\| \leq 2^{-2}$$

for $g \in G$, since $g^2 - 1 \in I_2 \subseteq J^2$, by Lemma 7.1. However, it is useful in this case to define a slightly different norm on $\mathbb{Q}_p[G]$, so that $\|g - 1\| \leq 2^{-2}$ for all $g \in G$; how to do this is explained in Exercise 10.

The proof of Theorem 7.4 depends on the next two lemmas; the first of these is a pair of simple identities which will also be important in Chapter 9.

7.8 Lemma *Let $u_1, \ldots, u_r \in G$ and put $v_i = u_i - 1$ for each i. Then for $\beta \in \mathbb{N}^r$,*

$$\mathbf{u}^\beta = \sum_{\alpha \in \mathbb{N}^r} \binom{\beta_1}{\alpha_1} \cdots \binom{\beta_r}{\alpha_r} \mathbf{v}^\alpha,$$

$$\mathbf{v}^\beta = \sum_{\alpha \in \mathbb{N}^r} (-1)^{\langle b \rangle - \langle \alpha \rangle} \binom{\beta_1}{\alpha_1} \cdots \binom{\beta_r}{\alpha_r} \mathbf{u}^\alpha.$$

(Each sum is finite since $\binom{\beta_i}{\alpha_i} = 0$ unless $\alpha_i \leq \beta_i$.)

Proof This is no more than the binomial theorem. We have

$$\mathbf{v}^\beta = (u_1 - 1)^{\beta_1}...(u_r - 1)^{\beta_r}$$

$$= \left(\sum_{\alpha_1=0}^{\beta_1} (-1)^{\beta_1-\alpha_1} \binom{\beta_1}{\alpha_1} u_1^{\alpha_1} \right) \cdots \left(\sum_{\alpha_r=0}^{\beta_r} (-1)^{\beta_r-\alpha_r} \binom{\beta_r}{\alpha_r} u_r^{\alpha_r} \right)$$

$$= \sum_{\alpha \in \mathbb{N}^r} (-1)^{\langle b \rangle - \langle \alpha \rangle} \binom{\beta_1}{\alpha_1} ... \binom{\beta_r}{\alpha_r} \mathbf{u}^\alpha.$$

The other formula is obtained similarly from

$$\mathbf{u}^\beta = (v_1 + 1)^{\beta_1}...(v_r + 1)^{\beta_r}.$$

The close relation between the $\mathbf{a}^\alpha$ and the $\mathbf{b}^\alpha$ is the key to the next lemma, where for $k \geq 1$ we write

$$T_k = \{\alpha \in \mathbb{N}^d \mid \alpha_i < p^{k-1} \text{ for } i = 1,...,d\}.$$

7.9 Lemma *Assume that G is powerful, and let $k \geq 1$.*
(i)
$$R = I_k + \sum_{\alpha \in T_k} \mathbb{Z}_p \mathbf{b}^\alpha.$$

(ii) *If G is uniform then*
$$R = I_k \oplus \bigoplus_{\alpha \in T_k} \mathbb{Z}_p \mathbf{b}^\alpha.$$

(iii) $\mathbf{b}^\alpha \in I_k + pR$ *for each* $\alpha \in \mathbb{N}^d \setminus T_k$.

Proof (i) Recall that I_k is the kernel of the natural epimorphism $\pi : R \to \mathbb{Z}_p[G/G_k]$. It follows from Proposition 3.7 that each element of G/G_k can be written in the form $a_1^{\alpha_1} ... a_d^{\alpha_d} G_k$ with $0 \leq \alpha_i < p^{k-1}$ for $i = 1,...,d$; hence $\{\mathbf{a}^\alpha \pi \mid \alpha \in T_k\}$ generates $\mathbb{Z}_p[G/G_k]$ as a $\mathbb{Z}_p$-module. Lemma 7.8 shows that $\{\mathbf{b}^\alpha \pi \mid \alpha \in T_k\}$ generates exactly the same module, so $R\pi = \sum_{\alpha \in T_k} \mathbb{Z}_p \mathbf{b}^\alpha \pi$. This implies (i).

(ii) Suppose now that G is uniform. Then $|G/G_k| = p^{(k-1)d}$, so $R\pi = \mathbb{Z}_p[G/G_k]$ is a free $\mathbb{Z}_p$-module of rank $p^{(k-1)d}$. Since $p^{(k-1)d} = |T_k|$ it follows that the generating set $\{\mathbf{b}^\alpha \pi \mid \alpha \in T_k\}$ is now actually a free basis for this module: this is equivalent to the assertion of (ii).

(iii) Let $\alpha \in \mathbb{N}^d \setminus T_k$. Then $\alpha_i \geq p^{k-1}$ for some i, so $\mathbf{b}^\alpha$ has a factor of the form

$$b_i^{p^{k-1}} = (a_i - 1)^{p^{k-1}} \equiv a_i^{p^{k-1}} - 1 \pmod{pR}.$$

As $a_i^{p^{k-1}} \in G_k$ it follows that $b_i^{p^{k-1}} \in (G_k - 1)R + pR = I_k + pR$. This gives (iii).

Proof of Theorem 7.4 (i) Let $c \in R$, and for $k \geq 1$ put

$$\Lambda_k = \left\{ (\lambda_\alpha)_{\alpha \in T_k} \mid c \equiv \sum_{\alpha \in T_k} \lambda_\alpha \mathbf{b}^\alpha \pmod{J^k} \right\} \subseteq \mathbb{Z}_p^{T_k}.$$

Since $J^k \supseteq I_k$ by Lemma 7.1, it follows by Lemma 7.9(i) that Λ_k is non-empty for each k. Also the natural projection mapping $\mathbb{Z}_p^{T_{k+1}} \to \mathbb{Z}_p^{T_k}$ sends Λ_{k+1} into Λ_k; for if $(\lambda_\alpha)_{\alpha \in T_{k+1}} \in \Lambda_{k+1}$ then

$$\sum_{\alpha \in T_k} \lambda_\alpha \mathbf{b}^\alpha = \sum_{\alpha \in T_{k+1}} \lambda_\alpha \mathbf{b}^\alpha - \sum_{\alpha \in T_{k+1} \setminus T_k} \lambda_\alpha \mathbf{b}^\alpha \equiv c \pmod{J^k},$$

since for each $\alpha \in \mathbb{N}^d \setminus T_k$ we have

$$\mathbf{b}^\alpha \in J^{p^{k-1}} \subseteq J^k \tag{4}$$

as $b_i = a_i - 1 \in J$ for each i. It follows by Proposition 1.4 that the inverse system of compact sets $(\Lambda_k)_{k \in \mathbb{N}}$ has a non-empty inverse limit; hence there exists a family $(\lambda_\alpha)_{\alpha \in \mathbb{N}^d}$ such that for each k, the subfamily $(\lambda_\alpha)_{\alpha \in T_k}$ belongs to Λ_k.

We complete the proof by showing that the series $\sum_{\alpha \in \mathbb{N}^d} \lambda_\alpha \mathbf{b}^\alpha$ converges to c. Let $\varepsilon > 0$, and let k be so large that $p^{-k} < \varepsilon$. It will suffice to show that for every finite subset S of $\mathbb{N}^d$ containing T_k we have $\|c - \sum_{\alpha \in S} \lambda_\alpha \mathbf{b}^\alpha\| < \varepsilon$; but this is clear from the above, since $c - \sum_{\alpha \in T_k} \lambda_\alpha \mathbf{b}^\alpha \in J^k$ by the choice of (λ_α), while $\sum_{\alpha \in S \setminus T_k} \lambda_\alpha \mathbf{b}^\alpha \in J^k$ by (4).

(ii) Assume now that G is uniform. To show that the series representation given in (i) is unique, it will be enough to establish that if $\sum_{\alpha \in \mathbb{N}^d} \mu_\alpha \mathbf{b}^\alpha = 0$ then $\mu_\alpha = 0$ for each $\alpha \in \mathbb{N}^d$. If this is not the case, we may suppose without loss of generality that μ_α is not divisible by p for at least one value of α.

Now fix a large positive integer k, and put $m = |G/G_k|$ According to Definition 6.8, the hypothesis $\sum_{\alpha \in \mathbb{N}^d} \mu_\alpha \mathbf{b}^\alpha = 0$ implies that $\|\sum_{\alpha \in S} \mu_\alpha \mathbf{b}^\alpha\| < p^{-m}$ for some finite set $S \supseteq T_k$. It follows that

$$\sum_{\alpha \in T_k} \mu_\alpha \mathbf{b}^\alpha = \sum_{\alpha \in S} \mu_\alpha \mathbf{b}^\alpha - \sum_{\alpha \in S \setminus T_k} \mu_\alpha \mathbf{b}^\alpha$$

$$\in J^m + (I_k + pR) = I_k + pR,$$

by Lemmas 7.1(ii) and 7.9(iii). But it follows from Lemma 7.9(ii) that

$$\sum_{\alpha \in T_k} \mathbb{Z}_p \mathbf{b}^\alpha \cap (I_k + pR) = \bigoplus_{\alpha \in T_k} p\mathbb{Z}_p \mathbf{b}^\alpha;$$

so equating coefficients we deduce that $\mu_\alpha \in p\mathbb{Z}_p$ for each $\alpha \in T_k$. As k was arbitrary it follows that μ_α is divisible by p for every α, contradicting our assumption. Thus $\mu_\alpha = 0$ for all $\alpha \in \mathbb{N}^d$ as required.

Before proving Theorem 7.5, we need to refine Lemma 7.9. The result, Lemma 7.11 below, depends on the following lemma, which expresses the ring-theoretic significance of G being powerful. For each $k \geq 1$ we write

$$J_{k,1} = pJ^{k-1} + J^{k+1}.$$

7.10 Lemma *Assume that G is powerful. Let $u \in J^k$ and $x \in G$. Then*

$$ux - xu \in J_{k+1,1}.$$

Proof Let y be any element of G. Then

$$yx - xy = xy([y,x] - 1) = xy(z^p - 1)$$

for some $z \in G$, since G is powerful (Theorem 3.6). Formula (1) above shows that $z^p - 1 \in J^p + pJ$. If $p \geq 3$ this ideal is contained in $J_{2,1}$. If $p = 2$ then G/G_3 is abelian, so $[y,x] - 1 \in I_3 \subseteq J^3$, by Lemma 7.1. In either case it follows that $yx - xy \in J_{2,1}$. As R is additively spanned by G this proves the lemma in the case $k = 1$.

Now let $k > 1$, and assume inductively that $vy - yv \in J_{k,1}$ for all $v \in J^{k-1}$ and $y \in G$. As J^k is additively spanned by elements of the form vw with $v \in J^{k-1}$ and $w \in J$, it will suffice to show that for any such v and w we have $vwx - xvw \in J_{k+1,1}$. But

$$vwx - xvw = v(wx - xw) + (vx - xv)w$$
$$\in J^{k-1}J_{2,1} + J_{k,1}J$$

by the first paragraph and the inductive hypothesis. The result follows, since each of the terms in the last line is obviously equal to $J_{k+1,1}$.

7.11 Lemma *Assume that G is powerful. Let $k \geq 1$. Then*

$$J^k = J^{k+1} + \sum_{\langle \alpha \rangle \leq k} p^{k - \langle \alpha \rangle} \mathbb{Z}_p \mathbf{b}^\alpha.$$

Proof Let us write

$$W_k = \sum_{\langle \alpha \rangle \le k} p^{k - \langle \alpha \rangle} \mathbb{Z}_p \mathbf{b}^\alpha.$$

Since $p \in J$ and $b_i = a_i - 1 \in J$ for each i, it is clear that $p^{k - \langle \alpha \rangle} \mathbf{b}^\alpha \in J^k$ for each α; so $J^k \supseteq J^{k+1} + W_k$.

The reverse inclusion is proved by induction on k. From Lemma 7.9(i) we have

$$J \subseteq I_2 + \sum_{\alpha \in T_2} \mathbb{Z}_p \mathbf{b}^\alpha \subseteq J^2 + \sum_{\langle \alpha \rangle = 1} \mathbb{Z}_p \mathbf{b}^\alpha + \mathbb{Z}_p 1_G,$$

since $I_2 \subseteq J^2$ and $\mathbf{b}^\alpha \in J^2$ when $\langle \alpha \rangle \ge 2$. It follows by the modular law that

$$J = J^2 + \sum_{\langle \alpha \rangle = 1} \mathbb{Z}_p \mathbf{b}^\alpha + (J \cap \mathbb{Z}_p 1_G) = J^2 + W_1,$$

since $J \cap \mathbb{Z}_p 1_G = p\mathbb{Z}_p 1_G = p\mathbb{Z}_p \mathbf{b}^0$. This establishes the case $k = 1$.

Now take $k > 1$ and suppose inductively that $J^\ell = J^{\ell+1} + W_\ell$ for all $\ell < k$. Then

$$J^k = J^{k-1} J = (J^k + W_{k-1})(J^2 + W_1) \subseteq J^{k+1} + W_{k-1} W_1.$$

As $W_1 = p\mathbb{Z}_p + \sum_{i=1}^d \mathbb{Z}_p b_i$, and $pW_{k-1} \subseteq W_k$, we see that J^k is additively spanned modulo $J^{k+1} + W_k$ by elements of the form $p^{k-1-\langle \alpha \rangle} \mathbf{b}^\alpha b_i$ with $i \in \{1, \dots, d\}$ and $\langle \alpha \rangle \le k - 1$. It will therefore suffice to show that for each such i and α we have $p^{k-1-\langle \alpha \rangle} \mathbf{b}^\alpha b_i \in J^{k+1} + W_k$.

Put

$$\mathbf{u} = b^{\alpha_1} \dots b^{\alpha_{i-1}}, \quad \mathbf{v} = b^{\alpha_i} b^{\alpha_{i+1}} \dots b^{\alpha_d}.$$

Then

$$\mathbf{b}^\alpha b_i = \mathbf{uv} b_i = \mathbf{u} b_i \mathbf{v} + \mathbf{u}(\mathbf{v} b_i - b_i \mathbf{v})$$
$$= \mathbf{b}^\beta + \mathbf{u} w,$$

say, where $w = \mathbf{v} b_i - b_i \mathbf{v}$ and $\beta_i = 1 + \alpha_i$, $\beta_j = \alpha_j$ for $j \ne i$. Now $\mathbf{v} \in J^n$ where $n = \alpha_i + \dots + \alpha_d$, so $w \in J_{n+1,1}$ by Lemma 7.10. As $\mathbf{u} \in J^{\langle \alpha \rangle - n}$ it follows that

$$\mathbf{u} w \in J^{\langle \alpha \rangle - n} J_{n+1,1} = J_{\langle \alpha \rangle + 1, 1} = pJ^{\langle \alpha \rangle} + J^{\langle \alpha \rangle + 2}.$$

Thus

$$p^{k-1-\langle \alpha \rangle} \mathbf{b}^\alpha b_i \in p^{k-1-\langle \alpha \rangle} \mathbb{Z}_p \mathbf{b}^\beta + p^{k-\langle \alpha \rangle} J^{\langle \alpha \rangle} + p^{k-1-\langle \alpha \rangle} J^{\langle \alpha \rangle + 2}.$$

The final term on the right lies in J^{k+1}. Since $\langle \alpha \rangle \leq k - 1$, we may apply the inductive hypothesis to see that the middle term lies in

$$p^{k-\langle \alpha \rangle} J^{\langle \alpha \rangle + 1} + p^{k-\langle \alpha \rangle} W_{\langle \alpha \rangle} \subseteq J^{k+1} + W_k.$$

Since $\langle \beta \rangle = \langle \alpha \rangle + 1$, the first term belongs to W_k. It follows that $p^{k-1-\langle \alpha \rangle} \mathbf{b}^\alpha b_i \in J^{k+1} + W_k$ as required.

Proof of Theorem 7.5 Let $c = \sum_{\alpha \in \mathbb{N}^d} \lambda_\alpha \mathbf{b}^\alpha \in \mathbb{Z}_p[G]$, and suppose that $\|c\| = p^{-k}$. Then $c \in J^k$, so by Lemma 7.11 we can write

$$c = \sum_{\langle \alpha \rangle \leq k} p^{k-\langle \alpha \rangle} \mu_\alpha(k) \mathbf{b}^\alpha + c_{k+1}$$

where $\mu_\alpha(k) \in \mathbb{Z}_p$ for each α and $c_{k+1} \in J^{k+1}$. Repeating this step we construct a sequence $(c_j)_{j \geq k}$ such that for each j, $c_j \in J^j$ and

$$c_j - c_{j+1} = \sum_{\langle \alpha \rangle \leq j} p^{j-\langle \alpha \rangle} \mu_\alpha(j) \mathbf{b}^\alpha = w_j$$

say. Then $c \equiv w_k + w_{k+1} + \cdots + w_n \pmod{J^{n+1}}$, so $\left\| c - \sum_{j=k}^n w_j \right\| \leq p^{-n-1}$ for all $n > k$; hence

$$c = \sum_{j=k}^\infty w_j.$$

Now let $T = \{(\alpha, i) \mid i \geq k, \langle \alpha \rangle \leq i\}$. Since $p^{i-\langle \alpha \rangle} \mathbf{b}^\alpha \in J^i$, we have $\|p^{i-\langle \alpha \rangle} \mu_\alpha(i) \mathbf{b}^\alpha\| \leq p^{-i}$ for each i, which implies that

$$\lim_{(\alpha, i) \in T} p^{i-\langle \alpha \rangle} \mu_\alpha(i) \mathbf{b}^\alpha = 0,$$

in the sense of §6.2. Then Proposition 6.9(ii) shows that the series $\sum_{(\alpha, i) \in T} p^{i-\langle \alpha \rangle} \mu_\alpha(i) \mathbf{b}^\alpha$ converges in the complete ring $\widehat{R}$. In view of Corollary 6.11 we may re-arrange this series: putting

$$\mu_\alpha = \sum_{i=\max\{k, \langle \alpha \rangle\}}^\infty p^{i-\langle \alpha \rangle} \mu_\alpha(i)$$

we obtain

$$\sum_{\alpha \in \mathbb{N}^d} \mu_\alpha \mathbf{b}^\alpha = \sum_{(\alpha, i) \in T} p^{i-\langle \alpha \rangle} \mu_\alpha(i) \mathbf{b}^\alpha = \sum_{i=k}^\infty w_i = c.$$

(As a matter of fact, the argument so far provides an alternative proof for Theorem 7.4(i).)

Now Theorem 7.4(ii) shows that $\lambda_\alpha = \mu_\alpha$ for all α. It follows that

$$p^{\langle\alpha\rangle}\lambda_\alpha = \sum_{i=\max\{k,\langle\alpha\rangle\}}^{\infty} p^i \mu_\alpha(i) \in p^k \mathbb{Z}_p$$

for each α, and hence that $\sup_{\alpha\in\mathbb{N}^d} p^{-\langle\alpha\rangle}|\lambda_\alpha| \leq p^{-k} = \|c\|$.

The reverse inequality follows from Proposition 6.9(iv). For if $|\lambda_\alpha| = p^{-n}$ then $\lambda_\alpha \mathbf{b}^\alpha \in p^n J^{\langle\alpha\rangle} \subseteq J^{n+\langle\alpha\rangle}$, showing that $\|\lambda_\alpha \mathbf{b}^\alpha\| \leq p^{-n-\langle\alpha\rangle} = p^{-\langle\alpha\rangle}|\lambda_\alpha|$; so

$$\|c\| \leq \sup_\alpha \|\lambda_\alpha \mathbf{b}^\alpha\| \leq \sup_\alpha p^{-\langle\alpha\rangle}|\lambda_\alpha|$$

as required.

7.2 The Lie algebra

Throughout this section, G denotes a uniform pro-p group of dimension d, generated topologically by elements $a_1, \ldots, a_d$, and $G_i = P_i(G)$ for each $i \geq 1$. We assume that G is contained as a subgroup of the unit group in a normed $\mathbb{Q}_p$-algebra A, and that the topology on G is induced by the norm. Let $\widehat{A}$ denote the completion of A. We put

$$\widehat{A}_0 = \begin{cases} \{x \in \widehat{A} \mid \|x\| \leq p^{-1}\} & \text{if } p \neq 2 \\ \{x \in \widehat{A} \mid \|x\| \leq 2^{-2}\} & \text{if } p = 2. \end{cases}$$

Finally, we make the hypothesis that

$$G - 1 \subseteq \widehat{A}_0.$$

If p is odd, Theorem 7.7 ensures that all our hypotheses are fulfilled on taking $A = (\mathbb{Q}_p[G], \|\cdot\|)$. If $p = 2$, they are fulfilled if $G = P_2(H)$ for some uniform pro-2 group H, and $A = (\mathbb{Q}_p[H], \|\cdot\|)$; however, Exercise 10 shows that for *any* uniform pro-2 group G, we can define a norm $\|\cdot\|_2$ on $\mathbb{Q}_p[G]$ in such a way that $A = (\mathbb{Q}_p[G], \|\cdot\|_2)$ satisfies all our requirements.

In Proposition 6.22 we defined a mapping $\log : 1 + \widehat{A}_0 \to \widehat{A}_0$. We now put

$$\Lambda = \log G \subseteq \widehat{A}.$$

Now the associative algebra $\widehat{A}$ has a natural Lie algebra structure, with Lie bracket operation $(x, y) = xy - yx$. On the other hand, Definitions 4.12 and 4.29 give an intrinsic addition and bracket operation on the uniform pro-p group G; to avoid confusion with the operations on $\widehat{A}$, we shall denote these by $+_G$ and $(\,,\,)_G$ respectively.

The operations are related as follows:

7.12 Lemma. *Let* $g, h \in G$ *and* $\lambda \in \mathbb{Z}_p$. *Then*
  (i) $\log g + \log h = \log(g +_G h)$;
  (ii) $\lambda \log g = \log g^\lambda$;
  (iii) $(\log g, \log h) = \log(g, h)_G$.

Proof Put $\gamma = \log g$ and $\eta = \log h$.

(i) Let $\Phi(X, Y) = X + Y + \sum_{n \geq 2} u_n(X, Y)$ be the Campbell–Hausdorff series given in Definition 6.26; here $u_n(X, Y)$ is a homogeneous polynomial of degree n. By Corollary 6.25(iii), for each $i \in \mathbb{N}$, $\log g^{p^i} = p^i \gamma$ and $\log h^{p^i} = p^i \eta$. Combining this with Proposition 6.27 gives

$$\log(g^{p^i} h^{p^i}) = p^i(\gamma + \eta) + \sum_{n \geq 2} p^{ni} u_n(\gamma, \eta).$$

Thus

$$\log(g^{p^i} h^{p^i})^{p^{-i}} = \gamma + \eta + p^i \sum_{n \geq 2} p^{(n-2)i} u_n(\gamma, \eta). \tag{5}$$

By Proposition 6.19, log is a continuous function, so

$$\lim_{i \to \infty} \log(g^{p^i} h^{p^i})^{p^{-i}} = \log(\lim_{i \to \infty} (g^{p^i} h^{p^i})^{p^{-i}}).$$

Since the norm $\|\cdot\|$ on $\widehat{A}$ is compatible with the topology on G, the definition of $+_G$ gives

$$\lim_{i \to \infty} (g^{p^i} h^{p^i})^{p^{-i}} = g +_G h \in G.$$

Since $\sum_{n \geq 2} u_n(\gamma, \eta)$ converges, its terms are bounded, and so the sum $\sum_{n \geq 2} p^{(n-2)i} u_n(\gamma, \eta)$ is bounded for all i. Hence

$$\lim_{i \to \infty} p^i \sum_{n \geq 2} p^{(n-2)i} u_n(\gamma, \eta) = 0.$$

Thus taking limits in (5) we obtain

$$\log g + \log h = \log(g +_G h).$$

(ii) Let $\lambda \in \mathbb{Z}_p$, so $\lambda = \lim_{i \to \infty} a_i$ for some $a_i \in \mathbb{N}$. It follows from the definition of p-adic exponentiation in G and the fact that the norm $\|\cdot\|$ on $\widehat{A}$ is compatible with the topology on G that

$$g^\lambda = \lim_{i \to \infty} g^{a_i} \in G,$$

where the limit is taken with respect to the topology of $\widehat{A}$. Since the mapping log is continuous, using Corollary 6.25(iii) we get

$$\lambda \log g = \lim_{i \to \infty} a_i \log g$$
$$= \lim_{i \to \infty} \log g^{a_i}$$
$$= \log(\lim_{i \to \infty} g^{a_i})$$
$$= \log g^\lambda.$$

(iii) Let $\Psi(X, Y) = XY - YX + \sum_{n \geq 3} v_n(X, Y)$ be the Commutator Campbell–Hausdorff series given in Definition 6.26; here $v_n(X, Y)$ is a homogeneous polynomial of degree n. Then, by Corollary 6.25(iii) and Proposition 6.27, we have

$$\log[g^{p^i}, h^{p^i}]^{p^{-2i}} = \gamma\eta - \eta\gamma + p^i \sum_{n \geq 3} p^{(n-3)i} v_n(\gamma, \eta). \qquad (6)$$

By a similar argument to (i), we find that

$$\lim_{i \to \infty} \log[g^{p^i}, h^{p^i}]^{p^{-2i}} = \log(\lim_{i \to \infty} [g^{p^i}, h^{p^i}]^{p^{-2i}})$$
$$= \log(g, h)_G$$

and $\lim_{i \to \infty}(p^i \sum_{n \geq 3} p^{(n-3)i} v_n(\gamma, \eta)) = 0$. Thus taking limits in (6) gives

$$(\log g, \log h) = \gamma\eta - \eta\gamma = \log(g, h)_G,$$

completing the proof.

Now G is closed under the operations $+_G$ and $(,)_G$; and Theorem 4.17 shows that $(G, +_G)$ is a free $\mathbb{Z}_p$-module of rank d, the action of $\mathbb{Z}_p$ being p-adic exponentiation. The main result of this section is thus an immediate consequence of Lemma 7.12:

7.13 Theorem $(\Lambda, +, (,))$ *is a $\mathbb{Z}_p$-Lie subalgebra of the Lie algebra* $(\widehat{A}, +, (,))$, *and is free of rank d as a $\mathbb{Z}_p$-module.*

Arguing in the reverse direction, we can now also deduce

7.14 Corollary $(G, +_G, (,)_G)$ *is a Lie algebra over $\mathbb{Z}_p$, and the mapping* $\log \colon (G, +_G, (,)_G) \to (\Lambda, +, (,))$ *is a $\mathbb{Z}_p$-Lie algebra isomorphism.*

Let us write L_G for the $\mathbb{Z}_p$-Lie algebra $(G, +_G, (,)_G)$. We have seen in Proposition 4.31 how suitable subgroups and quotients of G correspond to subalgebras and quotients of L_G. We are now ready to establish the converse:

7.15 Proposition *Let N be a $\mathbb{Z}_p$-Lie subalgebra of L_G such that L_G/N is torsion-free. Then*

(i) *N is a closed uniform subgroup of G;*

(ii) *if N is an ideal of L_G then N is normal in G and G/N is uniform.*

Proof Write $I = \log(N)$, and $\mathbb{Q}_p I$ for the $\mathbb{Q}_p$-subspace it spans in $\widehat{A}$. In view of Corollary 7.14, I is a $\mathbb{Z}_p$-Lie subalgebra of $\Lambda = \log G$, Λ/I is torsion-free, and in case (ii) I is an ideal. Write $\epsilon = 1$ if $p \neq 2$, $\epsilon = 2$ if $p = 2$. Since $[G, G] \leq G^{p^\epsilon}$ it follows from the definition (see Exercise 4.2(ii)) that $(L_G, L_G) \subseteq p^\epsilon L_G$, and hence that $(\Lambda, \Lambda) \subseteq p^\epsilon \Lambda$. Since Λ/I is torsion-free this implies that $(I, I) \subseteq p^\epsilon I$.

Now let $x, y \in N$ and put $u = \log x$, $v = \log y$. According to Proposition 6.27 we have

$$\log(xy^{-1}) = \sum_{n \geq 1} u_n(u, -v),$$

where $\sum_{n \geq 1} u_n(X, Y) = \Phi(X, Y)$ is the Campbell–Hausdorff series. According to Corollary 6.38, the series $\sum_{n \geq 1} u_n(u, -v)$ converges to an element of I. Hence $xy^{-1} \in \exp(I) = N$. Thus N is a subgroup of G, and (being compact) it is closed.

If $y \in G$ and $y^p \in N$ then $\log y \in \Lambda$ and $p \log y \in I$, so $\log y \in I$ and $y \in N$. Therefore $G^p \cap N = N^p$, whence N/N^p is abelian and N is powerful (if p is odd; replace p by 4 to reach the same conclusion for $p = 2$). As G is torsion-free it follows that N is uniform, by Theorem 4.5. This completes the proof of (i).

The argument just given shows that if N is also normal in G, then G/N is torsion-free, hence uniform. So to prove (ii) it remains only to show that if I is an ideal of Λ then N is indeed normal in G. Let $x \in N$ and $y \in G$, and let u, v be as above. Put

$$w = \sum_{n \geq 2} u_n(u, -v).$$

Corollary 6.38 shows that each term in this sum lies in Λ; as $\mathbb{Q}_p I$ is an ideal in $\mathbb{Q}_p\Lambda$, each term also lies in $\mathbb{Q}_p I$, hence belongs to $\Lambda \cap \mathbb{Q}_p I = I$. Since I is closed in Λ and the series converges in Λ it follows that $w \in I$.

As above, we see that $\log(xy^{-1}) = u - v + w$. Then

$$\log(yxy^{-1}) = v + (u - v + w) + \sum_{n \geq 2} u_n(v, u - v + w).$$

Since $(v, u - v + w) = (v, u + w) \in I$, we can repeat the argument with $u + w$ in place of w to deduce that $\log(yxy^{-1}) \in I$. Hence $yxy^{-1} \in \exp(I) = N$, as required.

Results like this illustrate how the Campbell–Hausdorff formula establishes a tight parallel between the structure of a uniform group and that of its Lie algebra. This correspondence will be developed further in Section 9.4. We conclude this section with another illustration:

7.16 Corollary *Let G be a uniform group and $L_G = (G, +_G, (,)_G)$ the corresponding Lie algebra.*
  (i) *G is abelian if and only if L_G is abelian;*
  (ii) *G is soluble if and only if L_G is soluble.*

Proof (i) If G is abelian then $(x, y)_G = 0$ for all $x, y \in G$, directly from the definition, so L_G is abelian. Suppose conversely that L_G is abelian. Let G and Λ be as above. Corollary 7.14 shows that Λ is abelian, so for $u, v \in \Lambda$ we have

$$\exp(u)\exp(v) = \exp(u + v) = \exp(v + u) = \exp(v)\exp(u),$$

by Corollary 6.36. Hence $G = \exp(\Lambda)$ is abelian.
  (ii) Suppose that G is soluble. Let A be a maximal abelian normal subgroup of G. We claim that G/A is uniform. Indeed, by Theorem 4.20 there is a finite normal subgroup T/A of G/A such that G/T is uniform; then T is uniform by 4.31(ii), and consequently abelian, by Exercise 4.9. Hence $T = A$. Arguing by induction on the derived length, we may suppose that the Lie algebra $L_{G/A}$ is soluble. Now Proposition 4.31(iii) shows that A is an ideal of L_G and that $L_G/A \cong L_{G/A}$; and A is abelian as an ideal, by (i). It follows that L_G is soluble.
  Suppose conversely that L_G is soluble. Let A be a maximal abelian ideal of L_G. It is easy to see that L_G/A must be torsion-free. Now Proposition 7.15 shows that A is a uniform normal subgroup of G and that G/A is uniform; and A is abelian as a subgroup, by (i). As above, we have $L_G/A \cong L_{G/A}$; arguing by induction on the derived length of L_G, we may therefore suppose that G/A is a soluble group, and it follows that G is soluble.

7.3 Linear representations

In this section we prove that every pro-p group of finite rank has a faithful linear representation. This depends on the analogous property of Lie algebras, a classic result due to Ado (for the proof, see e.g. Jacobson (1962), Chapter VI):

Ado's Theorem *Let L be a finite dimensional Lie algebra over a field k of characteristic zero. Then L admits a faithful finite dimensional linear representation*

$$\phi : L \to \mathrm{M}_n(k).$$

Here, $\mathrm{M}_n(k)$ is the algebra of all $n \times n$ matrices over k, considered as a Lie algebra with the Lie bracket operation given by $(x, y) = xy - yx$, and ϕ is a Lie algebra monomorphism.

We keep the notation of the previous section, so G is a uniform pro-p group, $\widehat{A}$ is the completion of $\mathbb{Q}_p[G]$, and $\Lambda = \log G$ (if $p = 2$, we can avoid having to rely on Exercise 10 by taking $\Lambda = \log G_2$ in this section). We write $\mathbb{Q}_p\Lambda$ for the $\mathbb{Q}_p$-vector subspace of $\widehat{A}$ spanned by Λ. Then $\mathbb{Q}_p\Lambda$ is a finite dimensional Lie algebra over $\mathbb{Q}_p$. Let

$$\phi : \mathbb{Q}_p\Lambda \to \mathrm{M}_n(\mathbb{Q}_p) = B, \text{ say,}$$

be the faithful linear representation of $\mathbb{Q}_p\Lambda$ provided by Ado's Theorem. Recall from Example 6.7 that $\mathrm{M}_n(\mathbb{Q}_p)$ is a normed $\mathbb{Q}_p$-algebra, and that Proposition 6.22 defines the mapping exp on the subset

$$B_0 = \begin{cases} \mathrm{M}_n(p\mathbb{Z}_p) & \text{if } p \neq 2 \\ \mathrm{M}_n(4\mathbb{Z}_2) & \text{if } p = 2, \end{cases}$$

with

$$\exp B_0 \subseteq 1 + B_0 \subseteq \mathrm{GL}_n(\mathbb{Z}_p). \tag{7}$$

Putting $\Lambda_0 = \phi^{-1}(\phi\Lambda \cap B_0)$, we have $\exp \Lambda_0 \subseteq G$.

7.17 Lemma. *There exists $m \geq 1$ such that $G_m \subseteq \exp \Lambda_0$.*

Proof Since Λ is finitely generated as a $\mathbb{Z}_p$-module, $p^i\phi\Lambda \subseteq \mathrm{M}_n(\mathbb{Z}_p)$ for some positive integer i. We take $m = i + 2$ if p is odd, $m = i + 4$ if $p = 2$. Now Lemma 4.14 shows that $G_m = p^{m-1} \cdot (G, +_G)$. So if p is odd we have $G_m = p^{i+1} \cdot (G, +_G)$, and Corollary 7.15 shows that this is mapped by log into $p^{i+1}\Lambda$. Thus

$$\phi(\log G_m) \subseteq \phi(p^{i+1}\Lambda) \subseteq p\mathrm{M}_n(\mathbb{Z}_p) = B_0;$$

hence $G_m \subseteq \exp \Lambda_0$. The argument is similar when $p = 2$.

We now define $\psi = \exp \circ \phi \circ \log : G_m \to \mathrm{M}_n(\mathbb{Q}_p)$, where $m \in \mathbb{N}$ is such that $G_m \subseteq \exp \Lambda_0$. Then (7) implies that $\psi(G_m) \subseteq \mathrm{GL}_n(\mathbb{Z}_p)$, and we have

7.18 Proposition. $\psi : G_m \to \mathrm{GL}_n(\mathbb{Z}_p)$ *is a faithful linear representation of* G_m.

Proof The mapping ψ is clearly injective. It remains therefore to prove that if $g, h \in G_m$ then $\psi(gh) = \psi(g) \cdot \psi(h)$. Suppose that $g = \exp x$ and $h = \exp y$ where $x, y \in \Lambda_0$. Let $\Phi(X, Y) = \sum_{n \in \mathbb{N}} u_n(X, Y)$ denote the Campbell–Hausdorff series. Then by Proposition 6.27

$$\log(gh) = \sum_{n \in \mathbb{N}} u_n(x, y).$$

Now Theorem 6.28 shows that each $u_n(X, Y)$ can be written as a $\mathbb{Q}$-linear combination of compound Lie brackets in X and Y. So $u_n(x, y) \in \mathbb{Q}_p\Lambda$ for all $n \in \mathbb{N}$. Hence $\phi(u_n(x, y))$ is defined and

$$\phi(u_n(x, y)) = u_n(\phi(x), \phi(y))$$

since ϕ is a Lie algebra homomorphism. By Exercise 6.3, the map $\phi \colon \mathbb{Q}_p\Lambda \to \mathrm{M}_n(\mathbb{Q}_p)$ is continuous. So

$$\phi(\log(gh)) = \phi(\sum_{n \in \mathbb{N}} u_n(x, y)) = \sum_{n \in \mathbb{N}} \phi(u_n(x, y))$$

$$= \sum_{n \in \mathbb{N}} u_n(\phi(x), \phi(y)).$$

Now $\phi(x)$ and $\phi(y)$ are in B_0, so applying Proposition 6.27 to the normed $\mathbb{Q}_p$-algebra $B = \mathrm{M}_n(\mathbb{Q}_p)$ we obtain

$$\sum_{n \in \mathbb{N}} u_n(\phi(x), \phi(y)) = \log(\exp(\phi(x)) \cdot \exp(\phi(y)))$$

$$= \log((\exp \circ \phi \circ \log(g)) \cdot (\exp \circ \phi \circ \log(h)))$$

$$= \log(\psi(g) \cdot \psi(h)).$$

Thus $\psi(g) \cdot \psi(h) = \exp(\phi \log(gh)) = \psi(gh)$ as required.

Now suppose that Γ is a pro-p group of finite rank. Corollary 4.3 shows that Γ contains a uniform subgroup G of finite index. Then G_m also has finite index in Γ, and the representation of Γ induced from $\psi : G_m \to \mathrm{GL}_n(\mathbb{Z}_p)$ is again faithful. We may now therefore deduce the main result:

7.19 Theorem. *Every pro-p group of finite rank admits a faithful linear representation over $\mathbb{Z}_p$.*

Remark The method we have used here – translating a Lie algebra homomorphism into a group homomorphism via exp and log, with the help of the Campbell–Hausdorff formula – is actually quite general; in §9.5 we construct a functor from Lie algebras to p-adic analytic groups, and Proposition 7.18 will appear in retrospect as an application of this functor.

7.4 The completed group algebra

We saw in Chapter 4 that a uniform pro-p group resembles in many ways an abelian group. It is thus not unreasonable to expect that its group algebra will bear some similarity to a commutative algebra, and this indeed turns out to be the case: although $\mathbb{Z}_p[G]$ is of course not in general commutative, we shall show that a naturally associated graded algebra is isomorphic to a commutative polynomial ring, and thereby derive some ring-theoretic properties of the completion $\mathbb{Z}_p[[G]]$. Similar considerations will also be applied to $\mathbb{F}_p[G]$.

Throughout, G will denote a uniform pro-p group, $R = \mathbb{Z}_p[G]$ its group algebra and $\widehat{R} = \mathbb{Z}_p[[G]]$ the completion of R defined in §7.1. As before, we fix a topological generating set $\{a_1, \ldots, a_d\}$ for G, where $d = \dim(G)$, and write $b_i = a_i - 1$ for $i = 1, \ldots, d$. Recall the notation

$$\langle \alpha \rangle = \alpha_1 + \cdots + \alpha_d, \quad \mathbf{b}^\alpha = b_1^{\alpha_1} \ldots b_d^{\alpha_d}$$

for $\alpha = (\alpha_1, \ldots, \alpha_d) \in \mathbb{N}^d$.

The first result is a direct consequence of Theorem 7.4:

7.20 Theorem *Each element of $\mathbb{Z}_p[[G]]$ is equal to the sum of a uniquely determined convergent series*

$$\sum_{\alpha \in \mathbb{N}^d} \lambda_\alpha \mathbf{b}^\alpha \tag{8}$$

with $\lambda_\alpha \in \mathbb{Z}_p$ for each $\alpha \in \mathbb{N}^d$; conversely, every such series converges in $\mathbb{Z}_p[[G]]$.

Proof The convergence of the series (8) is immediate from Proposition 6.9(ii), since $\|\mathbf{b}_\alpha\| \leq p^{-\langle\alpha\rangle}$ for each $\alpha \in \mathbb{N}^d$. The uniqueness statement follows from Theorem 7.4(ii), which shows that if the series (8) converges to 0 then $\lambda_\alpha = 0$ for each α.

Now let S denote the subset of $\widehat{R}$ consisting of all the elements that are equal to the sum of a series like (8). Then S contains R, by Theorem 7.4, so S is dense in $\widehat{R}$. To prove that $S = \widehat{R}$ it will therefore suffice to show that S is closed. We do this by showing that S is *compact*. Write $X = \mathbb{Z}_p^{\mathbb{N}^d}$, and define the map $\phi : X \to S$ by

$$\phi(\lambda) = \sum_{\alpha \in \mathbb{N}^d} \lambda_\alpha \mathbf{b}^\alpha.$$

Certainly ϕ is surjective; as X, with the product topology, is compact, by Tychonoff's Theorem, it suffices now to check that ϕ is continuous. So let $r = \phi(\lambda) \in S$ and let $\varepsilon > 0$. Choose $n \in \mathbb{N}$ such that $p^{-n} < \varepsilon$, and put

$$U = \{(\mu) \in X \mid \mu_\alpha \equiv \lambda_\alpha \pmod{p^n \mathbb{Z}_p} \text{ for all } \alpha \text{ with } \langle\alpha\rangle < n\}.$$

Then U is an open subset of X containing (λ); and for each $(\mu) \in U$ we have

$$\|\phi(\mu) - r\| = \left\| \sum_{\alpha \in \mathbb{N}^d} (\mu_\alpha - \lambda_\alpha)\mathbf{b}^\alpha \right\|$$
$$\leq p^{-n} < \varepsilon,$$

by Proposition 6.9(iv), since each term in the infinite sum is either divisible by p^n or divisible by some $\mathbf{b}^\alpha$ with $\langle\alpha\rangle \geq n$. This shows that ϕ is continuous, and completes the proof.

Corresponding to Theorem 7.5 we have

7.21 Theorem *If* $c = \sum_{\alpha \in \mathbb{N}^d} \lambda_\alpha \mathbf{b}^\alpha \in \mathbb{Z}_p[[G]]$, *where* $\lambda_\alpha \in \mathbb{Z}_p$ *for each* α, *then*

$$\|c\| = \sup_{\alpha \in \mathbb{N}^d} p^{-\langle\alpha\rangle} |\lambda_\alpha|.$$

Proof Say $\|c\| = p^{-k}$. Putting

$$r = \sum_{\langle \alpha \rangle \leq k} \lambda_\alpha \mathbf{b}^\alpha,$$

we have $\|c - r\| \leq p^{-k-1}$, by Proposition 6.9(iv). Therefore

$$\|c\| = \|r\| = \sup_{\langle \alpha \rangle \leq k} p^{-\langle \alpha \rangle} |\lambda_\alpha|$$

by Theorem 7.5, since $r \in R$. The result follows, since for each α with $\langle \alpha \rangle > k$ we have

$$p^{-\langle \alpha \rangle} |\lambda_\alpha| < p^{-k} = \|c\|.$$

Next, we proceed to construct the graded ring associated to $\widehat{R}$. There is a natural filtration on $\widehat{R}$ given by setting

$$R_k = \{c \in \widehat{R} \mid \|c\| \leq p^{-k}\}$$

for each $k \in \mathbb{N}$. Thus $R_0 = \widehat{R}$, $R_k \supseteq R_{k+1}$ for each k, and each of the subsets R_k is an ideal in $\widehat{R}$. Using property (N2) of the norm, we see that in fact

$$R_i R_j \subseteq R_{i+j} \tag{9}$$

for all i and j. However, in order to obtain a nice graded ring we have to refine this filtration; this is done as follows. For each $k, m \geq 0$ put

$$R_{k,m} = (R_k \cap p^m \widehat{R}) + R_{k+1} = R_k \cap (p^m \widehat{R} + R_{k+1}) = p^m R_{k-m} + R_{k+1};$$

the second equality here is a consequence of the modular law, and the third one follows from Theorem 7.21 which implies that $R_k \cap p^m \widehat{R} = p^m R_{k-m}$. Then

$$R_k = R_{k,0} \supseteq R_{k,1} \supseteq \cdots \supseteq R_{k,k} \supseteq R_{k,k+1} = R_{k+1},$$

since $\|p^{k+1} \cdot 1\| = p^{-(k+1)}$. A simple calculation using (9) shows that

$$R_{i,m} R_{j,n} \subseteq R_{i+j,m+n} \tag{10}$$

for all relevant values of the subscripts. Now for $k \geq 0$ and $0 \leq m \leq k$ we put

$$E_{k,m} = R_{k,m}/R_{k,m+1}.$$

The *associated graded ring* R^* is constructed as follows. The additive group of R^* is the direct sum

$$R^* = \bigoplus_{k=0}^{\infty} \bigoplus_{m=0}^{k} E_{k,m} = E_{0,0} \oplus E_{1,0} \oplus E_{1,1} \oplus E_{2,0} \oplus \cdots.$$

Multiplication is defined on homogeneous elements (those belonging to a single summand $E_{*,*}$) by the formula

$$(a + R_{i,m+1})(b + R_{j,n+1}) = ab + R_{i+j,m+n+1} \quad (a \in R_{i,m}, b \in R_{j,n});$$

it follows from (10) that this is well defined. Finally, the multiplication is extended to the whole of R^* by distributivity. It is a simple exercise, left to the reader, to verify that all ring axioms are satisfied.

In passing from a filtered ring to the associated graded ring, one loses a certain amount of information. The advantage is that the graded ring is usually easier to understand:

7.22 Theorem *Let* $x_0 = p1_G + R_{1,2} \in E_{1,1}$, *and for* $i = 1, \dots, d$ *put* $x_i = b_i + R_{1,1} \in E_{1,0}$. *Then the mapping* $X_i \mapsto x_i$ *(* $i = 0, \dots, d$ *) defines a ring isomorphism*

$$\mathbb{F}_p[X_0, X_1, \dots, X_d] \overset{\sim}{\to} R^*.$$

Proof It is clear from the definition that $pE_{k,m} = 0$ for all k and m, so R^* is an $\mathbb{F}_p$-algebra. Next, let us verify that the elements x_j commute among themselves. It is obvious that x_0 commutes with everything. Now let $1 \le i, j \le d$, and recall Lemma 7.10: this implies that

$$b_i b_j - b_j b_i \in pJ + J^3 \subseteq pR_1 + R_3 \subseteq R_{2,1}.$$

As $x_i x_j - x_j x_i = (b_i b_j - b_j b_i) + R_{2,1} \in E_{2,0}$ this shows that $x_i x_j - x_j x_i = 0$.

Thus the mapping $X_i \mapsto x_i$ $(0 \le i \le d)$ extends to a ring homomorphism $\psi : \mathbb{F}_p[X_0, X_1, \dots, X_d] \to R^*$. To prove that ψ is an isomorphism, it will suffice to show that the monomials

$$w(m, \alpha) = x_0^m x_1^{\alpha_1} \dots x_d^{\alpha_d}$$

with $\langle a \rangle + m = k$ form a vector-space basis for $E_{k,m}$.

For each $n \ge 0$ put

$$B_n = \sum_{\langle \alpha \rangle = n} \mathbb{Z}_p b^{\alpha}.$$

Then Theorem 7.21 can be interpreted as asserting that for each $k \geq 0$,

$$R_k = \sum_{m+n=k} p^m B_n + R_{k+1}.$$

From this it follows that $R_{k,m} = p^m B_{k-m} + R_{k,m+1}$ Now when $\langle a \rangle + m = k$ we have

$$w(m, \alpha) = p^m \mathbf{b}^\alpha + R_{k,m+1} \in E_{k,m};$$

thus

$$E_{k,m} = R_{k,m}/R_{k,m+1} = \left(\sum_{\langle \alpha \rangle = k-m} \mathbb{Z}_p p^m \mathbf{b}^\alpha + R_{k,m+1} \right) / R_{k,m+1}$$

$$= \sum_{\langle \alpha \rangle = k-m} \mathbb{F}_p w(m, \alpha).$$

This shows that the given monomials span $E_{k,m}$.

To see that they are linearly independent, we reverse the argument. Suppose that $\sum_{\langle a \rangle = k-m} z_\alpha w(m, \alpha) = 0$, with each $z_\alpha \in \mathbb{F}_p$. Interpreting each z_α as an element of $\{0, 1, \dots, p-1\} \subseteq \mathbb{Z}_p$, our hypothesis is equivalent to the statement: $\sum z_\alpha p^m \mathbf{b}^\alpha \in R_{k,m+1}$. Since $R_{k,m+1} \subseteq p^{m+1}\widehat{R} + R_{k+1}$, Theorems 7.20 and 7.21 show that there exist $\lambda_\alpha, \mu_\alpha \in \mathbb{Z}_p$ such that

$$\sum_{\langle a \rangle = k-m} z_\alpha p^m \mathbf{b}^\alpha = p^{m+1} \sum \lambda_\alpha \mathbf{b}^\alpha + \sum \mu_\alpha \mathbf{b}^\alpha$$

and $p^{-\langle \alpha \rangle} |\mu_\alpha| \leq p^{-(k+1)}$ for each α. Equating coefficients, as we may by Theorem 7.20, we find that when $\langle a \rangle = k - m$,

$$p^{-m} |z_\alpha - p\lambda_\alpha| = |p^m z_\alpha - p^{m+1}\lambda_\alpha| = |\mu_\alpha| \leq p^{-(m+1)}.$$

It follows that each z_α is divisible by p, hence equal to 0. Thus the elements $w(m, \alpha)$ with $\langle a \rangle = k - m$ are indeed linearly independent, and the proof is complete.

Also of interest is the group algebra $\mathbb{F}_p[G] = S$, say. This will be studied in greater generality in Chapter 12; here we consider the case where G is uniform. As we did for $\mathbb{Z}_p[G]$, we define the completed group algebra

$$\mathbb{F}_p[[G]] = \varprojlim_{N \triangleleft_o G} \mathbb{F}_p[G/N] = \varprojlim_{k \in \mathbb{N}} \mathbb{F}_p[G/G_k].$$

Let $I = (G - 1)\mathbb{F}_p[G]$ be the augmentation ideal of $\mathbb{F}_p[G]$. Writing

$-: \mathbb{Z}_p[G] \to \mathbb{F}_p[G]$ for the natural map (reducing coefficients modulo p), we have

$$I = \overline{I_1} = \overline{J}.$$

It follows by Lemma 7.1 that the chain $(I^k) = (\overline{J^k})$ is cofinal with the chain $(\overline{I_k})$, where $\overline{I_k} = (G_k - 1)\mathbb{F}_p[G]$ is the kernel of the natural map $\mathbb{F}_p[G] \to \mathbb{F}_p[G/G_k]$. A simple modification of the proof of Corollary 7.2 shows that

$$\bigcap_{n=1}^{\infty} I^n = 0,$$

and as in §7.1 we may define a norm on $\mathbb{F}_p[G]$ by putting

$$\|c\| = p^{-k} \quad \text{if } c \in I^k \backslash I^{k+1}$$
$$\|0\| = 0.$$

The completion of S with respect to this norm is denoted $\widehat{S}$; as before, $\widehat{S}$ can be identified with $\mathbb{F}_p[[G]]$, and it is clear that the natural mapping $-: \mathbb{Z}_p[G] \to \mathbb{F}_p[G]$ induces an isomorphism $\widehat{R}/p\widehat{R} \cong \widehat{S}$. From Theorems 7.20 and 7.21 it is now easy to infer

7.23 Theorem (i) *Each element c of $\mathbb{F}_p[[G]]$ is equal to the sum of a uniquely determined convergent series*

$$\sum_{\alpha \in \mathbb{N}^d} \lambda_\alpha \mathbf{b}^\alpha$$

with $\lambda_\alpha \in \mathbb{F}_p$ for each $\alpha \in \mathbb{N}^d$; conversely, every such series converges in $\mathbb{F}_p[[G]]$.
   (ii) *For c as above,*

$$\|c\| = \sup_{\alpha \in \mathbb{N}^d} p^{-\langle \alpha \rangle}.$$

Writing

$$S_k = \{c \in \widehat{S} \mid \|c\| \leq p^{-k}\},$$

we form the graded $\mathbb{F}_p$-algebra

$$S^* = \bigoplus_{k=0}^{\infty} S_k/S_{k+1};$$

multiplication of homogeneous elements is defined by

$$(a + S_{i+1})(b + S_{j+1}) = ab + S_{i+j+1} \quad (a \in S_i, b \in S_j),$$

and extended by linearity to the whole of S^*. The following structure theorem is now proved in the same way as Theorem 7.22, the argument however being considerably simpler (alternatively, it can be deduced from Theorem 7.22, as outlined in Exercise 7):

7.24 Theorem *Let* $y_i = b_i + S_2 \in S_1/S_2$, *for* $i = 1, \ldots, d$. *Then the mapping* $X_i \mapsto y_i$ ($i = 1, \ldots, d$) *defines a ring isomorphism*

$$\mathbb{F}_p[X_1, \ldots, X_d] \xrightarrow{\sim} S^*.$$

Although beautiful in themselves, these graded rings would be of limited interest if they conveyed no information at all about the original group algebras. The point of the 'method of graded rings' is that certain features of a complete filtered ring *are* reflected in the associated graded ring. As a typical application of the method we shall deduce

7.25 Corollary *The rings* $\mathbb{Z}_p[[G]]$ *and* $\mathbb{F}_p[[G]]$ *are right and left Noetherian, and have no zero divisors.*

7.26 Corollary *The group algebras* $\mathbb{Z}_p[G]$ *and* $\mathbb{F}_p[G]$ *have no zero divisors.*

It follows from the last corollary that if Γ is any group that can be embedded in a uniform pro-p group, then the group rings $\mathbb{Z}[\Gamma]$ and $\mathbb{F}_p[\Gamma]$ are without zero divisors. This applies, for example, when Γ is the principal congruence subgroup modulo p (p odd), or modulo 4 (when $p = 2$) in $\mathrm{GL}_n(\mathbb{Z})$, by Theorem 5.2. More general results along these lines can be deduced from stronger forms of the above theorems, in which G may be *any torsion-free pro-p group of finite rank*. We do not prove these here, as they require a number of new ingredients from ring theory and homological algebra, and refer the reader to Neumann (1988). (See also Exercise 6.)

We know from Theorems 7.22 and 7.24 that the graded rings R^* and S^* are Noetherian (by Hilbert's Basis Theorem) and without zero divisors. That the corresponding complete rings $\widehat{R}$ and $\widehat{S}$ also have these properties will follow by Proposition 7.27 which we prove below. Before doing so, we must set up some terminology.

Let A be a ring with a descending chain of ideals (A_i) satisfying the conditions of Lemma 6.5. Put $E_i = A_i/A_{i+1}$ for each $i \geq 0$. The *associated graded ring* is then

$$A^* = \bigoplus_{i=0}^{\infty} E_i.$$

This formula specifies the additive structure; the multiplicative structure depends on the choice of a 'grading monoid'. In the present context, this comes down to specifying a commutative, associative binary operation $*$ on $\mathbb{N}$, which satisfies the further conditions

$$i * 0 = i$$
$$i * j = i * k \Rightarrow j = k$$
$$j > k \Rightarrow i * j > i * k$$

for all i, j and k. It follows from the first and third conditions that $i * j \geq i + j$ for all i and j; returning to our ring A, we therefore have

$$A_i A_j \subseteq A_{i*j}, \quad A_i A_{j+1} + A_{i+1} A_j \subseteq A_{(i*j)+1}.$$

Thus the multiplication in A induces a well-defined product $E_i \times E_j \to E_{i*j}$. This extends uniquely to an associative, distributive product on A^*; the ring axioms are easily verified (here the associative property of $*$ is essential). If A is an algebra, over $\mathbb{F}_p$ or $\mathbb{Z}_p$, say, then A^* inherits an algebra structure provided each A_i is an algebra ideal; this will be the case in our applications. (In many cases one simply takes $*$ to be $+$; this works for $A = \widehat{S}$ but not for $A = \widehat{R}$.)

Finally, we endow A with a norm according to the recipe of Lemma 6.5.

7.27 Proposition *Let A be a ring with a chain of ideals (A_i) as above. Assume that A is complete relative to the given norm, and let A^* be the associated graded ring.*

(i) *If A^* has no zero divisors then A has no zero divisors.*

(ii) *If A^* is right (or left) Noetherian then so is A.*

Proof If $0 \neq a \in A$ then $a \in A_k \backslash A_{k+1}$ for some k; we call k the *degree* of a, and write

$$a^\natural = a + A_{k+1} \in E_k;$$

this is the *leading term* of a. Note that $a^\natural \neq 0$, by definition; putting

$0^\natural = 0$, we get a mapping $\natural : A \to A^*$, and it has the following useful property: for any $a, b \in A$,

$$a^\natural b^\natural \neq 0 \Rightarrow a^\natural b^\natural = (ab)^\natural. \tag{11}$$

Part (i) now follows quickly: if A^* has no zero divisors and $a, b \in A\backslash 0$ then $a^\natural$ and $b^\natural$ are non-zero, so $a^\natural b^\natural \neq 0$, and (11) shows that $(ab)^\natural \neq 0$. Hence $ab \neq 0$. (Of course, this is just the familiar proof that the ring of polynomials over an integral domain is again an integral domain.)

Suppose now that A^* is right Noetherian, and let L be a non-zero right ideal of A. For each $k \geq 0$, put

$$L(k) = ((L \cap A_k) + A_{k+1})/A_{k+1} \leq E_k$$

and let

$$L^* = \bigoplus_{k=0}^{\infty} L(k).$$

Thus the non-zero elements of $L(k)$ are just the leading terms of the degree-k elements of L. It follows, using (11), that $L(k)E_j \subseteq L(k * j)$ for all k and j, and hence that L^* is a right ideal of A^*. Then L^* is finitely generated as a right ideal, and splitting each generator into its homogeneous pieces we can find a generating set of the form $\{s_1, \ldots, s_d\}$ with $0 \neq s_i \in L(k_i)$, say, for each i.

Now for each i there exists $b_i \in L$ such that $s_i = b_i^\natural$. Our claim is that *the finite set* $\{b_1, \ldots, b_d\}$ *generates the right ideal* L; part (ii) will be proved once this is established. To this end, we start by defining partial functions $g_1, \ldots, g_d$ on $\mathbb{N}$ as follows: if $n = m * k_i$ then $g_i(n) = m$; otherwise $g_i(n)$ is undefined. Note that if $g_i(n_1) < g_i(n_2)$ then $n_1 < n_2$.

Now suppose $0 \neq a \in L$; we need to show that a can be written as a linear combination of $b_1, \ldots, b_d$. Say a has degree n. Then $a^\natural \in L(n) \subseteq L^*$, so there exist $x_1, \ldots, x_d \in A^*$ such that

$$a^\natural = \sum s_i x_i;$$

we may assume that $x_i \in E_{g(n_i)}$ for each i, since any homogeneous component of x_i other than that of degree $g(n_i)$ contributes nothing to the E_n-component of $s_i x_i$; we may also take $x_i = 0$ whenever $s_i x_i = 0$. Thus for each i we have $x_i = y_{i1}^\natural$, where $y_{i1} \in A$ has degree $g_i(n)$ if $s_i x_i \neq 0$ and $y_{i1} = 0$ if $s_i x_i = 0$. Using (11), we see that $\left(\sum_{i=1}^{d} b_i y_{i1}\right)^\natural = a^\natural$, so putting

$$a_1 = a - \sum_{i=1}^{d} b_i y_{i1},$$

we have $a_1 \in A_{n+1}$. If $a_1 = 0$ we are done. If not, we repeat the process with a_1 in place of a; thus we obtain a strictly increasing sequence $n = n_0 < n_1 < n_2 < \ldots$, and for each $i = 1, \ldots, d$ a sequence $(y_{i1}, y_{i2}, \ldots)$ in A, such that $y_{ij} \in A_{g_i(n_{j-1})}$ and

$$a_j = a - \sum_{i=1}^{d} b_i \sum_{\ell=1}^{j} y_{i\ell} \in A_{n_j} \cap L$$

for each j. If $a_j = 0$ for some j, we see that $a \in \sum_{i=1}^{d} b_i A$, and we are done. If not, then consider the infinite series $\sum_{\ell=1}^{\infty} y_{i\ell}$; since $g_i(n_{j-1}) \to \infty$ as $j \to \infty$, the partial sums of this series form a Cauchy sequence in A. Having made the assumption that A is *complete*, we may infer that the series converges to an element $z_i \in A$. Then

$$a - \sum_{i=1}^{d} b_i z_i = \lim_{j \to \infty} a_j = 0,$$

since $\|a_j\| = c^{-n_j} \to 0$ as $j \to \infty$. Thus $a \in \sum_{i=1}^{d} b_i A$ in this case too, and we have proved (ii) (essentially, Hilbert's Basis Theorem for power-series rings).

To deduce Corollary 7.25, it remains to show how the above procedure gives rise to R^* and S^*. The case of S^* is immediate: taking $A = \widehat{S}$ we put $A_i = S_i$ for each i, and define $m * n = m + n$ for all m and n.

When $A = \widehat{R}$, we take A_n to be the $(n+1)$th term in the sequence

$$\widehat{R} = R_{0,0} \supseteq R_{1,0} \supseteq \cdots \supseteq R_{k,0} \supseteq R_{k,1} \supseteq \cdots \supseteq R_{k,k} \supseteq R_{k+1,0} \supseteq \cdots .$$

Thus $A_n = R_{k,m}$ where

$$n = \frac{1}{2}k(k+1) + m = n(k, m),$$

say. Note that $n(k, m)$ uniquely determines k and m subject to $0 \le m \le k$. For $n = n(k, m)$ and $n' = n(k', m')$, we define

$$n * n' = n(k + k', m + m').$$

It is easy to verify that $n * n' \ge n + n'$, and that the operation $*$ satisfies the requirements listed above for making $\mathbb{N}$ into a 'grading monoid' (in effect, we are embedding $\mathbb{N}$ into $(\mathbb{N} \oplus \mathbb{N}, +)$ by $n(k,j) \mapsto (k,j)$).

The graded ring R^* constructed above is now seen to be identical to A^*. The norm on A defined by the chain (A_i) is *not* the same as our original norm on $\widehat{R}$; however it defines the same topology, since the chain (A_i) is merely a refinement of (R_i), and this is enough to ensure that A is complete. Thus all conditions of Proposition 7.27 are fulfilled, and Corollary 7.25 follows.

Notes

The relationship between the filtration on a pro-p group and the norm on its group algebra is explored from a different point of view in [L], Chapter II §2. Theorems 7.4 and 7.20, on the unique representation of elements by power series, follow from [L] Chapter III, Cor. 2.29. Theorem 7.22, on the graded algebra associated to $\mathbb{Z}_p[[G]]$, follows from [L] Chapter III, Theorem 2.3.3.

Exercises

G denotes a finitely generated pro-p group, and $R = \mathbb{Z}_p[G]$.

1. Using the universal property of inverse limits, show that there is a natural epimorphism

$$\pi : \mathbb{Z}_p[[G]] \longrightarrow \varprojlim_k (\mathbb{Z}/p^k\mathbb{Z})[G/G_k].$$

Prove that π is an isomorphism. Deduce that $\mathbb{Z}_p[[G]] \cong \widehat{R}$. [For the second claim use Lemma 7.1 and Exercise 6.2.]

2. Let $x \in G$. Show that if $x \in G_k$ then $\|x - 1\| \leq p^{-k}$, and that if $x \in I_k + pR$ then $x \in G_k$. Deduce that for each $k \geq 1$ there exists $m \geq 1$ such that

$$G \cap \overline{B}(1; p^{-m}) \subseteq G_k \subseteq G \cap \overline{B}(1; p^{-k}),$$

where $\overline{B}(1; \delta)$ denotes the closed ball of radius δ about 1 in R. (Thus the norm topology on R induces the original topology on G as a subspace of R.)

  Show that if G is *uniform* then $G_k = G \cap \overline{B}(1; p^{-k})$.

  Show that the norm topology on $\mathbb{F}_p[G]$, defined in §7.4, also induces the original topology on G.

3. (i) For $\lambda \in \mathbb{Z}_p$ and $1 \leq r \in \mathbb{N}$, prove that

$$\binom{\lambda}{r} = \frac{\lambda(\lambda - 1) \ldots (\lambda - r + 1)}{r!}$$

belongs to $\mathbb{Z}_p$.

  (ii) Suppose that the pro-p group G is *uniform*, with minimal generating set $\{a_1, \ldots, a_d\}$, and put $b_i = a_i - 1$. Show that for $\lambda_1, \ldots, \lambda_d \in \mathbb{Z}_p$,

$$a_1^{\lambda_1} \ldots a_d^{\lambda_d} = \sum_{\alpha \in \mathbb{N}^d} \binom{\lambda_1}{\alpha_1} \ldots \binom{\lambda_d}{\alpha_d} \mathbf{b}^\alpha$$

(where $\binom{\lambda}{0} = 1$)

  [*Hint:* Approximate $\lambda_1, \ldots, \lambda_d$ by integers, and use Lemma 7.8. The proof is given in full in §8.3.]

(iii) Deduce that each element of R is the sum of a convergent series

$$\sum_{\alpha \in \mathbb{N}^d} \lambda_\alpha \mathbf{b}^\alpha$$

with $\lambda_\alpha \in \mathbb{Z}_p$ for all α. (This gives a constructive proof for Theorem 7.4(i) when G is uniform.)

(iv) Let $\widehat{A}$ be the completion of $\mathbb{Q}_p[G] = A$. Prove that each element of $\widehat{A}$ is the sum of a convergent series

$$\sum_{\alpha \in \mathbb{N}^d} \lambda_\alpha \mathbf{b}^\alpha$$

with $\lambda_\alpha \in \mathbb{Q}_p$ for all α.

[*Hint*: Suppose that (w_i) is a Cauchy sequence in A. Show that $w_i = \sum_{\alpha \in \mathbb{N}^d} \lambda_{i\alpha} \mathbf{b}^\alpha$ with $\lambda_{i\alpha} \in \mathbb{Q}_p$ for all α. Prove that given $\varepsilon > 0$ we have

$$\|\lambda_{i\alpha} \mathbf{b}^\alpha - \lambda_{j\alpha} \mathbf{b}^\alpha\| < \varepsilon$$

for all sufficiently large i and j, uniformly in α. Deduce that for each α the sequence $(\lambda_{i\alpha})$ converges to $\lambda_\alpha \in \mathbb{Q}_p$, say, and show that then $\sum_{\alpha \in \mathbb{N}^d} \lambda_\alpha \mathbf{b}^\alpha$ converges to the sum $\lim_{i \to \infty} w_i$. (The proof depends on Theorem 7.5.)]

4. Assume that G is uniform, and keep the notation of Sections 7.1 and 7.4.

(i) Prove that $J_{k,1} = (pR \cap J^k) + J^{k+1} = R_{k,1} \cap R$.

(ii) Show that

$$J^k + R_{k+1} = R_k, \quad J^k \cap R_{k+1} = J^{k+1}.$$

Deduce that the images of the elements $p^{k-\langle\alpha\rangle} \mathbf{b}^\alpha$ with $\langle\alpha\rangle \le k$ form a basis for J^k/J^{k+1} over $\mathbb{F}_p$, and that the images of the elements $\mathbf{b}^\alpha$ with $\langle\alpha\rangle = k$ form a basis for $J^k/J_{k,1}$.

(iii) Write

$$f_d(k) = \binom{k+d-1}{d-1}.$$

Show that $f_d(k) = \sum_{i=0}^k f_{d-1}(i)$. Deduce that $f_d(k)$ is the number of $\alpha \in \mathbb{N}^d$ such that $\langle\alpha\rangle = k$.

(iv) Show that

$$\dim_{\mathbb{F}_p}(J^k/J^{k+1}) = \dim_{\mathbb{F}_p}(R_k/R_{k+1}) = f_{d+1}(k),$$
$$\dim_{\mathbb{F}_p}(I^k/I^{k+1}) = \dim_{\mathbb{F}_p}(S_k/S_{k+1}) = f_d(k)$$

and that $\dim_{\mathbb{F}_p}(\mathbb{F}_p[G]/I^{k+1}) = f_{d+1}(k)$ (recall that I denotes the augmentation ideal of $\mathbb{F}_p[G]$.)

5. Let H be a pro-p group and G an open normal subgroup. Suppose that G is uniform, and keep the notation of Exercise 4. Let I_H be the augmentation ideal of $\mathbb{F}_p[H]$, and put $I_0 = I\mathbb{F}_p[H]$. Show that $\mathbb{F}_p[H]/I_0^n$ is a free $\mathbb{F}_p[G]/I^n$-module on $|H:G|$ generators.

Show that $\mathbb{F}_p[H]/I_H^n$ is a homomorphic image of $\mathbb{F}_p[H]/I_0^n$; deduce that

$$\dim_{\mathbb{F}_p}(\mathbb{F}_p[H]/I_H^n) \leq |H:G|f_{d+1}(n-1) \leq Cn^d,$$

where C is a constant depending only on $|H:G|$ and d.

6. Let H and G be as in Exercise 5, and suppose that $G \geq H_q = P_q(H)$.
(i) Show that

$$H_{q+k-1} \leq G_k \leq H_k$$

for all $k \geq 1$. Deduce that

$$\varprojlim \mathbb{Z}_p[H/H_k] = \varprojlim \mathbb{Z}_p[H/G_k].$$

(ii) Show that $\varprojlim \mathbb{Z}_p[H/G_k]$ is generated as a module for the ring $\varprojlim \mathbb{Z}_p[G/G_k]$ by $|H:G|$ elements. Hence show that the ring $\mathbb{Z}_p[[H]]$ is (right and left) Noetherian.

7. We keep the notation of §7.4. Show that the natural map $\overline{} : \mathbb{Z}_p[G] \to \mathbb{F}_p[G]$ induces an epimorphism $\overline{} : \mathbb{Z}_p[[G]] \to \mathbb{F}_p[[G]]$ with kernel exactly $p\mathbb{Z}_p[[G]]$. Show that

$$\overline{R_{k,0}} = S_k$$
$$\overline{R_{k,m}} = 0 \quad \text{for } m \geq 1.$$

Deduce that the mapping $x_i \mapsto y_i$ $(1 \leq i \leq d)$, $x_0 \mapsto 0$ induces an isomorphism $R^*/pR^* \to S^*$. [*Hint:* use Exercise 4.]

8. (i) Show that the $\mathbb{Q}_p$-algebra $M_n(\mathbb{Q}_p)$ of all $n \times n$ matrices over $\mathbb{Q}_p$ becomes a complete normed $\mathbb{Q}_p$-algebra if we define

$$\|(g_{ij})\| = \max_{i,j} |g_{ij}|.$$

(ii) Let

$$\Gamma = \{g \in \mathrm{GL}_n(\mathbb{Z}_p) \mid g \equiv 1_n \pmod{p^\epsilon}\},$$

where $\epsilon = 1$ if p is odd, $\epsilon = 2$ if $p = 2$. Prove that $L_\Gamma \cong p^\epsilon M_n(\mathbb{Z}_p)$.

[Use Corollary 7.14, and Proposition 6.22 to show that $p^\epsilon M_n(\mathbb{Z}_p) = \log(\Gamma)$.]

9. Let H be a uniform pro-2 group and put $G = P_2(H)$. Show that for x and $y \in H$ we have

$$4(x, y) = (2x, 2y)$$

(in the sense of Definition 4.29). Hence show that if G is a Lie algebra over $\mathbb{Z}_p$, with the operations $+_G$ and $(\ ,\ )_G$, then so is H, with the operations $+_H$ and $(\ ,\ )_H$. (Thus we may deduce Theorem 4.30 from Corollary 7.14 without appealing to Exercise 10.)

[*Hint*: For the first bit, compare $(x^2, y^2)_{n-1}$ and $(x, y)_n^4$. Then note that $G = 2H$.]

10. In this exercise, G is a uniform pro-2 group. We are going to define a new norm on $\mathbb{Z}_2[G] = R$. The notation is as in §7.1, except that we'll write $I = I_1$ for the augmentation ideal of R, and $I^0 = R$.

(*i*) Let $J_0 = R$ and put

$$J_{2n+1} = 2J_{2n} + I^{n+1} \quad (n \geq 0)$$
$$J_{2n} = 2J_{2n-1} + I^n \quad (n \geq 1).$$

Verify that the chain (J_k) satisfies the conditions of Lemma 6.5, and that it is cofinal with the chain (J^k). Deduce that R is a normed ring with norm

$$\|r\|_2 = 2^{-k} \quad \text{if } r \in J_k \setminus J_{k+1},$$

and that this norm induces the original topology on G as a subspace of R.

(ii) For each $n \in \mathbb{N}$, put

$$L_n = \sum_{\langle \alpha \rangle = n} \mathbb{Z}_2 \mathbf{b}^\alpha.$$

Show that $I = L_1 + I^2$. Deduce that $J_1 = 2L_0 + J_2$ and that $J_2 = 4L_0 + L_1 + J_3$. [*Hint*: Use Theorem 7.4.]

(iii) Show that if $v \in I^n$ and $b \in R$ then $vb - vb \in 4I^n + 2I^{n+1} + I^{n+2}$. [Compare Lemma 7.10.]

(iv) Using (ii) and (iii), prove that for each $k \geq 1$,

$$J_k = J_{k+1} + \sum_{2n < k} 2^{k-2n} L_n.$$

[*Hint*: Compare the proof of Lemma 7.11; it may be easier to deal with the cases of even and odd k separately.]

(v) Deduce from (iv) that if $c = \sum_{\alpha \in \mathbb{N}^d} \lambda_\alpha \mathbf{b}^\alpha \in \mathbb{Z}_2[G]$, where $\lambda_\alpha \in \mathbb{Z}_2$ for each α, then

$$\|c\|_2 = \sup_{\alpha \in \mathbb{N}^d} 2^{-2\langle \alpha \rangle} |\lambda_\alpha|.$$

Hence show that $\|\cdot\|_2$ extends to a norm on $\mathbb{Q}_2[G]$ so that all the hypotheses of §7.2 are fulfilled.

Interlude B

Linearity criteria

A long-standing problem in group theory has been to formulate a set of conditions on an abstract group Γ which are necessary and sufficient for Γ to be a *linear group*, that is, for Γ to have a faithful linear representation (of finite dimension) over some field. Alex Lubotzky had the insight to realise that the theory of p-adic analytic groups provides an almost ready-made solution to this problem, under certain restrictions: namely, when the *group is assumed to be finitely generated* and the *field is restricted to having characteristic zero*.

The basic idea is as follows. One defines a 'p-congruence system', in an abstract group Γ, to be a family of subgroups which models the behaviour of the system of all principal congruence subgroups in $\mathrm{GL}_d(\mathbb{Z}_p)$. This latter system satisfies various finiteness conditions, each of which essentially expresses the fact that $\mathrm{GL}_d(\mathbb{Z}_p)$ is virtually a pro-p group of finite rank. If we now *assume* that Γ has a congruence system $\mathcal{C}$ which satisfies one of the appropriate finiteness conditions, we can hope to deduce that the completion

$$\widetilde{\Gamma} = \varprojlim(\Gamma/N)_{N \in \mathcal{C}}$$

is itself virtually a pro-p group of finite rank. In this case, $\widetilde{\Gamma}$ has a faithful linear representation over $\mathbb{Z}_p$, and we may conclude that Γ itself has such a representation.

B.1 Definition A *p-congruence system* in a group Γ is a descending chain $\mathcal{C} = (N_i)_{i \in \mathbb{N}}$ of normal subgroups of Γ such that
(i) Γ/N_1 is finite;
(ii) N_1/N_i is a finite p-group for all $i \geq 1$; and
(iii) $\bigcap_{i=1}^{\infty} N_i = 1$.

Of course, the mere existence of a p-congruence system in Γ means little more than that Γ is virtually residually a finite p-group. The concept is useful because we can attach conditions to it. (The reader should be warned that the similar term 'p-congruence structure' introduced by Lubotzky (1988) carries a more restrictive meaning: namely the condition defined below in Definition B.2(ii) is also supposed to hold.)

Now let $\mathcal{C} = (N_i)$ be a p-congruence system in some group Γ.

B.2 Definition (i) $\mathcal{C}$ has *finite rank* if there exists $r \in \mathbb{N}$ such that $\mathrm{rk}(N_1/N_j) \leq r$ for all j.

(ii) $\mathcal{C}$ is *uniformly f.g.* if there exists $d \in \mathbb{N}$ such that $\mathrm{d}(N_i/N_j) \leq d$ for all $i < j$.

(iii) $\mathcal{C}$ has *PIG* if there exist positive real numbers c and s such that

$$|N_1 : N_1^{p^n} N_j| \leq cp^{ns}$$

for all $n, j \in \mathbb{N}$. (PIG stands for '*polynomial index growth*'.)

(iv) $\mathcal{C}$ has *PSG* if there exist positive real numbers c and s such that

$$\sigma_n(N_1/N_j) \leq cp^{ns}$$

for all $n, j \in \mathbb{N}$, where $\sigma_n(F)$ denotes the number of subgroups of index at most p^n in a finite p-group F. (PSG stands for '*polynomial subgroup growth*'.)

B.3 Proposition *Let Γ be a group with a p-congruence system $\mathcal{C}$. If $\mathcal{C}$ satisfies one of the conditions* (i), (ii), (iii) *or* (iv) *of Definition B.2, then Γ has a faithful linear representation over $\mathbb{Z}_p$.*

Proof Since Γ/N_1 is finite, it will suffice to show that N_1 has a faithful representation: the corresponding induced representation of Γ will then also be faithful.

Now let

$$G = \varprojlim_{i \to \infty} N_1/N_i.$$

Then G is a pro-p group, and the natural map of N_1 into G is injective, by condition (iii) of Definition B.1. So it will suffice to show that G has a faithful representation. This will follow by Theorem 7.19 once we have shown that G has finite rank.

We consider N_1 as being embedded in G: then $\overline{N_1} = G$, the family of subgroups $(\overline{N_j})_{j \in \mathbb{N}}$ forms a base for the neighbourhoods of 1 in G, and $\overline{N_i}/\overline{N_j} \cong N_i/N_j$ whenever $j \geq i \geq 1$ (see Exercise 1.4). Now we take the four alternative hypotheses in turn.

(i) Suppose $\mathrm{rk}(N_1/N_j) \leq r$ for all j. If $K \lhd_o G$ then $K \geq \overline{N_j}$ for some j, and then

$$\mathrm{rk}(G/K) \leq \mathrm{rk}(G/\overline{N_j}) = \mathrm{rk}(N_1/N_j) \leq r.$$

Thus G has rank at most r (see Definition 3.12).

(ii) Suppose $\mathrm{d}(N_i/N_j) \leq d$ for all $i < j$. Then $\mathrm{d}(\overline{N_i}) \leq d$ for each i, by Proposition 1.5(ii), and Corollary 3.14 shows that G has finite rank.

(iii) Suppose that $|N_1 : N_1^{p^n} N_j| \leq cp^{ns}$ for all j and all n. Since $G/\overline{N_j} \cong N_1/N_j$ for each j, we have $|G : G^{p^n} \overline{N_j}| = |N_1 : N_1^{p^n} N_j| \leq cp^{ns}$. But $\bigcap_{j=1}^\infty G^{p^n} \overline{N_j} = \overline{G^{p^n}}$, so $|G : \overline{G^{p^n}}| \leq cp^{ns}$. As this holds for each n, Theorem 3.16 shows that G has finite rank.

(iv) Suppose $\sigma_n(N_1/N_j) \leq cp^{ns}$ for all j and all n. We claim that for each n, G has at most cp^{ns} open subgroups of index $\leq p^n$. Indeed, if this is false, then for some n we can find $k > cp^{ns}$ open subgroups of index $\leq p^n$ in G, and we can then find j such that $\overline{N_j}$ is contained in their intersection; but then $N_1/N_j \cong G/\overline{N_j}$ contains k subgroups of index $\leq p^n$, contrary to hypothesis. That G has finite rank now follows by Theorem 3.19.

This completes the proof.

The converse depends on the following lemma:

B.4 Lemma *Let R be a finitely generated integral domain of characteristic zero. Then there exist positive integers k and ℓ such that, for every prime p not dividing ℓ, R is isomorphic to a subring of $M_k(\mathbb{Z}_p)$.*

It is also true that for infinitely many primes p, R is isomorphic to a subring of $\mathbb{Z}_p$. An elementary proof of this can be found in Cassels (1986), Chapter 5.

Proof Let $\widetilde{R} = \mathbb{Q} \cdot R$ be the $\mathbb{Q}$-subalgebra of the field of fractions of R generated by R. By Noether's normalisation lemma (see e.g. Atiyah and Macdonald (1969), Chapter 5, Ex. 16), $\widetilde{R}$ contains finitely many algebraically independent elements $x_1, \ldots, x_m$ such that $\widetilde{R}$ is integral over its subalgebra $S = \mathbb{Q}[x_1, \ldots, x_m]$. This means that each of the finitely many generators of R is a zero of some monic polynomial with coefficients in S; since only finitely many such polynomials are being considered, we can find a positive integer ℓ such that these polynomials all have their coefficients in the subring $S_1 = \mathbb{Z}[1/\ell, x_1, \ldots, x_m]$ of S. Let

$$R_1 = R[1/\ell, x_1, \ldots, x_m] \subseteq \widetilde{R}.$$

Then R_1 is integral over its subring S_1, hence finitely generated as a module over S_1. Let us suppose that k module generators suffice.

Now let p be any prime not dividing ℓ. Since $\mathbb{Z}_p$ is uncountable, it contains infinitely many algebraically independent elements (over $\mathbb{Q}$), and since $\mathbb{Z}[1/\ell] \subseteq \mathbb{Z}_p$ we can embed S_1 as a subring into $\mathbb{Z}_p$, by mapping $x_1, \ldots, x_m$ to m algebraically independent elements of $\mathbb{Z}_p$. Then the field of fractions F of S_1 gets embedded into $\mathbb{Q}_p$. Now let E be the field of fractions of R_1. Since R_1 is a k-generator module over its subring S_1 it follows that E is a finite extension field of F, with $(E : F) \leq k$. Hence there exists an extension field K of $\mathbb{Q}_p$ with $(K : \mathbb{Q}_p) \leq k$ and $E \subseteq K$. Let A be the integral closure of $\mathbb{Z}_p$ in K; A is a free $\mathbb{Z}_p$-module of rank at most k. Also $R \subseteq R_1 \subseteq E \subseteq K$, and each element of R_1 is integral over $S_1 \subseteq \mathbb{Z}_p$. Hence $R \subseteq A$, and letting R act on A by multiplication we see that R is isomorphic to a subring of $\mathrm{End}_{\mathbb{Z}_p}(A) \cong \mathrm{M}_k(\mathbb{Z}_p)$. This completes the proof.

B.5 Proposition *Let R be a finitely generated integral domain of characteristic zero. Then there exists a positive integer ℓ such that if Γ is any subgroup of $\mathrm{GL}_n(R)$, where $n \in \mathbb{N}$, then for each prime p not dividing ℓ, Γ has a p-congruence system satisfying all the conditions* (i), (ii), (iii) *and* (iv) *of Definition* B.2.

Proof By Lemma B.4, there exist k and ℓ such that R can be embedded in $\mathrm{M}_k(\mathbb{Z}_p)$ for every prime $p \nmid \ell$. Then $\mathrm{GL}_n(R)$ embeds into $\mathrm{GL}_d(\mathbb{Z}_p)$, where $d = nk$. Thus it will suffice to show that if $p \nmid \ell$ and $\Gamma \leq \mathrm{GL}_d(\mathbb{Z}_p)$ then Γ has a p-congruence system satisfying (i)–(iv) of Definition B.2.

For each i, let Γ_i denote the principal congruence subgroup modulo p^i in $\mathrm{GL}_d(\mathbb{Z}_p)$, as in §5.1. Put $\varepsilon = 0$ if p is odd, $\varepsilon = 1$ if $p = 2$, and for each i let

$$N_i = \Gamma \cap \Gamma_{i+\varepsilon}.$$

Our claim is that $(N_i)_{i \in \mathbb{N}}$ is a p-congruence system in Γ with the required properties.

Now by Theorem 5.2, $\Gamma_{1+\varepsilon}$ is a uniform pro-p group of dimension d^2, and $P_i(\Gamma_{1+\varepsilon}) = \Gamma_{i+\varepsilon}$ for each i. For $j > i \geq 1$,

$$N_i/N_j \cong N_i \Gamma_{j+\varepsilon}/\Gamma_{j+\varepsilon} \leq \Gamma_{i+\varepsilon}/\Gamma_{j+\varepsilon},$$

so

$$\mathrm{d}(N_i/N_j) \leq \mathrm{rk}(N_i/N_j) \leq \mathrm{rk}(\Gamma_{i+\varepsilon}/\Gamma_{j+\varepsilon}) = d^2,$$

showing that the system (N_i) satisfies (i) and (ii) of Definition B.2. To establish properties (iii) and (iv), consider the group

$$G = \varprojlim_{j \to \infty} (N_1/N_j).$$

If $N \triangleleft_o G$ then G/N is a homomorphic image of N_1/N_j for some $j \geq 1$, so $\mathrm{rk}(G/N) \leq d^2$. Hence G is a pro-p group of rank at most d^2, and

Theorem 3.16 shows that there exist c and r such that $|G : G^{p^n}| \leq cp^{nr}$ for all n (indeed the proof of Theorem 3.16 gives $r = \mathrm{rk}(G) \leq d^2$). Since N_1/N_j is a homomorphic image of G, it follows that

$$|N_1 : N_1^{p^n} N_j| \leq cp^{nr} \quad \text{for all } n.$$

Thus the system (N_i) has PIG. A similar argument using Theorem 3.19 shows that the system also has PSG.

Propositions B.3 and B.5 now combine to yield the main result:

B.6 Theorem *Let Γ be a finitely generated group. The following are equivalent*

(i) *Γ is isomorphic to a linear group of finite degree over a field of characteristic zero.*

(ii) *For some prime p, Γ has a p-congruence system of finite rank.*

(iii) *For some prime p, Γ has a p-congruence system which is uniformly f.g.*

(iv) *For some prime p, Γ has a p-congruence system with PIG.*

(v) *For some prime p, Γ has a p-congruence system with PSG.*

(vi) *For all but finitely many primes p, Γ has a p-congruence system which satisfies all four of the above conditions.*

Proof Clearly, (vi) implies each of (ii), (iii), (iv) and (v). Each of these hypotheses, in turn, implies (i), by Proposition B.3. Finally, suppose (i) holds. We may suppose that $\Gamma = \langle x_1, \dots, x_m \rangle \leq \mathrm{GL}_n(F)$, where F is a field of characteristic zero. Take R to be the subring of F generated by the entries of the matrices x_i, x_i^{-1} ($1 \leq i \leq m$). Then $\Gamma \leq \mathrm{GL}_n(R)$, and (vi) follows by Proposition B.5.

The 'linearity criteria' have found a number of applications in problems involving the classification of residually finite groups subject to additional finiteness conditions. Typically, Proposition B.3 is used to reduce the problem to the special case of linear groups, whereupon an extensive range of powerful techniques can be brought into play. Details of the method are described in the survey articles Mann (1990) and Segal (1990). For an alternative approach, independent of pro-p groups, see Segal (1996).

The method can usefully be applied to groups that are *residually nilpotent*. This depends on the following observation; here, $\widehat{\Gamma}_p$ denotes the pro-p completion of a group Γ.

B7 Proposition *Let Γ be a finitely generated residually nilpotent group.*

Suppose that, for some prime p, the pro-p completion $\widehat{\Gamma}_p$ has finite rank. Then there exists a finite set π of primes such that the natural map

$$\Gamma \to \prod_{\ell \in \pi} \widehat{\Gamma}_\ell$$

is injective.

Proof Suppose that Λ is a torsion-free nilpotent quotient of Γ. Then $\widehat{\Lambda}_p$ is an image of $\widehat{\Gamma}_p$ (Exercise 1.21), so $\dim(\widehat{\Lambda}_p) \le \dim(\widehat{\Gamma}_p)$, by Theorem 4.8. It also follows from 4.8 (see Exercise 1.22) that $\dim(\widehat{\Lambda}_p)$ is equal to the Hirsch length $h(\Lambda)$ of Λ, so we have

$$h(\Lambda) \le \dim(\widehat{\Gamma}_p).$$

We may therefore choose a normal subgroup T of Γ such that Γ/T is torsion-free and nilpotent of maximal possible Hirsch length. Then $\gamma_n(\Gamma) \le T$ and $T/\gamma_n(\Gamma)$ is finite, for every n exceeding the nilpotency class of Γ/T (see Segal (1983), Chapter 1, for elementary properties of the Hirsch length). It follows that $T/[T, \Gamma]$ is finite, of order m, say.

Let $q \nmid m$ be a prime, and suppose that $Q \lhd \Gamma$ has finite index a power of q. Then $T = (T \cap Q)[T, \Gamma]$, and as $\Gamma/(T \cap Q)$ is nilpotent it follows that $T = (T \cap Q) \le Q$. Hence T is contained in the kernel $\Gamma(q)$, say, of the natural homomorphism $\Gamma \to \widehat{\Gamma}_q$. On the other hand, $T \ge \Gamma(q)$ because Γ/T is residually a finite q-group (Gruenberg's theorem, see Segal (1983), Chapter 1).

Thus $\Gamma(q) = T$ for every prime $q \nmid m$. Choose one such prime q and let π be the set of prime divisors of m together with q. The kernel of the natural map $\Gamma \to \prod_{\ell \in \pi} \widehat{\Gamma}_\ell$ is then

$$\bigcap_{\ell \in \pi} \Gamma(\ell) = T \cap \bigcap_{\ell \mid m} \Gamma(\ell) = \bigcap_{\text{all primes } \ell} \Gamma(\ell) = 1,$$

since Γ is residually nilpotent. The result follows.

The next corollary illustrates how this may be applied. Let us say that a group Γ satisfies an *upper finiteness condition* if one of the following holds:

(a) Γ has finite *upper rank*: i.e. $\mathrm{rk}(\Gamma^*)$ is bounded as Γ^* ranges over all finite quotients of Γ;

(b) Γ has *PIG*: i.e. there exist constants c and s such that $|\Gamma^* : \Gamma^{*^n}| \le cn^s$ for all positive integers n and all finite quotients Γ^* of Γ;

(c) Γ has *PSG*: i.e. there exist constants c and s such that $s_n(\Gamma) \le cn^s$

for all n, where $s_n(\Gamma)$ is the number of subgroups of index at most n in Γ.

B.8 Corollary *Let Γ be a finitely generated residually nilpotent group. If Γ satisfies an upper finiteness condition then Γ is isomorphic to a linear group over $\mathbb{C}$.*

Proof For each prime p let $\Gamma(p)$, as above, denote the finite-p residual of Γ, i.e. the kernel of the natural map $\Gamma \to \widehat{\Gamma}_p$. Now fix a prime p, let $N_1 = \Gamma$, and for $i \geq 1$ put $N_{i+1} = [N_i, \Gamma]N_i^p$. We establish below that each N_i has finite index in Γ. Since every normal subgroup of p-power index in Γ contains some N_i, it follows that $\bigcap_{i=1}^{\infty} N_i = \Gamma(p)$, and that $(N_i/\Gamma(p))$ is a p-congruence system in $\Gamma/\Gamma(p)$; moreover, each of the conditions (a)–(c) implies one of the alternative hypotheses of Proposition B.3, which therefore shows that $\Gamma/\Gamma(p)$ is isomorphic to a linear group over $\mathbb{Z}_p$. In fact the proof of B3 shows that $\varprojlim_{i \to \infty} N_1/N_i$ is a pro-p group of finite rank; but it is clear in the present context that this pro-p group is precisely $\widehat{\Gamma}_p$, so we see that $\widehat{\Gamma}_p$ has finite rank.

It follows by Proposition B7 that Γ can be embedded in the direct product of finitely many linear groups, over various rings $\mathbb{Z}_p$. Embedding all of these rings into $\mathbb{C}$ gives the required faithful linear representation of Γ.

It remains to show that each N_i has finite index in Γ. Let $i \geq 1$ and suppose we have shown that $|\Gamma : N_i|$ is finite. If N_i/N_{i+1} is infinite, then we can find a normal subgroup K of finite index in Γ such that $N_{i+1} < K < N_i$ and N_i/K is elementary abelian of arbitrarily large rank. It is easy to see that this contradicts (a), (b) and (c). So $|N_i : N_{i+1}|$ must be finite. The claim follows by induction.

Corollary B.8 provides a novel technique for 'lifting' linear representations from characteristic p to characteristic zero:

B.9 Corollary *Every finitely generated linear group that satisfies an upper finiteness condition is isomorphic to a linear group over $\mathbb{C}$.*

For such a group has a normal subgroup of finite index which is residually nilpotent (see Wehrfritz (1973), Theorem 4.7); and it is easy to see that upper finiteness conditions are inherited by subgroups of finite index.

8

p-adic analytic groups

A *p-adic analytic group* is a p-adic analytic manifold which is also a group, the group operations being given by analytic functions. This chapter has two main purposes: the first is to explain what these terms mean; the second is to establish Lazard's group-theoretic characterisation of p-adic analytic groups:

8.1 Theorem *A topological group G has the structure of a p-adic analytic group if and only if G has an open subgroup which is a powerful finitely generated pro-p group.*

This is the main result of this part of the book; it will be proved, along with some refinements and corollaries, in Sections 8.3 and 8.4. The first two sections are devoted to the definition and basic properties of p-adic analytic groups.

Further properties of these groups are developed in the following chapter.

8.1 p-adic analytic manifolds

We begin by considering functions on $\mathbb{Z}_p^r$. As usual, we write $\mathbf{X}$ for the r-tuple $(X_1, \ldots, X_r)$, etc. For $\mathbf{y} \in \mathbb{Z}_p^r$ and $h \in \mathbb{N}$ we put

$$B(\mathbf{y}, p^{-h}) = \{\mathbf{z} \in \mathbb{Z}_p^r \mid |z_i - y_i| \le p^{-h} \text{ for } i = 1, \ldots, r\}$$
$$= \{\mathbf{y} + p^h \mathbf{x} \mid \mathbf{x} \in \mathbb{Z}_p^r\}.$$

8.2 Definition Let V be a non-empty open subset of $\mathbb{Z}_p^r$ and let

$$\mathbf{f} = (f_1, \ldots, f_s)$$

be a function from V into $\mathbb{Z}_p^s$.

178

(i) Let $\mathbf{y} \in V$. Then $\mathbf{f}$ is *analytic* at $\mathbf{y}$ if there exist $h \in \mathbb{N}$ with $B(\mathbf{y}, p^{-h}) \subseteq V$ and formal power series $F_i(\mathbf{X}) \in \mathbb{Q}_p[[\mathbf{X}]]$ $(i = 1, \ldots, s)$ such that

$$f_i(\mathbf{y} + p^h\mathbf{x}) = F_i(\mathbf{x}) \quad \text{for all } \mathbf{x} \in \mathbb{Z}_p^r.$$

(ii) The function $\mathbf{f}$ is *analytic on* V if it is analytic at each point of V.

Equivalently, by Proposition 6.34, $\mathbf{f}$ is analytic at $\mathbf{y} \in V$ if there exists $h \in \mathbb{N}$, with $B(\mathbf{y}, p^{-h}) \subseteq V$, such that the function $\mathbf{x} \mapsto f_i(\mathbf{y} + p^h\mathbf{x})$ is strictly analytic on $\mathbb{Z}_p^r$, for each $i = 1, \ldots, s$, in the sense of Definition 6.17; here we are considering $\mathbb{Z}_p^r$ as an open set in A^r where A is the complete normed $\mathbb{Q}_p$-algebra $\mathbb{Q}_p$. Note that if $\{U_i\}_{i \in I}$ is a covering of V by open subsets, then $\mathbf{f}$ is analytic on V if and only if $\mathbf{f}|_{U_i}$ is analytic on U_i for each $i \in I$ (i.e. being analytic is a 'local' property).

8.3 Lemma *Suppose that* $F(\mathbf{X}) \in \mathbb{Q}_p[[\mathbf{X}]]$ *can be evaluated at* $\mathbf{x}$ *for all* $\mathbf{x} \in \mathbb{Z}_p^r$. *Let* $\mathbf{a} \in \mathbb{Z}_p^r$. *Then there exists* $G(\mathbf{X}) \in \mathbb{Q}_p[[\mathbf{X}]]$ *such that* $F(\mathbf{x} + \mathbf{a}) = G(\mathbf{x})$ *for all* $\mathbf{x} \in \mathbb{Z}_p^r$.

Proof Say $F(\mathbf{X}) = \sum_{\alpha \in \mathbb{N}^r} d_\alpha X_1^{\alpha_1} \ldots X_r^{\alpha_r}$. Then for $\mathbf{x} \in \mathbb{Z}_p^r$ we have

$$F(\mathbf{x} + \mathbf{a}) = \sum_{\alpha \in \mathbb{N}^r} d_\alpha (x_1 + a_1)^{\alpha_1} \ldots (x_r + a_r)^{\alpha_r}$$

$$= \sum_{\alpha \in \mathbb{N}^r} \left(\sum_{\beta \in \mathbb{N}^r} d_\alpha \binom{\alpha_1}{\beta_1} \ldots \binom{\alpha_r}{\beta_r} a_1^{\alpha_1 - \beta_1} \ldots a_r^{\alpha_r - \beta_r} x_1^{\beta_1} \ldots x_r^{\beta_r} \right)$$

$$= \sum_{\alpha \in \mathbb{N}^r} \left(\sum_{\beta \in \mathbb{N}^r} \lambda_{\alpha\beta} \mathbf{x}^\beta \right)$$

say, where $\mathbf{x}^\beta$ is shorthand for $x_1^{\beta_1} \ldots x_r^{\beta_r}$; note that $\lambda_{\alpha\beta} = 0$ whenever $\beta_i > \alpha_i$ for some i. Since $F(\mathbf{X})$ can be evaluated at $(1, \ldots, 1) \in \mathbb{Z}_p^r$ we have $\lim_{\alpha \in \mathbb{N}^r} d_\alpha = 0$; as $|\lambda_{\alpha\beta}| \leq |d_\alpha|$ for each α and β it follows that $\lim_{(\alpha,\beta) \in \mathbb{N}^r \times \mathbb{N}^r} \lambda_{\alpha\beta} = 0$.

By Proposition 6.9(ii), each of the series $\sum_{\alpha \in \mathbb{N}^r} \lambda_{\alpha\beta}$ converges, with sum c_β say; invoking Corollary 6.11 we deduce that

$$F(\mathbf{x} + \mathbf{a}) = \sum_{\beta \in \mathbb{N}^r} \left(\sum_{\alpha \in \mathbb{N}^r} \lambda_{\alpha\beta} \mathbf{x}^\beta \right) = \sum_{\beta \in \mathbb{N}^r} c_\beta \mathbf{x}^\beta.$$

The result follows on taking $G(\mathbf{X}) = \sum_{\beta \in \mathbb{N}^r} c_\beta X_1^{\beta_1} \ldots X_r^{\beta_r}$. (Note that we can't apply Theorem 6.35 directly, here, since the 'power series' $X_i +$ a_i have non-zero constant term in general.)

8.4 Corollary *Suppose that $V \subseteq \mathbb{Z}_p^r$ can be written as a union $\bigcup\{B(\mathbf{y}(i), p^{-h(i)}) \mid i \in I\}$ of balls and that $\mathbf{f} = (f_1, \ldots, f_s)$ is a function from V into $\mathbb{Z}_p^s$ such that, for each $i \in I$, the functions $\mathbf{x} \mapsto f_j(\mathbf{y}(i) + p^{h(i)}\mathbf{x})$ are strictly analytic on $\mathbb{Z}_p^r$ for $j = 1, \ldots, s$. Then $\mathbf{f}$ is analytic on V.*

Proof Let $\mathbf{y} \in V$. Then there exists $i \in I$ such that $\mathbf{y} \in B(\mathbf{y}(i), p^{-h(i)})$, and for $j = 1, \ldots, s$ there exists $F_j(\mathbf{X}) \in \mathbb{Q}_p[[\mathbf{X}]]$ such that, for all $\mathbf{x} \in \mathbb{Z}_p^r$,

$$F_j(\mathbf{x}) = f_j(\mathbf{y}(i) + p^{h(i)}\mathbf{x}).$$

Now put $\mathbf{a} = p^{-h(i)}(\mathbf{y} - \mathbf{y}(i))$. Then $f_j(\mathbf{y} + p^{h(i)}\mathbf{x}) = F_j(\mathbf{x} + \mathbf{a})$. By Lemma 8.3, for each j there exists $G_j(\mathbf{X}) \in \mathbb{Q}_p[[\mathbf{X}]]$ such that $F_j(\mathbf{x} + \mathbf{a}) = G_j(\mathbf{x})$ for all $\mathbf{x} \in \mathbb{Z}_p^r$. Then

$$f_j(\mathbf{y} + p^{h(i)}\mathbf{x}) = G_j(\mathbf{x})$$

for all $\mathbf{x} \in \mathbb{Z}_p^r$ and $j = 1, \ldots, s$. Thus $\mathbf{f}$ is analytic at $\mathbf{y}$, and the result follows.

A special case of Corollary 8.4 is the reassuring proposition that if the function $f : \mathbb{Z}_p^r \to \mathbb{Z}_p$ is strictly analytic on $\mathbb{Z}_p^r$ then it is analytic on $\mathbb{Z}_p^r$.

8.5 Lemma *Let $\mathbf{f} : U \to V$ and $\mathbf{g} : V \to W$ be analytic functions, where $U \subseteq \mathbb{Z}_p^r$, $V \subseteq \mathbb{Z}_p^s$ and $W \subseteq \mathbb{Z}_p^t$ are non-empty open sets. Then $\mathbf{g} \circ \mathbf{f}$ is analytic on U.*

Proof Let $\mathbf{y} \in U$. We are required to find $h \in \mathbb{N}$ such that, for each $j = 1, \ldots, t$, there exists $H_j(\mathbf{X}) \in \mathbb{Q}_p[[\mathbf{X}]]$ with the property that $g_j(f_1(\mathbf{y} + p^h\mathbf{x}), \ldots, f_s(\mathbf{y} + p^h\mathbf{x})) = H_j(\mathbf{x})$ for all $\mathbf{x} \in \mathbb{Z}_p^r$.

Since $\mathbf{f}$ and $\mathbf{g}$ are analytic there exist h_1 and $h_2 \in \mathbb{N}$ such that

(i) for each $i = 1, \ldots, s$, there exists $F_i(\mathbf{X}) = \sum_{\alpha \in \mathbb{N}^r} b_\alpha(i) X_1^{\alpha_1} \ldots X_r^{\alpha_r} \in \mathbb{Q}_p[[\mathbf{X}]]$ such that $f_i(\mathbf{y} + p^{h_1}\mathbf{x}) = F_i(\mathbf{x})$ for all $\mathbf{x} \in \mathbb{Z}_p^r$; and

(ii) if we set $\mathbf{b} = (b_0(1), \ldots, b_0(s)) = (f_1(\mathbf{y}), \ldots, f_s(\mathbf{y}))$, then for each $j = 1, \ldots, t$, there exists $G_j(\mathbf{y}) = \sum_{\beta \in \mathbb{N}^s} c_\beta(j) Y_1^{\beta_1} \ldots Y_s^{\beta_s} \in \mathbb{Q}_p[[\mathbf{Y}]]$ such that $g_j(\mathbf{b} + p^{h_2}\mathbf{y}) = G_j(\mathbf{y})$ for all $\mathbf{y} \in \mathbb{Z}_p^s$.

By Proposition 6.19, the function $\mathbf{x} \mapsto F_i(\mathbf{x})$ is continuous for each $i = 1, \ldots, s$. So there exists $h_3 \in \mathbb{N}$ such that

$$(f_1(\mathbf{y} + p^{h_1+h_3}\mathbf{x}), \ldots, f_s(\mathbf{y} + p^{h_1+h_3}\mathbf{x})) \in B(\mathbf{b}, p^{-h_2})$$

for all $\mathbf{x} \in \mathbb{Z}_p^r$. Thus

$$g_j\left(f_1(\mathbf{y} + p^{h_1+h_3}\mathbf{x}), \ldots, f_s(\mathbf{y} + p^{h_1+h_3}\mathbf{x})\right) = G_j(z_1, \ldots, z_s)$$

where $z_i = p^{-h_2}\left(F_i(p^{h_3}\mathbf{x}) - b_0(i)\right)$ for each $i = 1, \ldots, s$. We now apply the work done in Chapter 6 on composition of power series. By Lemma 6.18, there exists $h_4 \in \mathbb{N}$ such that for $i = 1, \ldots, s$ and all $\alpha \in \mathbb{N}^r$

$$p^{\langle\alpha\rangle h_4} \cdot b_\alpha(i) \in \mathbb{Z}_p.$$

Let $h_5 = \max(h_3, h_4) + h_2$. For each $i = 1, \ldots, s$ let

$$E_i(\mathbf{X}) = \sum_{\alpha \in \mathbb{N}^r} e_\alpha(i) X_1^{\alpha_1} \ldots X_r^{\alpha_r} \in \mathbb{Q}_p[[\mathbf{X}]],$$

where $e_0(i) = 0$ and $e_\alpha(i) = p^{\langle\alpha\rangle h_5 - h_2} \cdot b_\alpha(i)$ if $\alpha \neq \mathbf{0}$. Put $H_j(\mathbf{X}) = (G_j \circ \mathbf{E})(\mathbf{X})$ for each $j = 1, \ldots, t$.

Now $\lim_{\beta \in \mathbb{N}^s} c_\beta(j) = 0$ and $|e_\alpha(i)| \leq 1$ for all $\alpha \in \mathbb{N}^r$ and $i = 1, \ldots, s$. So, by Theorem 6.35, taking $h = h_5 + h_1$ we have

$$\begin{aligned}
g_j\left(f_1(\mathbf{y} + p^h\mathbf{x}), \ldots, f_s(\mathbf{y} + p^h\mathbf{x})\right) &= G_j(E_1(\mathbf{x}), \ldots, E_s(\mathbf{x})) \\
&= (G_j \circ \mathbf{E})(\mathbf{x}) \\
&= H_j(\mathbf{x}).
\end{aligned}$$

Thus $\mathbf{g} \circ \mathbf{f}$ is analytic on U as claimed.

8.6 Definition (i) Let X be a topological space and U a non-empty open subset of X. A triple (U, ϕ, n) is a *chart* on X if ϕ is a homeomorphism from U onto an open subset of $\mathbb{Z}_p^n$ for some $n \in \mathbb{N}$. The *dimension* of the chart is n. The chart (U, ϕ, n) is a *global chart* if $U = X$.

(ii) Two charts (U, ϕ, n) and (V, ψ, m) on a topological space X are *compatible* if the maps $\psi \circ \phi^{-1}|_{\phi(U \cap V)}$ and $\phi \circ \psi^{-1}|_{\psi(U \cap V)}$ are analytic functions on $\phi(U \cap V)$ and $\psi(U \cap V)$ respectively.

(iii) An *atlas* on a topological space X is a set of pairwise compatible charts that covers X; i.e. it is a set of the form

$$A = \{(U_i, \phi_i, n_i) \mid i \in I\}$$

with the following properties:

- for each $i \in I$, (U_i, ϕ_i, n_i) is a chart on X;
- for all $i, j \in I$, (U_i, ϕ_i, n_i) and (U_j, ϕ_j, n_j) are compatible;
- $X = \bigcup_{i \in I} U_i$.

A is a *global atlas* if for some $i \in I$ the chart (U_i, ϕ_i, n_i) is global.

(iv) Let A and B be atlases on a topological space X. Then A and B are *compatible* if every chart in A is compatible with every chart in B; that is, if $A \cup B$ is an atlas on X.

As a temporary notation, let us write X_A to denote the topological space X endowed with the atlas A. A function $f : X_A \to Y_B$ is said to be *analytic* if for each pair of charts $(U, \phi, n) \in A$ and $(V, \psi, m) \in B$, the following hold:

(i) $f^{-1}(V)$ is open in X, and

(ii) the composition

$$\psi \circ f \circ \phi^{-1}|_{\phi(U \cap f^{-1}(V))}$$

is an analytic function from the open set $\phi(U \cap f^{-1}(V)) \subseteq \mathbb{Z}_p^n$ into $\mathbb{Z}_p^m$.

8.7 Lemma *Let X, Y and Z be topological spaces and A, B, C atlases on X, Y, Z respectively. If $f : X_A \to Y_B$ and $g : Y_B \to Z_C$ are analytic, then $g \circ f : X_A \to Z_C$ is analytic.*

Proof Write $h = g \circ f$. Let $(V, \psi, m) \in C$. Then $g^{-1}(V) \cap W$ is an open subset of W for each $(W, \theta, t) \in B$, and $g^{-1}(V)$ is covered by such subsets. So to show that $h^{-1}(V) = f^{-1}g^{-1}(V)$ is open in X it suffices to show that for each such W and each $(U, \phi, n) \in A$, the set $U \cap f^{-1}(g^{-1}(V) \cap W)$ is open in U. Now the map $\theta \circ f \circ \phi^{-1}|_{\phi(U \cap f^{-1}(W))}$ is analytic, hence continuous by Proposition 6.19; as both θ and ϕ are homeomorphisms it follows that the restriction of f to $U \cap f^{-1}(W)$ is also continuous. Therefore $U \cap f^{-1}(g^{-1}(V) \cap W)$ is open in $U \cap f^{-1}(W)$, and as $f^{-1}(W)$ is open in X it follows that $U \cap f^{-1}(g^{-1}(V) \cap W)$ is open in U, as required. Thus h satisfies condition (i).

To verify condition (ii), let $(U, \phi, n) \in A$. It will suffice to show that for each $y \in U \cap h^{-1}(V)$ there is a neighbourhood N of $\phi(y)$ in $\phi(U \cap h^{-1}(V))$ such that $\psi \circ h \circ \phi^{-1}|_N$ is analytic. Now the atlas B contains a chart (W, θ, t) such that $f(y) \in W$. We take $N = \phi(U \cap f^{-1}(g^{-1}(V) \cap W))$. Then $\psi \circ h \circ \phi^{-1}|_N$ is the composition of $\theta \circ f \circ \phi^{-1} : N \to \theta(g^{-1}(V) \cap W)$ and $\psi \circ g \circ \theta^{-1} : \theta(g^{-1}(V) \cap W) \to \psi(V) \subseteq \mathbb{Z}_p^m$. Each of these maps is analytic, by hypothesis, so it follows by Lemma 8.5 that $\psi \circ h \circ \phi^{-1}|_N$ is analytic as required.

Now according to Definition 8.6(iv), two atlases A and B on a space X are compatible if and only if the identity maps $X_A \to X_B$ and $X_B \to X_A$ are analytic. Lemma 8.7 therefore implies that compatibility of atlases is a *transitive* relation; as it is plainly both reflexive and symmetric, we see that compatibility is an equivalence relation on the class of all atlases on X. We can therefore make the following

8.8 Definition Let X be a topological space. A *p-adic analytic manifold structure* on X is an equivalence class of compatible atlases on X. If such a structure exists, X is a *p-adic analytic manifold*. Any atlas belonging to the given equivalence class is called an atlas of (the manifold) X; any chart belonging to such an atlas is called a chart of (the manifold) X.

In this chapter and the next, when we write 'analytic manifold', or just 'manifold', we always mean '*p*-adic analytic manifold'. Analytic manifolds of a more general type will be discussed in Chapter 13.

8.9 Examples

(i) Let X be a discrete topological space. Then X is a *p*-adic analytic manifold with structure determined by the atlas $\{(\{x\}, \phi_x, 0) \mid x \in X\}$ where $\phi_x : x \mapsto 0$.

(ii) Let $X = \mathbb{Q}_p^n$. For each $i \in \mathbb{N}$, let $\phi_i : p^{-i}\mathbb{Z}_p^n \to \mathbb{Z}_p^n$ be the map defined by $\mathbf{x} \mapsto p^i\mathbf{x}$. Then the set $A = \{(p^{-i}\mathbb{Z}_p^n, \phi_i, n) \mid i \in \mathbb{N}\}$ is an atlas on X since $\{p^{-i}\mathbb{Z}_p^n \mid i \in \mathbb{N}\}$ is an open covering of $\mathbb{Q}_p^n$. This atlas gives X the structure of a *p*-adic analytic manifold. [This implies that if we weakened the definition of a chart (U, ϕ, n) to require that ϕ should be a homeomorphism from U onto an open subset of $\mathbb{Q}_p^n$ rather than $\mathbb{Z}_p^n$, we would still get the same class of *p*-adic analytic manifolds.]

(iii) Let X be a manifold and let U be an open subset of X. Let $A = \{(V_i, \phi_i, n_i) \mid i \in I\}$ be an atlas of X. Then $B = \{V_i \cap U, \phi_i|_{V_i \cap U}, n_i) \mid i \in I\}$ is an atlas on U and the manifold structure determined by this atlas is called the *induced manifold structure* on U. If V and Y are *p*-adic analytic manifolds such that V is an open subset of Y and the manifold structure on V is the induced manifold structure from Y, then we say that Y *extends* the manifold structure on V.

(iv) Let $X = \mathrm{GL}_n(\mathbb{Q}_p)$ with the subspace topology induced from that on $\mathrm{M}_n(\mathbb{Q}_p) = \mathbb{Q}_p^{n^2}$.

(a) Let $A = \{(p^{-i}\mathrm{M}_n(\mathbb{Z}_p), \phi_i, n^2) \mid i \in \mathbb{N}\}$ be the atlas on the vector space $\mathrm{M}_n(\mathbb{Q}_p)$ given in Example (ii). Then, by Example (iii), $B = \{(X \cap p^{-i}\mathrm{M}_n(\mathbb{Z}_p), \phi_i|_{X \cap p^{-i}\mathrm{M}_n(\mathbb{Z}_p)}, n^2) \mid i \in \mathbb{N}\}$ is an atlas on X.

(b) Let $U = 1_n + p\mathrm{M}_n(\mathbb{Z}_p)$, an open subgroup of X. Define $\phi : U \to p\mathrm{M}_n(\mathbb{Z}_p)$ by $\phi(u) = u - 1_n$ for all $u \in U$. Then $\{(U, \phi, n^2)\}$ is a global atlas on U. For each $h \in X$, let $V_h = hU$, an open neighbourhood of h

in X. Define $\phi_h : V_h \to pM_n(\mathbb{Z}_p)$ by $\phi_h(x) = \phi(h^{-1}x)$. We leave it as an exercise to show that $\{(V_h, \phi_h, n^2) \mid h \in X\}$ is an atlas on X, and that it is compatible with the atlas B defined in (a).

The two atlases defined in (a) and (b) thus determine the *same* analytic manifold structure on X.

(v) Let G be a uniform pro-p group of dimension d, generated topologically by $\{a_1, \dots, a_d\}$, say.

(a) In Theorem 4.9 we defined a homeomorphism $\phi : G \to \mathbb{Z}_p^d$ by $\phi(x) = (\lambda_1, \dots, \lambda_d)$ where $x = a_1^{\lambda_1} \dots a_d^{\lambda_d}$. Then $A = \{(G, \phi, d)\}$ is a global atlas.

(b) In Theorem 4.17 we showed that G is a free $\mathbb{Z}_p$-module on the basis $\{a_1, \dots, a_d\}$, with respect to the module operations defined in §4.3. Define $\psi : G \to \mathbb{Z}_p^d$ by $\psi(x) = (\mu_1, \dots, \mu_d)$ where $x = \mu_1 a_1 + \cdots + \mu_d a_d$. Then Proposition 4.16 shows that ψ is a homeomorphism, so $B = \{(G, \psi, d)\}$ is a global atlas.

(c) In Corollary 7.14 we saw that $\log : G \to \hat{A}$ gives a homeomorphism of G onto $\Lambda = \log G$, a $\mathbb{Z}_p$-Lie subalgebra of $\hat{A}$ of dimension d, where $\hat{A}$ denotes the completed group algebra of G defined in Chapter 7. Let $\theta : \Lambda \to \mathbb{Z}_p^d$ be an isomorphism, obtained by choosing a basis for the $\mathbb{Z}_p$-module Λ. Then $C = \{(G, \theta \circ \log, d)\}$ is a global atlas on G.

Now Corollary 7.14 shows that $\log$ is a $\mathbb{Z}_p$-module isomorphism from G onto Λ, where G has the module structure mentioned above in (b); so we may choose $\{\log a_1, \dots, \log a_d\}$ to be our basis for Λ, and then we see that C is the *same* atlas as B. Since every invertible linear transformation on $\mathbb{Z}_p^d$ is an analytic homeomorphism, it follows that the atlases B and C are compatible whatever the choice of θ.

As a matter of fact, the atlases A and C are also compatible. A direct proof of this is outlined in Exercise 3. Alternatively, it can be deduced from a uniqueness theorem to be proved in §9.4. Thus all three atlases determine the same manifold structure on G.

(vi) Let X and Y be analytic manifolds determined by atlases A and B respectively. Then $Z = X \times Y$ has the structure of an analytic manifold defined by the atlas $C = \{(U \times V, \phi \times \psi, m + n) \mid (U, \phi, m) \in A$ and $(V, \psi, n) \in B\}$ where $\phi \times \psi : U \times V \to \mathbb{Z}_p^{m+n}$ is defined by $(u, v) \mapsto (\phi(u), \psi(v))$. This manifold Z is called the *product of X and Y*.

8.10 Definition Let X and Y be analytic manifolds and $f : X \to Y$ a function. f is *analytic* if there exist atlases A and B of X and Y respectively such that $f : X_A \to Y_B$ is analytic.

It follows from Lemma 8.7 that if $f : X \to Y$ satisfies the condition stated in this definition with respect to the atlases A and B, then it satisfies the corresponding condition with respect to *every* pair of atlases of the manifolds X and Y respectively. Another immediate consequence of Lemma 8.7 is

8.11 Lemma *Suppose that* $f : X \to Y$ *and* $g : Y \to Z$ *are analytic functions where* X, Y *and* Z *are analytic manifolds. Then* $g \circ f : X \to Z$ *is an analytic function.*

8.12 Lemma *Let* $f : X \to Y$ *be a function, where* X *and* Y *are manifolds. Suppose that* $X = \bigcup_{i \in I} X_i$ *where* X_i *is an open subset of* X *and that* $f|_{X_i} : X_i \to Y$ *is analytic with respect to the induced manifold structure on* X_i, *for each* $i \in I$. *Then* f *is an analytic function.*

Proof Let A be an atlas of the manifold X. For each $i \in I$ let A_i be the atlas on X_i determined by A as in Example 8.9(iii). Then $A^* = \bigcup_{i \in I} A_i$ is easily seen to be an atlas on X compatible with A. The lemma follows, since the hypothesis implies that $f : X_{A^*} \to Y$ is analytic.

The final result of this section follows from Proposition 6.19:

8.13 Lemma *Let* $f : X \to Y$ *be an analytic function. Then* f *is continuous.*

8.2 *p*-adic analytic groups

8.14 Definition A topological group G is a *p-adic analytic group* if G has the structure of a p-adic analytic manifold with the properties

(i) the function $f : G \times G \to G$ given by $(x, y) \mapsto xy$ is analytic;

(ii) the function $i : G \to G$ defined by $x \mapsto x^{-1}$ is analytic.

It follows from Lemma 8.11 that conditions (i) and (ii) may be replaced by the single condition:

(i)': the function $g : G \times G \to G$ defined by $(x, y) \mapsto xy^{-1}$ is analytic.

Using the Inverse Function Theorem 6.37, one can deduce condition (ii) from condition (i); however we prefer to state both in the definition. A p-adic analytic group is sometimes referred to as a *p-adic Lie group*.

The following proposition provides a 'local' criterion to determine whether a topological group G has the structure of a p-adic analytic group:

8.15 Proposition *Let G be a topological group containing an open subgroup H. Suppose that H has the structure of a p-adic analytic group, and that the following holds: for each $g \in G$, there exists an open neighbourhood V_g of the identity in H such that*

(i) $gV_g g^{-1} \subseteq H$ *and*

(ii) *the function $k_g : V_g \to H$ defined by $x \mapsto gxg^{-1}$ is analytic.*

Then there exists a unique analytic manifold structure on G extending the manifold structure on H and making G into a p-adic analytic group.

Proof Let T be a transversal to the left cosets of H in G. Let A be an atlas defining the manifold structure on H. For each $t \in T$, we define an atlas $A(t)$ on the coset tH, an open neighbourhood of t, as follows:

$$A(t) = \{(tU, \phi_t, m) \mid (U, \phi, m) \in A\}$$

where $\phi_t : tU \to \mathbb{Z}_p^m$ is defined by $\phi_t(x) = \phi(t^{-1}x)$. Since $G = \bigcup_{t \in T} tH$ and this union is disjoint, the set $\mathcal{A} = \bigcup_{t \in T} A(t)$ is an atlas on G. The manifold structure on G defined by the atlas $\mathcal{A}$ extends the manifold structure on H. To verify condition (i)$'$, it suffices to show that the function $f : G \times G \to G$ defined by $(x, y) \mapsto xy^{-1}$ is analytic on $sH \times tH$ for each s and t in T. If h_1 and h_2 are in H, then

$$(sh_1)(th_2)^{-1} = st^{-1}t(h_1 h_2^{-1})t^{-1}. \tag{1}$$

Since H is a p-adic analytic group, (1) together with Lemma 8.11 implies that f is analytic on $sH \times tH$, provided that

(a) for each $g \in G$, the function $f_g : G \to G$ defined by $x \mapsto gx$ is analytic; and

(b) for each $g \in G$, the function $k_g : H \to G$ defined by $x \mapsto gxg^{-1}$ is analytic.

To prove (a) it suffices to show that the function f_g is analytic on tH for each $t \in T$. Now $gt = sh$ for some $s \in T$ and $h \in H$. For each (U, ϕ, m) and $(W, \psi, n) \in A$, the function

$$\psi_s \circ f_g \circ \phi_t^{-1} : \phi_t(tU \cap f_g^{-1}(sW)) \to \psi_s(sW)$$

is precisely the function

$$\psi \circ f_h \circ \phi^{-1} : \phi(U \cap f_h^{-1}W) \to \psi W$$

which is analytic since H is an analytic group with respect to the manifold determined by A. Therefore f_g is analytic.

To prove (b) it suffices by Lemma 8.12 to show that for each $h \in H$, k_g is analytic on some open neighbourhood of h. By hypothesis there exists V, an open neighbourhood of the identity, with the property that $V \subseteq H \cap g^{-1}Hg$ and $k_g : V \to H$ is analytic. Consider the open neighbourhood hV of h. Then the function $k_g : hV \to ghg^{-1}H$ is a composition of functions

$$hV \xrightarrow{f_{h^{-1}}} V \xrightarrow{k_g} H \xrightarrow{f_{ghg^{-1}}} ghg^{-1}H.$$

By (a) and our hypothesis each of these functions is analytic. That $k_g|_{hV}$ is analytic therefore follows from Lemma 8.11, and this proves (b).

It follows that the manifold structure on G determined by A gives G the structure of a p-adic analytic group.

Finally, we have the question of uniqueness to settle. Suppose that B is an atlas defining a manifold structure on G extending the manifold structure on H and that G is a p-adic analytic group with respect to this structure. We want to prove that A and B are compatible. We may assume that $A \subseteq B$ since the manifold structure determined by B extends the structure on H. Let $(W, \psi, n) \in B$, $g \in G$ and $(U, \phi, m) \in A$. We are required to prove that the functions

$$\phi_g \circ \psi^{-1} : \psi(W \cap gU) \to \phi_g(W \cap gU)$$
$$\psi \circ \phi_g^{-1} : \phi_g(W \cap gU) \to \psi(W \cap gU)$$

are analytic.

G is a p-adic analytic group with respect to the manifold defined by B. So for each $g \in G$ the function $f_g : G_B \to G_B$ defined by $x \mapsto gx$ is analytic. Thus, for each $(W, \psi, n) \in B$ and $(U, \phi, m) \in A \subseteq B$, the functions

$$\phi \circ f_{g^{-1}} \circ \psi^{-1} : \psi(W \cap f_{g^{-1}}^{-1}U) \to \phi(f_{g^{-1}}W \cap U)$$
$$\psi \circ f_g \circ \phi^{-1} : \phi(U \cap f_g^{-1}W) \to \psi(f_gU \cap W)$$

are analytic. But these functions are precisely the functions detailed above. Thus A and B are compatible, and the result follows.

Remark In Chapter 9 we shall prove a far stronger uniqueness result: if G is a topological group then there is at most one p-adic manifold structure defined on G with respect to which G is a p-adic analytic group.

8.16 Lemma *Let G and G' be p-adic analytic groups and let $\phi : G \to G'$ be a group homomorphism. Suppose that $\phi|_H$ is analytic for some open subgroup H of G. Then ϕ is an analytic function.*

Proof By Lemma 8.12 it suffices to show that $\phi|_{gH}$ is analytic for each $g \in G$. Now $\phi|_{gH}$ is a composition of functions

$$gH \xrightarrow{f_{g^{-1}}} H \xrightarrow{\phi} G' \xrightarrow{f_{\phi(g)}} G'$$

where the functions $f_x : y \mapsto xy$ are analytic. By hypothesis $\phi|_H$ is analytic. The result follows by Lemma 8.11.

8.17 Examples

(i) Let G be a group with the discrete topology and manifold structure as defined in Example 8.9(i). Since any function on such a manifold is analytic, G is a p-adic analytic group.

(ii) Let $G = (\mathbb{Q}_p^n, +)$. With respect to the analytic manifold structure defined in Example 8.9(ii), the function $(x, y) \mapsto x - y$ is analytic. Thus G is a p-adic analytic group.

(iii) Let G be a p-adic analytic group and H an open subgroup of G. Then H is a p-adic analytic group with respect to the manifold structure induced from G, as detailed in Example 8.9(iii) .

(iv) Let $G = \mathrm{GL}_n(\mathbb{Q}_p)$. We begin by showing that the open subgroup $U = 1_n + p\mathrm{M}_n(\mathbb{Z}_p)$ has the structure of a p-adic analytic group with respect to the global atlas defined in Example 8.9(iv)(b). The function $f : U \times U \to U$ defined by $(x, y) \mapsto xy$ is clearly analytic since it is described by polynomials in the matrix entries of x and y. We leave it as an exercise to prove that the function $i : U \to U$ defined by $x \mapsto x^{-1}$ is analytic. Therefore U has the structure of a p-adic analytic group. For each $g \in G$, let $V_g = U \cap g^{-1}Ug$; then the function $k_g : V_g \to U$ defined by $x \mapsto gxg^{-1}$ is clearly analytic since it is described by linear polynomials in the entries of x. Thus, by Proposition 8.15, G has the structure of a p-adic analytic group. It is a straightforward exercise to show that the manifold structure constructed in Proposition 8.15 is precisely the manifold structure defined on G in Example 8.9(iv).

(v) In the following section we prove that if G is a uniform pro-p group then G is a p-adic analytic group with respect to the manifold structure given in Example 8.9(v)(a). In Chapter 9 we shall see how the Campbell–Hausdorff formula can be used to prove that G is a p-adic analytic group with respect to the manifold structure given in Example

8.9(v)(b). The uniqueness result to be proved in Chapter 9 will then imply that the atlases of Example 8.9(v) are compatible.

8.3 Uniform pro-p groups

In this section we prove our first major result:

8.18 Theorem *Let G be a topological group containing an open subgroup which is a uniform pro-p group. Then G is a p-adic analytic group.*

The proof leans heavily on the theory developed in §7.1. Until further notice, we assume that G is a uniform pro-p group, with minimal generating set $a_1, \ldots, a_d$, and write $b_i = a_i - 1$ for each i. We shall be working inside $\widehat{R}$, the completion of $(\mathbb{Z}_p[G], \|\cdot\|)$ constructed in §7.1.

Recall that for $\lambda \in \mathbb{Z}_p$ and $1 \leq r \in \mathbb{N}$,

$$\binom{\lambda}{r} = \frac{\lambda(\lambda - 1) \ldots (\lambda - r + 1)}{r!} \in \mathbb{Z}_p;$$

we also write $\binom{\lambda}{0} = 1$

8.19 Lemma *Let $u_1, \ldots, u_r \in G$ and put $v_i = u_i - 1$ for each i. Then for $\lambda_1, \ldots, \lambda_r \in \mathbb{Z}_p$ we have*

$$u_1^{\lambda_1} \ldots u_r^{\lambda_r} = \sum_{\alpha \in \mathbb{N}^r} \binom{\lambda_1}{\alpha_1} \ldots \binom{\lambda_r}{\alpha_r} \mathbf{v}^{\alpha}$$

where $\mathbf{v}^{\alpha} = v_1^{\alpha_1} \ldots v_r^{\alpha_r}$.

Proof Fix $\varepsilon > 0$, and choose $M \leq N \in \mathbb{N}$ so that $p^{-M} < \varepsilon$ and $p^{-N} < |M!| \, \varepsilon$. Now take $\beta_i \in \mathbb{N}$ such that $\beta_i \equiv \lambda_i \pmod{p^N}$. Then for each i we have

$$u_i^{\lambda_i} - u_i^{\beta_i} = u_i^{\beta_i}(u_i^{\lambda_i - \beta_i} - 1) \in \mathbb{Z}_p[G](G_{N+1} - 1) = I_{N+1},$$

in the notation of §7.1. It follows that

$$\left\| u_1^{\lambda_1} \ldots u_r^{\lambda_r} - u_1^{\beta_1} \ldots u_r^{\beta_r} \right\| \leq p^{-(N+1)} < \varepsilon, \tag{2}$$

as the expression inside the norm may be written as a sum of terms each of which has a factor of the form $u_i^{\lambda_i} - u_i^{\beta_i}$.

Next, suppose that $\alpha \in \mathbb{N}^r$ satisfies $\langle \alpha \rangle \leq M$. Then $\prod_{i=1}^r \alpha_i! \mid \langle \alpha \rangle! \mid M!$ (Exercise 6.5), and a simple argument then shows that

$$\left| \binom{\lambda_1}{\alpha_1} \ldots \binom{\lambda_r}{\alpha_r} - \binom{\beta_1}{\alpha_1} \ldots \binom{\beta_r}{\alpha_r} \right| \leq \frac{p^{-N}}{|M!|}.$$

On the other hand, if $\alpha \in \mathbb{N}^r$ and $\langle \alpha \rangle > M$ then

$$\|\mathbf{v}^\alpha\| \leq \prod_{i=1}^r \|v_i^{\alpha_i}\| \leq p^{-M}$$

since $\|v_i\| \leq p^{-1}$ for each i. It follows by Proposition 6.9(iv) that

$$\left\| \sum_{\alpha \in \mathbb{N}^r} \left(\binom{\lambda_1}{\alpha_1} \cdots \binom{\lambda_r}{\alpha_r} - \binom{\beta_1}{\alpha_1} \cdots \binom{\beta_r}{\alpha_r} \right) \mathbf{v}^\alpha \right\|$$
$$\leq \max\{p^{-N}/|M!|, p^{-M}\} < \varepsilon.$$

Finally, Lemma 7.8 shows that

$$u_1^{\beta_1} \cdots u_r^{\beta_r} = \sum_{\alpha \in \mathbb{N}^r} \binom{\beta_1}{\alpha_1} \cdots \binom{\beta_r}{\alpha_r} \mathbf{v}^\alpha.$$

Combining this with the previous line and (2) we infer that

$$\left\| u_1^{\lambda_1} \cdots u_r^{\lambda_r} - \sum_{\alpha \in \mathbb{N}^r} \binom{\lambda_1}{\alpha_1} \cdots \binom{\lambda_r}{\alpha_r} \mathbf{v}^\alpha \right\| < \varepsilon.$$

As ε was arbitrary, the lemma follows.

8.20 Lemma *Let $g : \mathbb{Z}_p^r \to \mathbb{Z}_p$ be a function such that*

$$g(\mathbf{x}) = \sum_{\alpha \in \mathbb{N}^r} c_\alpha \binom{x_1}{\alpha_1} \cdots \binom{x_r}{\alpha_r}$$

for all $\mathbf{x} \in \mathbb{Z}_p^r$, where $c_\alpha \in \mathbb{Q}_p$. Suppose that $|c_\alpha| \leq c p^{-e\langle \alpha \rangle}$ for each α, where $c > 0$ and $e > (p-1)^{-1}$. Then g is strictly analytic on $\mathbb{Z}_p^r$.

Proof Define $\nu_{\alpha\beta} \in \mathbb{Q}_p$ by

$$\binom{X_1}{\alpha_1} \cdots \binom{X_r}{\alpha_r} = \sum_{\beta \in \mathbb{N}^r} \nu_{\alpha\beta} \mathbf{X}^\beta.$$

Note that $\nu_{\alpha\beta} = 0$ whenever $\beta_i > \alpha_i$ for some $i = 1, \ldots, r$, and that for all $(\alpha, \beta) \in \mathbb{N}^r \times \mathbb{N}^r$ we have

$$|\nu_{\alpha\beta}| \leq |(\alpha_1! \ldots \alpha_r!)^{-1}|$$
$$\leq p^{(\langle \alpha \rangle - r)/(p-1)},$$

by Lemma 6.20. It follows that

$$|c_\alpha \nu_{\alpha\beta}| \leq c' p^{-\rho\langle \alpha \rangle}$$

where $c' = c p^{-r/(p-1)}$ and $\rho = e - (p-1)^{-1} > 0$.

We deduce that $\lim_{(\alpha,\beta)\in\mathbb{N}^r\times\mathbb{N}^r} c_\alpha\nu_{\alpha\beta}\mathbf{x}^\beta = 0$ for each $\mathbf{x}\in\mathbb{Z}_p^r$. Applying Proposition 6.9(ii) and Corollary 6.11 as in the proof of Lemma 8.3, we conclude that for all $\mathbf{x}\in\mathbb{Z}_p^r$,

$$g(\mathbf{x}) = \sum_{\alpha\in\mathbb{N}^r} c_\alpha\Big(\sum_{\beta\in\mathbb{N}^r}\nu_{\alpha\beta}\mathbf{x}^\beta\Big) = \sum_{\beta\in\mathbb{N}^r} d_\beta\mathbf{x}^\beta$$

where $d_\beta = \sum_{\alpha\in\mathbb{N}^r} c_\alpha\nu_{\alpha\beta}$ for each $\beta\in\mathbb{N}^r$. The result follows.

8.21 Proposition *Let $u_1,\dots,u_r$ be arbitrary elements of G (if $p > 2$) or of $P_2(G)$ (if $p = 2$). Define $\mathbf{g} = (g_1,\dots,g_d) : \mathbb{Z}_p^r \to \mathbb{Z}_p^d$ by $g_i(\lambda_1\dots,\lambda_r) = \mu_i$ where*

$$u_1^{\lambda_1}\dots u_r^{\lambda_r} = a_1^{\mu_1}\dots a_d^{\mu_d}.$$

Then $\mathbf{g}$ is an analytic function on $\mathbb{Z}_p^r$.

Proof Let $v_i = u_i - 1$ for $i = 1,\dots,r$. Lemma 8.19 shows that

$$u_1^{\lambda_1}\dots u_r^{\lambda_r} = \sum_{\alpha\in\mathbb{N}^r} \binom{\lambda_1}{\alpha_1}\dots\binom{\lambda_r}{\alpha_r}\mathbf{v}^\alpha. \tag{3}$$

According to Theorem 7.4, for each $\alpha\in\mathbb{N}^r$ we can write

$$\mathbf{v}^\alpha = \sum_{\beta\in\mathbb{N}^d} c_{\alpha\beta}\mathbf{b}^\beta \tag{4}$$

where $c_{\alpha\beta}\in\mathbb{Z}_p$ for each $\beta\in\mathbb{N}^d$. Put $\epsilon = 1$ if $p > 2$, $\epsilon = 2$ if $p = 2$. Then by hypothesis $\|v_i\| = \|u_i - 1\| \le p^{-\epsilon}$ for each i, so $\|\mathbf{v}^\alpha\| \le p^{-\epsilon\langle\alpha\rangle}$. It follows by Theorem 7.5 that

$$|c_{\alpha\beta}| \le p^{\langle\beta\rangle}\|\mathbf{v}^\alpha\| \le \min\{1, p^{-\epsilon\langle\alpha\rangle+\langle\beta\rangle}\} \tag{5}$$

for all $\alpha\in\mathbb{N}^r$ and $\beta\in\mathbb{N}^d$. Since $\binom{\lambda}{\alpha}\in\mathbb{Z}_p$ and $\|\mathbf{b}^\beta\| \le p^{-\langle\beta\rangle}$, we may deduce that

$$\lim_{(\alpha,\beta)\in\mathbb{N}^r\times\mathbb{N}^d} \binom{\lambda_1}{\alpha_1}\dots\binom{\lambda_r}{\alpha_r}c_{\alpha\beta}\mathbf{b}^\beta = 0.$$

Substituting (4) into (3) and applying Corollary 6.11, we get

$$u_1^{\lambda_1}\dots u_r^{\lambda_r} = \sum_{\beta\in\mathbb{N}^d}\Big(\sum_{\alpha\in\mathbb{N}^r}\binom{\lambda_1}{\alpha_1}\dots\binom{\lambda_r}{\alpha_r}c_{\alpha\beta}\Big)\mathbf{b}^\beta.$$

On the other hand, Lemma 8.19 also gives

$$u_1^{\lambda_1}\dots u_r^{\lambda_r} = a_1^{\mu_1}\dots a_d^{\mu_d} = \sum_{\beta\in\mathbb{N}^d}\binom{\mu_1}{\beta_1}\dots\binom{\mu_d}{\beta_d}\mathbf{b}^\beta,$$

where $\mu_i = g_i(\lambda_1,\dots,\lambda_r)$ for each i. Equating coefficients as we may by Theorem 7.4(ii) we see that

$$\binom{\mu_1}{\beta_1} \cdots \binom{\mu_d}{\beta_d} = \sum_{\alpha \in \mathbb{N}^r} \binom{\lambda_1}{\alpha_1} \cdots \binom{\lambda_r}{\alpha_r} c_{\alpha\beta}$$

for all $\beta \in \mathbb{N}^d$.

Denote by ϵ_i the d-tuple with entry δ_{ij} (Kronecker δ) in the jth place for $j = 1, \ldots, d$.

Choosing $\beta = \epsilon_i$ gives

$$g_i(\lambda_1, \ldots, \lambda_r) = \mu_i = \sum_{\alpha \in \mathbb{N}^r} \binom{\lambda_1}{\alpha_1} \cdots \binom{\lambda_r}{\alpha_r} c_{\alpha\epsilon_i}.$$

Since $\epsilon > (p-1)^{-1}$, Lemma 8.20 with (5) shows that each g_i is a strictly analytic function on $\mathbb{Z}_p^r$. It follows by Corollary 8.4 that $\mathbf{g}$ is analytic on $\mathbb{Z}_p^r$.

We can now prove Theorem 8.18. The group G is no longer assumed to be uniform; instead, G is a topological group containing an open subgroup N which is a uniform pro-p group. Let $H = N$ if $p > 2$, $H = P_2(N)$ if $p = 2$. Suppose that N is generated topologically by $a_1, \ldots, a_d$. For $i = 1, \ldots, d$, let $u_i = a_i$ (if $p > 2$) or $u_i = a_i^2$ (if $p = 2$). Then the uniform pro-p group H is generated topologically by $u_1, \ldots, u_d$. Define $\phi : H \to \mathbb{Z}_p^d$ by $\phi(x) = (\lambda_1, \ldots, \lambda_r)$ where $x = u_1^{\lambda_1} \ldots u_d^{\lambda_d}$. As we observed in Example 8.9(v)(a), H has the structure of a p-adic analytic manifold defined by the global atlas $\{(H, \phi, d)\}$. By Proposition 8.15, to prove that G has the structure of a p-adic analytic group it suffices to prove the following two claims:

(i) the function $H \times H \to H$ given by $(x, y) \mapsto xy^{-1}$ is analytic with respect to the manifold structure defined on H; and

(ii) for each $g \in G$, there exists an open neighbourhood $V = V_g$ of the identity in H such that $gVg^{-1} \subset H$ and the function $k_g : V \to H$ defined by $x \mapsto gxg^{-1}$ is analytic.

Let $\epsilon = 1$ if $p > 2$, $\epsilon = 2$ if $p = 2$, so $u_i = a_i^\epsilon$ for each i.

To prove (i), we have to show that the function $\mathbf{f} : \mathbb{Z}_p^d \times \mathbb{Z}_p^d \to \mathbb{Z}_p^d$ given by $\mathbf{f}(\tau, \sigma) = \nu$, where

$$(u_1^{\tau_1} \ldots u_d^{\tau_d}) \cdot (u_1^{\sigma_1} \ldots u_d^{\sigma_d})^{-1} = u_1^{\nu_1} \ldots u_d^{\nu_d},$$

is analytic. For $i = 1, \ldots, d$ put $u_{d+i} = u_{d-i+1}^{-1}$. Then $\mathbf{f} = \epsilon^{-1}\mathbf{f}^*$, where $\mathbf{f}^* : \mathbb{Z}_p^{2d} \to \mathbb{Z}_p^d$ is the function defined by $\mathbf{f}^*(\lambda_1, \ldots, \lambda_{2d}) = \mu$ when

$$u_1^{\lambda_1} \ldots u_{2d}^{\lambda_{2d}} = a_1^{\mu_1} \ldots a_d^{\mu_d}.$$

Proposition 8.21 shows that $\mathbf{f}^*$ is analytic; and (i) follows.

Now we prove (ii). Let $g \in G$; then $H \cap g^{-1}Hg$ is an open subgroup of H, so there exists $m \in \mathbb{N}$ such that $P_{m+1}(H) \subseteq H \cap g^{-1}Hg$. We put $V = P_{m+1}(H)$. Then the induced manifold structure on V is defined by the atlas $\{(V, \phi|_V, d)\}$ where $\phi|_V$ maps V onto $p^m\mathbb{Z}_p^d$. It will suffice to show that the function $\mathbf{k}^* : p^m\mathbb{Z}_p^d \to \mathbb{Z}_p^d$ defined by $\mathbf{k}^*(\tau) = \nu$ where

$$g(u_1^{\tau_1} \ldots u_d^{\tau_d})g^{-1} = u_1^{\nu_1} \ldots u_d^{\nu_d}$$

is analytic on $p^m\mathbb{Z}_p^d$. Write $w_i = gu_i^{p^m}g^{-1}$ $(i = 1, \ldots, d)$, and consider the function $\mathbf{h} : \mathbb{Z}_p^d \to \mathbb{Z}_p^d$ defined by $\mathbf{h}(\lambda_1, \ldots, \lambda_d) = \mu$ where $w_1^{\lambda_1} \ldots w_d^{\lambda_d} = a_1^{\mu_1} \ldots a_d^{\mu_d}$; this function is analytic by Proposition 8.21. Now let $\lambda \in \mathbb{Z}_p^d$. Then

$$g(u_1^{p^m\lambda_1} \ldots u_d^{p_d^m\lambda_d})g^{-1} = w_1^{\lambda_1} \ldots w_d^{\lambda_d}$$
$$= a_1^{\mu_1} \ldots a_d^{\mu_d} = u_1^{\epsilon^{-1}\mu_1} \ldots u_d^{\epsilon^{-1}\mu_d}$$

where $\mu = \mathbf{h}(\lambda)$. Thus $\mathbf{k}^*(p^m\lambda) = \epsilon^{-1}\mathbf{h}(\lambda)$ for all $\lambda \in \mathbb{Z}_p^d$, and it follows that $\mathbf{k}^*$ is analytic on $p^m\mathbb{Z}_p^d$. This completes the proof.

Remark What we have shown is that the 'co-ordinates of the second kind' on a uniform pro-p group H endow it with the structure of an analytic group; the proof was based on the generalised binomial expansion of Lemma 8.19. The 'co-ordinates of the first kind', given by the additive structure of H, also make H into an analytic group, where the group multiplication is expressed by the Campbell–Hausdorff series. This approach is spelt out in §9.4.

8.4 Standard groups

In this section we establish the converse of Theorem 8.18. This is done in two stages. The first is to show that every p-adic analytic group has an open subgroup whose analytic-group structure is of a particularly simple form, that of a *standard group*; the second is to show that every standard group is a uniform pro-p group.

Throughout, $X_1, X_2, \ldots, Y_1, Y_2, \ldots$ will denote commuting indeterminates, and we write

$$\mathbb{Z}_p[[X_1, \ldots, X_n]] = \mathbb{Z}_p[[\mathbf{X}]]$$

to denote the subring of $\mathbb{Q}_p[[\mathbf{X}]]$ consisting of formal power series

$$F(\mathbf{X}) = \sum_{\alpha \in \mathbb{N}^n} a_\alpha X_1^{\alpha_1} \ldots X_n^{\alpha_n} = \sum_{\alpha \in \mathbb{N}^n} a_\alpha \mathbf{X}^\alpha$$

where $a_\alpha \in \mathbb{Z}_p$ for each $\alpha \in \mathbb{N}^n$. Note that if $F(\mathbf{X}) \in \mathbb{Z}_p[[\mathbf{X}]]$ then $F(\mathbf{x})$ exists for all $\mathbf{x} \in p\mathbb{Z}_p^n$, by Proposition 6.9(ii).

As before, we use ϵ_i to denote a tuple (of suitable length) with δ_{ij} in the jth place for each j.

8.22 Definition Let G be a p-adic analytic group. Then G is a *standard group* (of dimension r over $\mathbb{Q}_p$) if

(i) the analytic manifold structure on G is defined by a global atlas of the form $\{(G, \psi, r)\}$ where ψ is a homeomorphism of G onto $p\mathbb{Z}_p^r$ (if $p > 2$) or onto $4\mathbb{Z}_2^r$ (if $p = 2$), with $\psi(1) = \mathbf{0}$; and

(ii) for $j = 1, \ldots, r$ there exists $P_j(\mathbf{X}, \mathbf{Y}) \in \mathbb{Z}_p[[\mathbf{X}, \mathbf{Y}]]$ such that

$$\psi_j(xy^{-1}) = P_j(\psi(x), \psi(y))$$

for all $x, y \in G$, where $\psi = (\psi_1, \ldots, \psi_r)$.

Remark Our definition of a standard group differs from Bourbaki's definition at the prime $p = 2$ (see Bourbaki (1989b), Chapter III §7.3). The definition will be generalised in Chapter 13.

8.23 Examples
(i) Let $G = (p\mathbb{Z}_p^r, +)$ if $p > 2$, or $G = (4\mathbb{Z}_2^r, +)$ if $p = 2$. Then G is a standard group of dimension r over $\mathbb{Q}_p$.

(ii) Let $G = 1_n + p\mathrm{M}_n(\mathbb{Z}_p)$ if $p > 2$, or $G = 1_n + 4\mathrm{M}_n(\mathbb{Z}_2)$ if $p = 2$. Let $\{(G, \phi, n^2)\}$ be the atlas defined in Example 8.9(iv)(b). We leave it as an exercise to verify that this atlas makes G a standard group over $\mathbb{Q}_p$. (We saw in Chapter 5 that G is a uniform pro-p group: this will be generalised in Theorem 8.31).

(iii) In §9.4 we shall see that if G is a uniform pro-p group, then the subgroup $H = P_2(G)$ (or $H = P_3(G)$ if $p = 2$) is a standard group with respect to the chart

$$\log|_H : H \to \Lambda \subseteq \mathbb{Z}_p^d$$

obtained by fixing a $\mathbb{Z}_p$-basis for the Lie algebra $\Lambda = \log(G) \subseteq \widehat{A}$, where $A = \mathbb{Q}_p[G]$.

8.24 Lemma *Let* $G[Y_1, \ldots, Y_m] \in \mathbb{Z}_p[\mathbf{Y}]$ *and let* $F_i[\mathbf{X}] \in \mathbb{Z}_p[\mathbf{X}]$ *for* $i = 1, \ldots, m$. *Suppose that each of the series* $F_i[\mathbf{X}]$ *has constant term* 0. *Then* $G \circ \mathbf{F} \in \mathbb{Z}_p[\mathbf{X}]$ *and* $(G \circ \mathbf{F})(\mathbf{x}) = G(F_1(\mathbf{x}), \ldots, F_1(\mathbf{x}))$ *for all* $\mathbf{x} \in p\mathbb{Z}_p^n$.

Proof The first claim is clear. The second is an application of Theorem 6.35.

Using Lemma 8.24, it is a simple exercise to verify that condition (ii) in Definition 8.22 is equivalent to

(ii)′ for $j = 1, \ldots, r$, there exist $M_j(X_1, \ldots, X_r, Y_1, \ldots, Y_r) \in \mathbb{Z}_p[[\mathbf{X}, \mathbf{Y}]]$ and $I_j(X_1, \ldots, X_r) \in \mathbb{Z}_p[[\mathbf{X}]]$ such that for all $x, y \in G$

$$\psi_j(xy) = M_j(\psi(x), \psi(y))$$
$$\psi_j(x^{-1}) = I_j(\psi(x)).$$

We can generalise this remark as follows:

8.25 Lemma *Let G be a standard group of dimension r. Let $w(x_1, \ldots, x_n)$ be a group word in the variables $x_1, \ldots, x_n$. Then there exist*

$$F_j[X_{11}, \ldots, X_{1r}, \ldots, X_{n1}, \ldots, X_{nr}] \in \mathbb{Z}_p[[\mathbf{X}_1, \ldots, \mathbf{X}_n]]$$

($j = 1, \ldots, r$) such that for all $x_1, \ldots, x_n \in G$

$$\psi_j(w(x_1, \ldots, x_n)) = F_j(\psi(x_1), \ldots, \psi(x_n)).$$

Proof This is by induction on the length of the word w. Suppose $w(x_1, \ldots, x_n) = v(x_1, \ldots, x_n)x_k$. We may assume that

$$\psi_j(v(x_1, \ldots, x_n)) = H_j(\psi(x_1), \ldots, \psi(x_n))$$

for suitable power series $H_1, \ldots, H_r$. Then writing $\mathbf{x}_i = \psi(x_i)$ for $i = 1, \ldots, n$ we have

$$\psi_j(w(x_1, \ldots, x_n)) = M_j(H_1(\mathbf{x}_1, \ldots, \mathbf{x}_n), \ldots, H_r(\mathbf{x}_1, \ldots, \mathbf{x}_n), \mathbf{x}_k).$$

Now put $F_j = M_j \circ \mathbf{H}$ where $\mathbf{H} = (H_1, \ldots, H_r, X_{k1}, \ldots, X_{kr})$. It follows by Lemma 8.24 that $F_j \in \mathbb{Z}_p[[\mathbf{X}_1, \ldots, \mathbf{X}_n]]$ and that $\psi_j(w(x_1, \ldots, x_n)) = F_j(\psi(x_1), \ldots, \psi(x_n))$.

The same argument gives the result when $w(x_1, \ldots, x_n) = v(x_1, \ldots, x_n)x_k^{-1}$, using P_j in place of M_j.

The following lemmas provide information about the coefficients of certain power series.

8.26 Lemma *Let $F(\mathbf{X}) = \sum_{\alpha \in \mathbb{N}^n} a_\alpha \mathbf{X}^\alpha \in \mathbb{Q}_p[[\mathbf{X}]]$. Suppose that there exists an open neighbourhood V of 0 in $\mathbb{Q}_p$ such that, for all*

$$\lambda_1, \ldots, \lambda_n \in V^n,$$

$$F(\lambda_1, \ldots, \lambda_n) = 0.$$

Then $a_\alpha = 0$ for all $\alpha \in \mathbb{N}^n$.

Proof This is by induction on n. The case $n = 1$ is dealt with in Proposition 6.13. Suppose that $n > 1$ and that the lemma is true for $n - 1$. For each $i \in \mathbb{N}$, put

$$F_i(X_1, \ldots, X_{n-1}) = \sum_{\substack{\alpha \in \mathbb{N}^n \\ \alpha_n = i}} a_\alpha X_1^{\alpha_1} \ldots X_{n-1}^{\alpha_{n-1}}.$$

Fix $(\lambda_1, \ldots, \lambda_{n-1}) \in V^{n-1}$. It follows from Proposition 6.10 that for any $\lambda_n \in V$, $F_i(\lambda_1, \ldots, \lambda_{n-1})\lambda_n^i = b_i\lambda_n^i$, say, exists for each i and that

$$\sum_{i \in \mathbb{N}} b_i\lambda_n^i = F(\lambda_1, \ldots, \lambda_n) = 0.$$

Proposition 6.13 now implies that $b_i = 0$ for each i. Taking $\lambda_n \neq 0$ we deduce that $F_i(\lambda_1, \ldots, \lambda_{n-1}) = 0$ for each i. It follows by the inductive hypothesis that $a_\alpha = 0$ for all $\alpha \in \mathbb{N}^n$.

8.27 Lemma *Let*

$$F(\mathbf{X}, \mathbf{Y}) = \sum a_{\alpha\beta} X^\alpha Y^\beta \in \mathbb{Q}_p[[\mathbf{X}, \mathbf{Y}]],$$

and suppose that $F(\lambda, \mu)$ exists for all $(\lambda, \mu) \in V \times V$, where V is some neighbourhood of 0 in $\mathbb{Q}_p^r$. Let $1 \leq i \leq r$.

(i) *If $F(\lambda, \lambda) = 0$ and $F(\lambda, 0) = \lambda_i$ for all $\lambda = (\lambda_1, \ldots, \lambda_r) \in V$, then*

$$a_{\epsilon_i 0} = 1, \qquad a_{\alpha 0} = 0 \quad \text{for } \alpha \neq \epsilon_i$$
$$a_{0\epsilon_i} = -1, \qquad a_{0\epsilon_j} = 0 \quad \text{for } j \neq i.$$

(ii) *If $F(\lambda, 0) = \lambda_i$ and $F(0, \mu) = \mu_i$ for all $\lambda = (\lambda_1, \ldots, \lambda_r)$, $\mu = (\mu_1, \ldots, \mu_r) \in V$, then*

$$a_{\epsilon_i 0} = a_{0\epsilon_i} = 1$$
$$a_{\alpha 0} = a_{0\beta} = 0 \quad \text{for all } \alpha \neq \epsilon_i, \beta \neq \epsilon_i.$$

Proof (i) Put $G(\mathbf{X}) = -X_i + \sum_{\alpha \in \mathbb{N}^r} a_{\alpha 0}\mathbf{X}^\alpha$. Then $G(\lambda) = F(\lambda, 0) - \lambda_i = 0$ for all $\lambda \in V$, so by Lemma 8.26 we have $a_{\epsilon_i 0} = 1$ and $a_{\alpha 0} = 0$ for all $\alpha \neq \epsilon_i$.

Now put $H(\mathbf{X}) = \sum_{\gamma \in \mathbb{N}^r} (\sum_{\alpha+\beta=\gamma} a_{\alpha\beta})\mathbf{X}^\gamma$. Proposition 6.10 shows that $H(\lambda) = F(\lambda, \lambda) = 0$ for all $\lambda \in V$. Applying Lemma 8.26 again, we deduce that for each $\gamma \in \mathbb{N}^r$

$$\sum_{\alpha+\beta=\gamma} a_{\alpha\beta} = 0.$$

In particular, setting $\gamma = \epsilon_j$ we get

$$a_{\mathbf{0}\epsilon_j} = -a_{\epsilon_j\mathbf{0}} = -\delta_{ij} \quad (\text{Kronecker } \delta).$$

(ii) As in case (i) we get $a_{\epsilon_i\mathbf{0}} = 1$, $a_\alpha = 0$ for all $\alpha \neq \epsilon_i$. The rest follows by symmetry.

8.28 Corollary. *Let G be a standard group over $\mathbb{Q}_p$.*

(i) *If $P_1, \dots, P_r \in \mathbb{Z}_p[[\mathbf{X}, \mathbf{Y}]]$ are power series satisfying condition*
(ii) *of Definition 8.22, then for each j we have*

$$P_j(\mathbf{X}, \mathbf{Y}) = X_j - Y_j + \sum_{(\alpha,\beta)\in I} a_{j,\alpha\beta} \mathbf{X}^\alpha \mathbf{Y}^\beta,$$

where $I = \{(\alpha, \beta) \in \mathbb{N}^r \times \mathbb{N}^r \mid \langle\alpha\rangle + \langle\beta\rangle \geq 2 \text{ and } \beta \neq \mathbf{0}\}$.

(ii) *If $M_1, \dots, M_r \in \mathbb{Z}_p[[\mathbf{X}, \mathbf{Y}]]$ are power series satisfying condition*
(ii)′, *then for each j we have*

$$M_j(\mathbf{X}, \mathbf{Y}) = X_j + Y_j + \sum_{(\alpha,\beta)\in J} b_{j,\alpha\beta} \mathbf{X}^\alpha \mathbf{Y}^\beta,$$

where $J = \{(\alpha, \beta) \in \mathbb{N}^r \times \mathbb{N}^r \mid \langle\alpha\rangle \geq 1 \text{ and } \langle\beta\rangle \geq 1\}$.

We are now ready to prove the first main result of this section:

8.29 Theorem. *Let G be a p-adic analytic group. Then G has an open subgroup H which is a standard group with respect to the manifold structure induced from G.*

Proof The subgroup H is going to be a small neighbourhood of the identity in G, carefully chosen so that the power series expressing the group operation have the desired form. The details are as follows.

Let A be an atlas defining the manifold structure on G and let $(U, \phi, n) \in A$ be a chart such that $1 \in U$. Then $(U \times U, \phi \times \phi, 2n)$ is a chart of the manifold $G \times G$, as detailed in Example 8.9(vi). Since G is an analytic group, the function $f : G \times G \to G$ given by $(x, y) \mapsto xy^{-1}$ is analytic, so putting $W = (U \times U) \cap f^{-1}(U)$ we see that the function

$$\phi \circ f \circ (\phi \times \phi)^{-1}\big|_{(\phi\times\phi)W} : (\phi \times \phi)W \to \mathbb{Z}_p^n$$

is analytic. Thus there exist $h \in \mathbb{N}$ and power series $F_j \in \mathbb{Q}_p[[\mathbf{X}, \mathbf{Y}]]$ $(j = 1, \ldots, n)$ such that

$$\phi_j \circ f \circ (\phi \times \phi)^{-1}(\phi(1) + p^h\lambda, \phi(1) + p^h\mu) = F_j(\lambda, \mu).$$

for all λ and $\mu \in \mathbb{Z}_p^n$. It is clear that

$$F_j(\lambda, \lambda) = \phi_j(1), \quad F_j(\lambda, \mathbf{0}) = \phi(1) + p^h\lambda_j \qquad (6)$$

for all $\lambda \in \mathbb{Z}_p^n$.

Now let $U_0 = \phi^{-1}(\phi(1) + p^h\mathbb{Z}_p^n)$, an open neighbourhood of 1 in U, and define $\psi : U_0 \to \mathbb{Z}_p^n$ by $\psi(x) = p^{-h}(\phi(x) - \phi(1))$. Then for λ and $\mu \in \mathbb{Z}_p^n$ we have

$$\psi_j \circ f \circ (\psi \times \psi)^{-1}(\lambda, \mu) = G_j(\lambda, \mu)$$

where

$$G_j(\mathbf{X}, \mathbf{Y}) = p^{-h}F_j(\mathbf{X}, \mathbf{Y}) - p^{-h}\phi_j(1) \in \mathbb{Q}_p[[\mathbf{X}, \mathbf{Y}]].$$

In view of (6), we may apply Lemma 8.27 and deduce that for each j,

$$G_j(\mathbf{X}, \mathbf{Y}) = X_j - Y_j + \sum_{(\alpha, \beta) \in I} a_{j, \alpha\beta} \mathbf{X}^\alpha \mathbf{Y}^\beta$$

where $I = \{(\alpha, \beta) \in \mathbb{N}^r \times \mathbb{N}^r \mid \langle\alpha\rangle + \langle\beta\rangle \geq 2 \text{ and } \beta \neq \mathbf{0}\}$.

By Lemma 6.18, there exists $k_0 \in \mathbb{N}$ such that, for all $(\alpha, \beta) \in I$ and $j = 1, \ldots, n$,

$$p^{k_0(\langle\alpha\rangle + \langle\beta\rangle)} a_{j, \alpha\beta} \in \mathbb{Z}_p.$$

Putting $k = 2k_0$, we then have

$$b_{j, \alpha\beta} = p^{k(\langle\alpha\rangle + \langle\beta\rangle - 1)} a_{j\alpha\beta} \in \mathbb{Z}_p$$

for all $(\alpha, \beta) \in I$ and $j = 1, \ldots, n$. We put

$$H = \psi^{-1}(p^{k+1}\mathbb{Z}_p^n) \quad \text{if } p > 2$$
$$H = \psi^{-1}(2^{k+2}\mathbb{Z}_2^n) \quad \text{if } p = 2.$$

Define $\theta : H \to p\mathbb{Z}_p^n$ (if $p > 2$) or $\theta : H \to 4\mathbb{Z}_2^n$ (if $p = 2$) by $\theta(x) = p^{-k}\psi(x)$. Then for all $x, y \in H$, if $\theta(x) = (\lambda_1, \ldots, \lambda_n)$ and $\theta(y) = (\mu_1, \ldots, \mu_n)$,

$$\theta_j(xy^{-1}) = \lambda_j - \mu_j + \sum_{(\alpha, \beta) \in I} b_{j, \alpha\beta} \lambda^\alpha \mu^\beta.$$

This shows that H is a standard group with respect to the global atlas

$\{(H, \theta, n)\}$. It is clear that this atlas is compatible with the manifold structure on H induced from G.

One further lemma is required before we can complete the proof of the main result:

8.30 Lemma *Let G be a standard group over $\mathbb{Q}_p$ with the global atlas $\{(G, \psi, r)\}$. Then there exist power series $F_1(\mathbf{X}), \dots, F_r(\mathbf{X}) \in \mathbb{Z}_p[[X_1, \dots, X_r]]$ such that*

$$\psi(x^p) = \mathbf{F}(\psi(x)) \quad \text{for all } x \in G$$

$$F_k(\mathbf{X}) = pX_k + \sum_{\langle \alpha \rangle > 1} c_{k,\alpha} \mathbf{X}^\alpha \quad \text{for each } k,$$

where $c_{k,\alpha} \in \mathbb{Z}_p$ for each α and k. Also each $c_{k,\alpha} \equiv 0 \,(\mathrm{mod}\, p)$ when $\langle \alpha \rangle = 2$, provided $p \neq 2$.

Proof Except for the final claim, this follows from Corollary 8.28(ii) and Lemma 8.24, by an obvious induction on p (where, just for the moment, we allow p to range over $\mathbb{N}$). For the final claim, we have to keep track of the quadratic terms: keeping the notation of Corollary 8.28(ii), we find that

$$c_{k, 2\epsilon i} = \frac{1}{2} p(p-1) a_{k, \epsilon_i \epsilon_i}$$

$$c_{k, \epsilon_i + \epsilon_j} = \frac{1}{2} p(p-1)(a_{k, \epsilon_i \epsilon_j} + a_{k, \epsilon_j \epsilon_i}) \quad \text{if } i \neq j,$$

for each k, giving the result.

8.31 Theorem *Let G be a standard group of dimension r over $\mathbb{Q}_p$. Then G is a uniform pro-p group of dimension r.*

Proof Let $\{(G, \psi, r)\}$ be the global atlas described in condition (i) of Definition 8.22. For $i \geq 1$ (if $p > 2$) or $i \geq 2$ (if $p = 2$) we put

$$G(i) = \psi^{-1}(p^i \mathbb{Z}_p^r).$$

Then $\{G(i) \mid i \geq 2\}$ is an open neighbourhood base of the identity in G. Corollary 8.28(i) shows that if $x \in G(i)$ and $y \in G(j)$ then, for $k = 1, \dots, r$,

$$\psi_k(xy^{-1}) \equiv \psi_k(x) - \psi_k(y) \,(\mathrm{mod}\, p^c) \tag{7}$$

where $c = j + \min(i, j)$. Taking $i = j$ we deduce that

(a) $G(i)$ is a subgroup of G and

(b) ψ induces an epimorphism $G(i) \rightarrow p^i \mathbb{Z}_p^r / p^{i+1} \mathbb{Z}_p^r$ with kernel $G(i+1)$.

It follows that $G(i+1) \lhd G(i)$ and $G(i)/G(i+1) \cong p^i \mathbb{Z}_p^r / p^{i+1} \mathbb{Z}_p^r \cong (\mathbb{Z}/p\mathbb{Z})^r$. Thus each $G(i)$ is a subnormal subgroup of p-power index in G. Since the intersection of the conjugates of a subnormal subgroup of p-power index is itself a normal subgroup of p-power index, it follows that G has a neighbourhood base at the identity consisting of normal subgroups of p-power index. Hence G is a pro-p group.

By Lemma 8.30, if $x \in G(i)$ then

$$\psi_j(x^p) \equiv p\psi_j(x) \,(\mathrm{mod}\,p^{i+2})$$

for each j. Since $\lambda \mapsto p\lambda$ induces an isomorphism from $p^i \mathbb{Z}_p^r / p^{i+1} \mathbb{Z}_p^r$ onto $p^{i+1} \mathbb{Z}_p^r / p^{i+2} \mathbb{Z}_p^r$, it follows that $x \mapsto x^p$ induces an isomorphism from $G(i)/G(i+1)$ onto $G(i+1)/G(i+2)$. It then follows by induction on n that $G(i+1) = G(i)^p G(i+n)$ for all $n \geq 1$, and hence that

$$G(i+1) = \overline{G(i)^p}. \tag{8}$$

Arguing by induction on i we deduce that $G(i) \lhd G$ for each i.

If $p > 2$ then

$$G/\overline{G^p} = G(1)/G(2) \cong (\mathbb{Z}/p\mathbb{Z})^r.$$

Thus $G/\overline{G^p}$ is abelian and hence G is powerful, with $d(G) = r$.

If $p = 2$, to show that G is powerful we are required to prove that $G/\overline{G^4}$ is abelian. Recall that now $G = G(2)$. Let $y \in G(4)$ and $n \geq 4$. Then there exist $u \in G(3)$ and $x \in G(2)$ such that $y \equiv u^2 \bmod G(n)$ and $u \equiv x^2 \bmod G(n)$. So $y \equiv x^4 \bmod G(n)$. Hence

$$G(4) \subseteq \bigcap_{n \geq 4} G^4 G(n) = \overline{G^4}.$$

But it follows from (7) that $G(2)/G(4)$ is isomorphic to $2^2 \mathbb{Z}_2^r / 2^4 \mathbb{Z}_2^r$. Hence $G/\overline{G^4}$ is abelian, again showing that G is powerful with $d(G) = r$.

Finally to prove that G is uniform it suffices to show that $P_i(G) = G(i+\varepsilon)$ for each i, where $\varepsilon = 0$ if $p \neq 2$, $\varepsilon = 1$ if $p = 2$. This follows from (8), by Theorem 3.6.

Combining Theorems 8.18, 8.29 and 8.31 we obtain

8.32 Theorem *Let G be a topological group. Then G has the structure of a p-adic analytic group if and only if G contains an open subgroup which is a uniform pro-p group.*

With the results of Chapter 4 this gives Theorem 8.1, and the following variations:

8.33 Corollary *A topological group G is p-adic analytic if and only if G has an open subgroup which is a pro-p group of finite rank.*

8.34 Corollary *The following are equivalent for a topological group G:*
   (i) *G is a compact p-adic analytic group;*
   (ii) *G contains an open normal uniform pro-p subgroup of finite index;*
   (iii) *G is a profinite group containing an open subgroup which is a pro-p group of finite rank.*

Combining this with Theorem 5.7 we deduce

8.35 Corollary *Let G be a compact p-adic analytic group. Then $\mathrm{Aut}(G)$ is a compact p-adic analytic group.*

We conclude this chapter with the following theorem which allows us to define the concept of *dimension* for a p-adic analytic group.

8.36 Theorem *Let G be a p-adic analytic group. Then there exists a unique non-negative integer n with the following properties:*

- *every chart belonging to an atlas defining the manifold structure on G has dimension n, in the sense of Definition 8.6;*
- *every open pro-p subgroup of G has finite rank and dimension n, in the sense of Definition 4.7.*

Proof Let A be an atlas giving the analytic manifold structure of G, and let $(U, \phi, n) \in A$ be a chart with $1 \in U$. In the proof of Theorem 8.29 we constructed from this chart a chart (H, θ, n) on G such that $H = H_U$, say, is a standard group of dimension n. Theorem 8.31 shows that then H_U is a uniform pro-p group of dimension n. Now let (V, ψ, m) be any other chart in A. We are required to prove that $m = n$. Let $g \in V$, put $W = g^{-1}V$ and define $\psi_g : W \to \mathbb{Z}_p^m$ by $\psi_g(x) = \psi(gx)$. Using the fact that the function $x \mapsto gx : G \to G$ is analytic (see Exercise 1), we see that (W, ψ_g, m) is a chart compatible with every chart in A, so $A \cup \{(W, \psi_g, m)\}$ is an atlas of G. Now $1 \in W$; as above we obtain an open uniform subgroup H_W in G with $\dim(H_W) = m$. Since $H_U \cap H_W$ is an open subgroup of both H_U and H_W, Lemma 4.6 shows that $m = n$.

Now let M be any open pro-p subgroup of G. We have seen that G contains a uniform open subgroup H of dimension n. Then $M \cap H$ is an open subgroup of H. So $M \cap H$ is a pro-p group of finite rank with $\dim(M \cap H) = n$. Since $M \cap H$ is also an open subgroup of M, it follows that M has finite rank and $\dim(M) = n$.

8.37 Definition Let G be a p-adic analytic group G. Then the *dimension*

$$\dim(G)$$

of G is the number n specified in Theorem 8.36.

Notes

The results and methods of this chapter are all essentially from [L]. Alternative accounts of much of this material may be found in Bourbaki (1989b) Chapter III and Serre (1965), Part II.

The main result, Theorem 8.32, combines Prop. 3.1.3 of [L], Chapter III, which implies that a p-adic analytic group of dimension r has a uniform open subgroup of dimension r, and **3.4.4.1** and **3.4.4.2** of [L], Chapter III, which state that every finitely generated powerful pro-p group is p-adic analytic.

Exercises

1. Let X and Y be p-adic analytic manifolds.

(i) Prove that the projection map $X \times Y \to Y$ is an analytic function.

(ii) Let $v \in X$. Prove that the map $y \mapsto (v, y) : Y \to X \times Y$ is an analytic function. Deduce that if G is a p-adic analytic group then for $g \in G$, the map $x \mapsto gx : G \to G$ is analytic.

2. Prove that condition (i) in Definition 8.14 implies condition (ii).

3. Let G be a uniform pro-p group of dimension d. Let $x_1, \ldots, x_d$ be a $\mathbb{Z}_p$-basis for the Lie algebra $\Lambda = \log G \subseteq \widehat{A}$, where $\widehat{A}$ is the completed group algebra of G (assume that $G \subseteq 1 + \widehat{A}_0$, as in §7.2).

(i) Show that $\{a_1 = \exp x_1, \ldots, a_d = \exp x_d\}$ is a set of topological generators for G.

Define $\phi : G \to \mathbb{Z}_p^d$ by $\phi(a_1^{\lambda_1} \ldots a_d^{\lambda_d}) = (\lambda_1, \ldots, \lambda_d)$ and define $\psi : G \to \mathbb{Z}_p^d$ by $\psi(x) = (\mu_1, \ldots, \mu_d)$ where $\log x = \mu_1 x_1 + \cdots + \mu_d x_d$.

(ii) Define $\Phi^{(d)}(X_1, \ldots, X_d) = (\mathcal{L} \circ H)(\mathbf{X}) \in \mathbb{Q}_p \langle\langle \mathbf{X} \rangle\rangle$, where $H(X_1, \ldots, X_d) = \mathcal{E}(X_1) \ldots \mathcal{E}(X_d) - 1$, $\mathcal{E}$ and $\mathcal{L}$ being the exponential and logarithmic series of Definition 6.21. Prove that, if $y_1, \ldots, y_d \in \widehat{A}_0$ then $\Phi^{(d)}(y_1, \ldots, y_d)$ converges to the sum $\log(\exp(y_1) \ldots \exp(y_d))$.

(iii) Show that for $\lambda = (\lambda_1, \ldots, \lambda_d) \in \mathbb{Z}_p^d$,

$$\Phi^{(d)}(\lambda_1 x_1, \ldots, \lambda_d x_d) = \mu_1 x_1 + \cdots + \mu_d x_d$$

where $\psi\phi^{-1}(\lambda) = (\mu_1, \ldots, \mu_d)$. Deduce that $\psi \circ \phi^{-1}$ is an analytic function on $\mathbb{Z}_p^d$.

(iv) Let $b_i = a_i - 1$ for $i = 1, \ldots, d$. Show that there exist $y_\beta \in \widehat{A}$ such that

$$(n!)^{-1}(\mu_1 x_1 + \cdots + \mu_d x_d)^n = \sum_{\langle \beta \rangle = n} \mu_1^{\beta_1} \ldots \mu_d^{\beta_d} y_\beta$$

for all $\mu_1, \ldots, \mu_d \in \mathbb{Z}_p$ and $n \in \mathbb{N}$, and that there exist $d_{\alpha\beta} \in \mathbb{Q}_p$ such that

$$y_\beta = \sum_{\alpha \in \mathbb{N}^d} d_{\alpha\beta} b_1^{\alpha_1} \ldots b_d^{\alpha_d}$$

for all $\alpha, \beta \in \mathbb{N}^d$.

[*Hint:* use Exercise 7.3(iv).]

Deduce that if $\psi(x) = (\mu_1, \ldots, \mu_d)$, then

$$x = \sum_{\alpha \in \mathbb{N}^d} \left(\sum_{\beta \in \mathbb{N}^d} d_{\alpha\beta} \mu_1^{\beta_1} \cdots \mu_d^{\beta_d} \right) b_1^{\alpha_1} \cdots b_d^{\alpha_d}.$$

Hence show that if $\phi(x) = (\lambda_1, \ldots, \lambda_d)$ then for each $\alpha \in \mathbb{N}^d$,

$$\binom{\lambda_1}{\alpha_1} \cdots \binom{\lambda_d}{\alpha_d} = \sum_{\beta \in \mathbb{N}^d} d_{\alpha\beta} \mu_1^{\beta_1} \cdots \mu_d^{\beta_d}.$$

[*Hint*: Use Lemma 8.19 and Theorem 7.4.]

Taking $\alpha = \epsilon_i$ for $i = 1, \ldots, d$, deduce that $\phi \circ \psi^{-1}$ is an analytic function on $\mathbb{Z}_p^d$.

(v) By (iii) and (iv), (G, ϕ, d) and (G, ψ, d) are compatible atlases. Suppose that $\{z_1, \ldots, z_d\}$ is another $\mathbb{Z}_p$-basis for Λ and define $\theta : G \to \mathbb{Z}_p$ by $\theta(x) = (\nu_1, \ldots, \nu_d)$ where $\log x = \nu_1 z_1 + \cdots + \nu_d z_d$. Prove that (G, ψ, d) and (G, θ, d) are compatible atlases.

4. (i) Show that if G is a p-adic analytic group then $\dim(G) = 0$ if and only if G is discrete.

(ii) Let $1 \to N \xrightarrow{f} G \to Q \to 1$ be an exact sequence of continuous homomorphisms of p-adic analytic groups. Show that

$$\dim(G) = \dim(N) + \dim(Q)$$

provided either (a) N is compact or (b) $f : N \to f(N)$ is an open mapping. Show by example that the proviso is necessary.

[*Hint*: use Theorem 4.8.]

5. Find the dimensions of the p-adic analytic groups in Examples 8.17.

6. Let $\epsilon = 1$ if $p \neq 2$, $\epsilon = 2$ if $p = 2$. Suppose that the function $f : \mathbb{Z}_p \to \mathbb{Z}_p$ satisfies

$$f(x) = \sum_{j=1}^{\infty} c_j \binom{x}{j}$$

for all $x \in \mathbb{Z}_p$, where $c_j \in \mathbb{Z}_p$ satisfies $|c_j| \leq p^{-\epsilon j}$ for all $j \geq 2$ and $|c_1| \leq p^{-2\epsilon}$. Show that there exists a power series

$$F(X) = \sum_{n=1}^{\infty} d_n X^n$$

such that $f(x) = F(x)$ for all $x \in \mathbb{Z}_p$, whose coefficients satisfy

$$d_1 \equiv 0 \pmod{p^{1+\epsilon}}$$
$$d_n \equiv 0 \pmod{p^{\epsilon}} \quad \text{for all } n.$$

[*Hint*: compare the proof of Lemma 8.20 (but d_1 needs to be done 'by hand').]

7. Let G be a compact subgroup of $\mathrm{GL}_n(\mathbb{Q}_p)$. Show that $G \subseteq \mathrm{GL}_n(\mathbb{Z}_p) \cdot X$ for some finite subset X of G. Deduce that there exists $y \in \mathrm{GL}_n(\mathbb{Q}_p)$ such that $yGy^{-1} \leq \mathrm{GL}_n(\mathbb{Z}_p)$. Hence prove that *every compact subgroup of* $\mathrm{GL}_n(\mathbb{Q}_p)$ *is contained in a maximal one, and every maximal compact subgroup of* $\mathrm{GL}_n(\mathbb{Q}_p)$ *is conjugate to* $\mathrm{GL}_n(\mathbb{Z}_p)$.

[*Hint*: Let X be a transversal to the right cosets of $G \cap \mathrm{GL}_n(\mathbb{Z}_p)$ in G. Show that the $\mathbb{Z}_p$-module $\sum_{x \in X} (\mathbb{Z}_p^d) \cdot x$ is equal to $(\mathbb{Z}_p^d) \cdot y$ for some $y \in \mathrm{GL}_n(\mathbb{Q}_p)$.]

Platonov and Rapinchuk (1994), §§3.3, 3.4 contains a discussion of compact subgroups in $\mathfrak{H}(\mathbb{Q}_p)$, where $\mathfrak{H}$ is an algebraic group defined over $\mathbb{Q}_p$; if $\mathfrak{H}$ is reductive, then the maximal compact subgroups of $\mathfrak{H}(\mathbb{Q}_p)$ lie in finitely many conjugacy classes.

Interlude C

Finitely generated groups, p-adic analytic groups and Poincaré series

Here we reproduce an announcement by one of the authors, concerning the detailed behaviour of the 'subgroup counting functions' whose growth properties have been discussed in Chapter 3. An adequate explanation of the results from p-adic model theory on which this depends is beyond the scope of this book; however, it may be of interest to the reader to see how information about the structure of uniform pro-p groups, referred to below as *p-saturable groups*, is exploited in a rather different context (the relation between uniform groups and p-saturable groups is discussed in the *Notes* to Chapter 4).

Unlike most of the other applications, this paper uses the full strength of Lazard's characterisation of compact p-adic analytic groups: on the one hand, the existence of a uniform pro-p structure in such a group; on the other hand, the fact that in a uniform pro-p group, the group operations are given by analytic functions. Thus the whole machinery developed in Part II of the book is relevant here.

For a full account see du Sautoy (1993).

BULLETIN (New Series) OF THE
AMERICAN MATHEMATICAL SOCIETY
Volume 23, Number 1, July 1990

FINITELY GENERATED GROUPS, p-ADIC ANALYTIC GROUPS, AND POINCARÉ SERIES

MARCUS P. F. DU SAUTOY

INTRODUCTION

Igusa [I 1, I 2] was the first to exploit p-adic integration with respect to the Haar measure on $\mathbf{Q}_p$ in the study of Poincaré series arising in number theory and developed a method using Hironaka's resolution of singularities to evaluate a limited class of such integrals. Denef [D 1, D 2] and, more recently, Denef and van den Dries [DvdD] have applied results from logic, profiting from the flexibility of the concept of definable, greatly to enlarge the class of integrals amenable to Igusa's method. In [DvdD] these results are employed to answer questions posed by Serre [S] and Oesterlé [O] concerning the rationality of various Poincaré series associated with the p-adic points of a closed analytic subset of $\mathbf{Z}_p^m$. In this note we apply these techniques to prove that various Poincaré series associated with finitely generated groups and p-adic analytic groups are rational in p^{-s}, extending results of [GSS].

RESULTS

Let G be a group and denote by $a_n(G)$ the number of subgroups of index n in G. We are interested in groups for which $a_n(G)$ is finite for every $n \in \mathbf{N}$. For each prime p, we can then associate the following Poincaré series with this arithmetical function:

$$(1) \qquad \zeta_{G,p}(s) = \sum_{n=0}^{\infty} a_{p^n}(G)p^{-ns} = \sum_{H \in \mathbf{X}_p} |G : H|^{-s}$$

where $\mathbf{X}_p = \{H \le G : H \text{ has finite } p\text{-power index in } G\}$.

Received by the editors June 22, 1989 and, in revised form, November 18, 1989.

1980 *Mathematics Subject Classification* (1985 *Revision*). Primary 22E20; Secondary 03C10, 11M41, 20E18, 20F99.

§1. *p*-ADIC ANALYTIC GROUPS

We consider firstly the case where G is a compact p-adic analytic group—that is, a compact topological group with the underlying structure of a p-adic analytic manifold with respect to which the group operations are analytic (see [Lz] and [DduSMS]). For such groups, $a_n(G)$ is finite for every n. We wish to announce the following:

Theorem 1. *If G is a compact p-adic analytic group, then $\zeta_{G,p}(s)$ is rational in p^{-s}.*

The philosophy behind the proof is to express our Poincaré series as a p-adic integral

$$(2) \qquad \int_M |f(\mathbf{x})|^s |g(\mathbf{x})| \, |d\mathbf{x}|,$$

where $|d\mathbf{x}|$ is the normalized Haar measure on $\mathbf{Z}_p^n$ and the functions f, $g: \mathbf{Z}_p^n \to \mathbf{Z}_p$ and the subset M are definable in the language describing the analytic theory of the p-adic numbers. We can then evaluate such definable integrals applying the techniques developed by Denef and van den Dries [DvdD] (which include quantifier elimination results for the analytic theory of $\mathbf{Z}_p$) to prove our theorem.

The translation from our Poincaré series to such a definable p-adic integral makes full use of Lazard's results on the close relationship between the structure of compact p-adic analytic groups and filtrations defined on such groups [Lz]. In answer to "Hilbert's 5th problem for p-adic analytic groups," Lazard has shown that a compact topological group has the structure of a p-adic analytic group if and only if there exists a normal subgroup G_1 of finite index in G which is *p-saturable*—that is, there exists a filtration on G_1

$$G_1 > G_2 > \cdots > G_i > \cdots$$

such that: (i) G_1 is a pro-p group with a fundamental system of neighborhoods of the identity given by $\{G_i : i \in \mathbf{N}\}$; (ii) for all $i \geq 1$, G_i/G_{i+1} is an elementary Abelian p-group of finite rank; and (iii) for all $i \geq 1$, the map $P_i : G_i/G_{i+1} \to G_{i+1}/G_{i+2}$ defined by $xG_{i+1} \to x^p G_{i+2}$ is an isomorphism of $\mathbf{F}_p$ vector spaces.

A p-saturable group has the underlying structure of a pro-p, p-adic analytic group with a global coordinate system $\mathbf{Z}_p^r$ given by p-adic powers of elements $x_1, \ldots, x_r$ where $x_1 G_2, \ldots, x_r G_2$ is an $\mathbf{F}_p$ vector space basis for G_1/G_2.

We first prove Theorem 1 in the case where G is a p-saturable group. Recall that subgroups of finite index in a pro-p group have p-power index. The idea is to associate with every subgroup H of finite index in G, a subset $M(H)$ of $r \times r$ matrices over $\mathbf{Z}_p$ whose rows form coordinates for a *good basis* for H. Every subgroup of finite index in a compact p-adic analytic group is open. So, there exists m such that $H \geq G_m$. We define the concept of a *good basis* for H as a set of elements $h_1, \ldots, h_r$ such that, for each $i = 1, \ldots, m$, if we let $v_j = h_j^{p^{e(i,j)}} \in H \cap G_i$ for $\omega(h_j) \leq i$, where $e(i, j) = i - \omega(h_j)$ and $\omega(g) = n$ if $g \in G_n \backslash G_{n+1}$, then

$$\{v_j G_{i+1} : j \text{ such that } \omega(h_j) \leq i\}$$

is an $\mathbf{F}_p$ vector space basis for $(H \cap G_i)G_{i+1}/G_{i+1}$. The index of H in G is encoded in the measure of the subset $M(H)$ and we identify functions $f, g : \mathbf{Z}_p^n \to \mathbf{Z}_p$ such that, for all subgroups H of finite index in G

$$|G : H|^{-s} = \int_{M(H)} |f(\mathbf{x})|^s |g(\mathbf{x})| \, |d\mathbf{x}|.$$

Summing over all subgroups of finite (necessarily p-power) index in G, we can express our Poincaré series $\zeta_{G,p}(s)$ as a p-adic integral of the form (2) where $M \subseteq \mathbf{Z}_p^{r^2}$ is the (disjoint) union of subsets $M(H)$ for all $H \in \mathbf{X}_p$.

The problem now is to show that this integral is definable in the sense of [DvdD]. The set of r-tuples of elements of G which form a good basis for some subgroup of finite index in G is definable by a filtered group theoretic statement. We show how to translate such statements into statements about coordinates of elements of G definable now in the language describing the analytic theory of the p-adic numbers. Using this translation we can show that the subset M is definable. With regards to the functions f and g, we show that there exists a finite partition of $\mathbf{Z}_p^{r^2}$ into definable subsets such that, on each subset, f and g are defined by polynomial functions. Thus f and g are definable functions. We are then in a position to apply the techniques of [DvdD] to this definable integral and thus prove Theorem 1 in the case where G is a p-saturable group.

We extend this to a proof of Theorem 1 using the following ideas. Let G be a compact p-adic analytic group and G_1 a normal p-saturable subgroup of finite index in G. If H is subgroup of

p-power index in G, then it is determined by $H_1 = H \cap G_1$ and a transversal for H_1 in H. We associate with each H a subset $N(H)$ consisting of coordinates both for a good basis for H_1 and coordinates for a transversal for H_1 in H. Extending our integral associated with the p-saturable group G_1, we can express $\zeta_{G,p}(s)$ as a definable p-adic integral over the union of the subsets $N(H)$. We can therefore apply [DvdD] to prove that our Poincaré series associated with the compact p-adic analytic group G is rational in p^{-s}.

§2. FINITELY GENERATED GROUPS

Theorem 1 has various corollaries for finitely generated groups. If Γ is a finitely generated group, then $a_n(\Gamma)$ is finite for all n. We can therefore consider the Poincaré series defined in (1). We consider first a variant of this Poincaré series. Define $a_n^{\lhd\lhd}(\Gamma)$ to be the number of subnormal subgroups of index n in Γ. For each prime p, we associate with the finitely generated group Γ, the following Poincaré series:

$$\zeta_{\Gamma,p}^{\lhd\lhd}(s) = \sum_{n=0}^{\infty} a_{p^n}^{\lhd\lhd}(\Gamma) p^{-ns} = \sum_{H \in Y_p} |\Gamma : H|^{-s},$$

where $Y_p = \{H \leq \Gamma : H \text{ is subnormal of } p\text{-power index in } \Gamma\}$.

We say that $a_{p^n}(\Gamma)$ grows polynomially if there exists $c \in \mathbf{N}$ such that $a_{p^n}(\Gamma) \leq p^{nc}$ for all n. Similarly for $a_{p^n}^{\lhd\lhd}(\Gamma)$. We then have the following:

Theorem 2. *Let* Γ *be a finitely generated group and* p *a prime. If* $a_{p^n}^{\lhd\lhd}(\Gamma)$ *grows polynomially, then* $\zeta_{\Gamma,p}^{\lhd\lhd}(s)$ *is rational in* p^{-s}.

There is a one-to-one correspondence between subnormal subgroups of finite p-power index in Γ and subgroups of finite index in the pro-p completion G of Γ. So $\zeta_{G,p}(s) = \zeta_{\Gamma,p}^{\lhd\lhd}(s)$. By Lubotzky and Mann's characterization of pro-p groups which have the underlying structure of a p-adic analytic group [LM], if $a_{p^n}(G) = a_{p^n}^{\lhd\lhd}(\Gamma)$ grows polynomially, then G is a p-adic analytic pro-p group. By Theorem 1, $\zeta_{\Gamma,p}^{\lhd\lhd}(s)$ is rational in p^{-s}.

We recall the definition of an *upper p-chief factor* of Γ—that is, a chief factor of some finite quotient of Γ whose order is divisible by p. We then have the following:

Theorem 3. *Let* Γ *be a finitely generated group and* p *a prime such that the order of all* p*-chief factors of* Γ *is bounded. If* $a_{p^n}(\Gamma)$ *grows polynomially, then* $\zeta_{\Gamma,p}(s)$ *is rational in* p^{-s}.

The bound on the order of p-chief factors in Γ implies that there exists a normal subgroup Γ_0 of finite index in Γ whose subgroups of p-power index are all subnormal. We construct a finite extension G of the pro-p completion of Γ_0 whose subgroups of finite index are in one-to-one correspondence with subgroups of p-power index in Γ. If $a_{p^n}(\Gamma)$ grows polynomially, then, by [LM], G is a finite extension of a p-adic analytic pro-p group. So G is a compact p-adic analytic group and by Theorem 1, $\zeta_{G,p}(s) = \zeta_{\Gamma,p}(s)$ is rational in p^{-s}.

Theorem 3 includes a large class of examples, some of which we collect together in the following corollary. We recall that the *upper p-rank* of Γ is the supremum of $r(P)$ as P ranges over all p-subgroups of finite quotients of Γ, where $r(P)$ is the rank of P.

Corollary 4. *If Γ is a finitely generated group of finite upper p-rank, then $\zeta_{\Gamma,p}(s)$ is rational in p^{-s}.*

This follows from a remark in [MS].

REFERENCES

[D 1]   J. Denef, *The rationality of the Poincaré series associated to the p-adic points on a variety*, Invent. Math. **77** (1984), 1–23.

[D 2]   _____, *On the evaluation of certain p-adic integrals*, Seminaire de Théorie des Nombres, Paris 1983-4, Progr. Math. **59** (1985), 25–47.

[DvdD] J. Denef and L. van den Dries, *P-adic and real subanalytic sets*, Ann. of Math. **128** (1988), 79–138.

[DduSMS] J. Dixon, M. P. F. du Sautoy, A. Mann and D. Segal, *Analytic pro-p groups*, London Math. Society Lecture Notes (to appear).

[GSS]   F. J. Grunewald, D. Segal, and G. C. Smith, *Subgroups of finite index in nilpotent groups*, Invent. Math. **93** (1988), 185–223.

[I 1]   J.-I. Igusa, *Complex powers and asymptotic expansions I*, J. Reine Angew. Math. **268/269** (1974), 110–130.

[I 2]   _____, *Some observations on higher degree characters*, Amer. J. Math. **99** (1977), 393–417.

[Lz]   M. Lazard, *Groupes analytiques p-adiques*, Publ. Math. Inst. Hautes Ètudes Sci. **26** (1965), 389–603.

[LM]   A. Lubotzky and A. Mann, *On groups of polynomial subgroup growth*, Invent. Math. (to appear).

[MS]   A. Mann and D. Segal, *Uniform finiteness conditions in residually finite groups*, Proc. London Math. Soc. (to appear).

[O]   J. Oesterlé, *Réduction modulo p^n des sous-ensembles analytiques fermés de $\mathbf{Z}_p^N$* , Invent. Math. **66** (1982), 325–341.

[S]   J.-P. Serre, *Quelques applications du théorème de densité de Chebotarev,* Publ. Math. Inst. Hautes Ètudes Sci. **54** (1981), 123–202.

MATHEMATICAL INSTITUTE, 24-29 ST. GILES, OXFORD, ENGLAND

Current address: School of Mathematical Sciences, Queen Mary and Westfield College, University of London, London E1 4N5, England

9

Lie theory

This concluding chapter of Part II is devoted to some of the more categorical aspects of p-adic analytic groups. In the first part of the chapter we establish the fundamental result that *every continuous homomorphism of analytic groups is analytic*; this implies that the analytic structure of a p-adic analytic group is uniquely determined by its structure as a topological group. Using this together with the group-theoretic characterisation given in the previous chapter, it is then easy to show that the category of p-adic analytic groups is closed with respect to taking closed subgroups, quotients and group extensions.

The second part of the chapter is devoted to the correspondence between analytic groups and Lie algebras. We show that there is an exact correspondence between uniform pro-p groups and powerful Lie algebras over $\mathbb{Z}_p$, and deduce that this correspondence gives rise to an equivalence between the 'local category' of p-adic analytic groups and the category of p-adic Lie algebras.

9.1 Powers

Recall the definition of standard group from §8.4. If G is a standard group and m is a positive integer, then the function $g \mapsto g^m$ is analytic on G. What we want to establish in this section is that for each fixed $g \in G$, the function $m \mapsto g^m$ is analytic (as a function from $\mathbb{Z}_p$ to G). More precisely, we shall prove

9.1 Proposition *Let G be a standard group, with respect to the global chart (G, ψ, r), and let $v \in G$. Then there exist power series*

$$K_i(X) = \sum_{j=1}^{\infty} c_{ij} X^j \in \mathbb{Z}_p[[X]],$$

213

for $i = 1, \ldots, r$, such that for all $\lambda \in \mathbb{Z}_p$

$$\psi(v^\lambda) = (K_1(\lambda), \ldots, K_r(\lambda)).$$

Moreover, putting $\epsilon = 1$ if p is odd and $\epsilon = 2$ if $p = 2$, we have, for each i,

$$c_{i1} \equiv \psi_i(v) \pmod{p^{1+\epsilon}}$$
$$c_{ij} \equiv 0 \pmod{p^\epsilon} \quad \text{for all } j \geq 2.$$

This depends on the following result about formal power series (we shall write $\mathbf{X}$ for $(X_1, \ldots, X_r)$ and $\mathbf{X}^\alpha = X_1^{\alpha_1} \ldots X_r^{\alpha_r}$):

9.2 Lemma *For $i = 1, \ldots, r$ let*

$$F_i(\mathbf{X}, \mathbf{Y}) \in \mathbb{Z}_p[[X_1, \ldots, X_r, Y_1, \ldots, Y_r]]$$

be a power series with zero constant term, and linear term $X_i + Y_i$. Define the power series G_i^m, for $i = 1, \ldots, r$ and $m \geq 1$, recursively, as follows: $G_i^1 = X_i$; for $m \geq 1$, $G_i^{m+1} = F_i \circ \mathbf{H}^m$ where $\mathbf{H}^m(\mathbf{X}) = (X_1, \ldots, X_r, G_1^m, \ldots, G_r^m)$. Then there exist $\xi_{ij\alpha} \in \mathbb{Z}_p$ such that, for $i = 1, \ldots, r$ and all $m \geq 1$,

$$G_i^m = \sum_{\langle \alpha \rangle \geq 1} \left(\sum_{j=1}^{\langle \alpha \rangle} \xi_{ij\alpha} \binom{m}{j} \right) X_1^{\alpha_1} \ldots X_r^{\alpha_r}.$$

Proof Note that each G_i^m is a well-defined power series, with zero constant term: this follows recursively from the hypothesis on the F_i. Thus we have

$$G_i^m = \sum_{\langle \alpha \rangle \geq 1} c_{i\alpha}(m) \mathbf{X}^\alpha$$

with each $c_{i\alpha}(m) \in \mathbb{Z}_p$, and what we have to show is that there exist $\xi_{ij\alpha} \in \mathbb{Z}_p$ such that

$$c_{i\alpha}(m) = \sum_{j=1}^{\langle \alpha \rangle} \xi_{ij\alpha} \binom{m}{j} \tag{1}$$

for $i = 1, \ldots, r$ and all $m \geq 1$. This we do by induction on $\langle \alpha \rangle$.

If $\langle \alpha \rangle = 1$ then $\alpha = \epsilon_k$ for some k. It is easy to see (by induction on m) that for all m,

$$c_{k\epsilon_k}(m) = m, \quad c_{j\epsilon_k}(m) = 0 \quad \text{if } j \neq k;$$

so (1) is satisfied for $\alpha = \epsilon_k$ if we put

$$\xi_{i1\epsilon_k} = \delta_{ik} \quad \text{(Kronecker } \delta\text{)}.$$

Now suppose $\langle \alpha \rangle > 1$, and assume that for all β with $\langle \beta \rangle < \langle \alpha \rangle$ we have found $\xi_{ij\beta} \in \mathbb{Z}_p$ ($1 \le i \le r$, $1 \le j \le \beta$) such that (1) holds with β in place of α, for all $m \ge 1$. Say

$$F_i(\mathbf{X}, \mathbf{Y}) = X_i + Y_i + \sum_{\substack{\langle \sigma \rangle \ge 1 \\ \langle \pi \rangle \ge 1}} a_{i\sigma\pi} \mathbf{X}^\sigma \cdot \mathbf{Y}^\pi,$$

and suppose that

$$X_1^{\sigma_1} \dots X_r^{\sigma_r} \cdot (G_1^m)^{\pi_1} \dots (G_r^m)^{\pi_r} = \sum b_{\sigma\pi\rho}(m) X_1^{\rho_1} \dots X_r^{\rho_r}.$$

Then by definition of G_i^{m+1} we have

$$c_{i\alpha}(m+1) = c_{i\alpha}(m) + \sum_{\substack{\langle \sigma \rangle \ge 1 \\ \langle \pi \rangle \ge 1}} a_{i\sigma\pi} b_{\sigma\pi\alpha}(m).$$

Now if $\langle \sigma \rangle + \langle \pi \rangle > \langle \alpha \rangle$ then $b_{\sigma\pi\alpha}(m) = 0$; otherwise, $b_{\sigma\pi\alpha}(m)$ is the sum of a finite number of terms of the form

$$c_{i_1\gamma(1)}(m) \dots c_{i_d\gamma(d)}(m)$$

where $d = \langle \pi \rangle$ and $\sigma + \gamma(1) + \dots + \gamma(d) = \alpha$. Since $\langle \sigma \rangle \ge 1$, each $c_{i_j\gamma(j)}(m)$ is given by an expression like (1); it follows that $b_{\sigma\pi\alpha}(m)$ is equal to a polynomial in m, of degree at most $\langle \alpha \rangle - \langle \sigma \rangle$, with coefficients in $\mathbb{Q}_p$ and zero constant term. Hence for each i there exists a polynomial $f_i(T) \in \mathbb{Q}_p[T]$, of degree at most $\langle \alpha \rangle - 1$, such that $f_i(0) = 0$ and $c_{i\alpha}(m+1) - c_{i\alpha}(m) = f_i(m)$ for all positive integers m. This implies that for each i there exists a polynomial $g_i(T) \in \mathbb{Q}_p[T]$, of degree at most $\langle \alpha \rangle$ and with $g_i(0) = 0$, such that $c_{i\alpha}(m) = g_i(m)$ for all positive integers m (see Exercise 1, noting that $c_{i\alpha}(1) = 0$ since $\langle \alpha \rangle > 1$).

Since $g_i(0) = 0$ and $c_{i\alpha}(m) \in \mathbb{Z}_p$ for each m, we can write

$$g_i(T) = \sum_{j=1}^{\langle \alpha \rangle} \xi_{ij\alpha} \binom{T}{j}$$

with each $\xi_{ij\alpha} \in \mathbb{Z}_p$ (see Exercise 2).

This establishes the inductive step, and completes the proof.

Proof of Proposition 9.1 We fix the element v in the standard group

G, with global chart (G, ψ, r); here $\psi(G) = p\mathbb{Z}_p$ if p is odd, $\psi(G) = 4\mathbb{Z}_2^r$ if $p = 2$. There exist power series $F_1, \dots, F_r$ over $\mathbb{Z}_p$ such that

$$\psi(xy) = (F_1(\psi(x), \psi(y)), \dots, F_r(\psi(x), \psi(y)))$$

for all $x, y \in G$, and Corollary 8.28(ii) shows that these power series satisfy the hypotheses of Lemma 9.2. Then for each positive integer m, we have

$$\psi(v^m) = (G_1^m(\psi(v)), \dots, G_r^m(\psi(v))), \qquad (2)$$

in the notation of Lemma 9.2; this follows by repeated applications of Lemma 8.24. Let $\xi_{ij\alpha} \in \mathbb{Z}_p$ be the numbers given in Lemma 9.2, and write

$$t_\alpha = \psi_1(v)^{\alpha_1} \dots \psi_r(v)^{\alpha_r}.$$

Then $|t_\alpha| \leq p^{-\langle \alpha \rangle}$ if p is odd, $|t_\alpha| \leq 2^{-2\langle \alpha \rangle}$ if $p = 2$, so for each $\lambda \in \mathbb{Z}_p$ we have

$$\lim_{\substack{(\alpha, j) \in \mathbb{N}^{r+1} \\ j \leq \langle \alpha \rangle}} \xi_{ij\alpha} \binom{\lambda}{j} t_\alpha = 0.$$

We can therefore define $f_i : \mathbb{Z}_p \to \mathbb{Q}_p$ by

$$f_i(\lambda) = \lambda \psi_i(v) + \sum_{\langle \alpha \rangle \geq 2} \sum_{j=1}^{\langle \alpha \rangle} \xi_{ij\alpha} \binom{\lambda}{j} t_\alpha$$

(see Proposition 6.9), and we see from (2) that for all positive integers m,

$$\psi(v^m) = (f_1(m), \dots, f_r(m)).$$

By Corollary 6.11, the series defining $f_i(\lambda)$ may be re-arranged, giving

$$f_i(\lambda) = \lambda \psi_i(v) + \sum_{j=1}^{\infty} \left(\sum_{\substack{\langle \alpha \rangle \geq j \\ \langle \alpha \rangle \geq 2}} \xi_{ij\alpha} t_\alpha \right) \binom{\lambda}{j}.$$

Proposition 6.9(iv) shows that for each j,

$$\left| \sum_{\substack{\langle \alpha \rangle \geq j \\ \langle \alpha \rangle \geq 2}} \xi_{ij\alpha} t_\alpha \right| \leq \min \left\{ p^{-j\epsilon}, p^{-2\epsilon} \right\},$$

where $\epsilon = 1$ if $p \neq 2$, $\epsilon = 2$ if $p = 2$. We may therefore apply Exercise 8.6, to infer that there exists $K_i(X) = \sum d_{ij} X^j \in \mathbb{Z}_p[[X]]$ with $f_i(\lambda) =$

$K_i(\lambda)$ for all $\lambda \in \mathbb{Z}_p$, such that $d_{i0} = 0$, $d_{i1} \equiv \psi_i(v)$ $(\mathrm{mod}\, p^{1+\epsilon})$ and $d_{ij} \equiv 0$ $(\mathrm{mod}\, p^\epsilon)$ for all $j \geq 2$.

It remains to show that $\psi_i(v^\lambda) = K_i(\lambda)$ for all $\lambda \in \mathbb{Z}_p$ and $1 \leq i \leq r$. The estimates given above for the coefficients c_{ij} show that the function f_i is strictly analytic on $\mathbb{Z}_p$, and therefore continuous, by Proposition 6.19. Since ψ_i is also continuous, we get

$$\psi_i(v^\lambda) = \lim_{k \to \infty} \psi_i(v^{m(k)}) = \lim_{k \to \infty} f_i(m(k)) = f_i(\lambda) = K_i(\lambda)$$

when $\lambda \in \mathbb{Z}_p$ is the limit of a sequence $(m(k))$ in $\mathbb{N}$. This completes the proof.

9.2 Analytic structures

It follows easily from Proposition 9.1 and Lemma 8.5 that the mapping $(\lambda_1, \ldots, \lambda_d) \mapsto u_1^{\lambda_1} \ldots u_d^{\lambda_d}$, from $\mathbb{Z}_p^d$ into a standard group G, is analytic $(u_1, \ldots, u_d$ being given elements of G). If $\{u_1, \ldots, u_d\}$ is a minimal topological generating set for G, the inverse of this mapping gives a global chart for G, as we saw already in Chapter 4: our main task in this section is to show that this chart is compatible with the given analytic structure on G as a standard group:

9.3 Lemma *Let G be a standard group, with respect to the global chart (G, ψ, d). Let $G_2 = P_2(G)$, let $\{u_1, \ldots, u_d\}$ be a set of topological generators for G, and define $\phi : G \to \mathbb{Z}_p^d$ by*

$$\phi(u_1^{\lambda_1} \ldots u_d^{\lambda_d}) = \lambda \quad \text{for each} \quad \lambda = (\lambda_1, \ldots, \lambda_d) \in \mathbb{Z}_p^d.$$

Then the two charts $(G_2, \psi|_{G_2}, d)$ and $(G_2, \phi|_{G_2}, d)$ are compatible.

Proof Note that ϕ is a well defined homeomorphism, by Theorem 8.31 and Theorem 4.9, and that $\phi(G_2) = p\mathbb{Z}$, $\psi(G_2) = p^{1+\epsilon}\mathbb{Z}$, where $\epsilon = 1$ if p is odd, $\epsilon = 2$ if $p = 2$ (see the proof of Theorem 8.31). We have to show that $\psi \circ \phi^{-1} : p\mathbb{Z} \to p^{1+\epsilon}\mathbb{Z}$ and $\phi \circ \psi^{-1} : p^{1+\epsilon}\mathbb{Z} \to p\mathbb{Z}$ are analytic functions. To this end, we shall construct power series $U_1, \ldots, U_d$ and $V_1, \ldots, V_d$ such that

$$\psi(\phi^{-1}(\lambda)) = \mathbf{U}(\lambda) \quad \text{for all} \quad \lambda \in p\mathbb{Z}$$
$$\phi(\psi^{-1}(\lambda)) = \mathbf{V}(\lambda) \quad \text{for all} \quad \lambda \in p^{1+\epsilon}\mathbb{Z}.$$

We start with the construction of $\mathbf{U}$. By Corollary 8.25, there exist $F_1, \ldots, F_d \in \mathbb{Z}_p[[\mathbf{X}(1), \ldots, \mathbf{X}(d)]]$ such that

$$\psi(w_1 \ldots w_d) = \mathbf{F}(\psi(w_1), \ldots, \psi(w_d))$$

for all $w_1, \ldots, w_d \in G$; it follows easily from Corollary 8.28(ii) that

$$\mathbf{F}^{[0]} = 0, \quad \mathbf{F}^{[1]}(\mathbf{X}(1), \ldots, \mathbf{X}(d)) = \mathbf{X}(1) + \cdots + \mathbf{X}(d).$$

Now Proposition 9.1 gives us power series $K_i^{(j)} \in \mathbb{Z}_p[[X]]$ such that

$$\psi(u_j^\lambda) = \mathbf{K}^{(j)}(\lambda) \quad \text{for all } \lambda \in \mathbb{Z}_p,$$

for $j = 1, \ldots, d$. For each j, the linear part of $\mathbf{K}^{(j)}$ is

$$(\psi(u_j) + \nu_j)\mathbf{X}$$

where $\nu_j \in p^{1+\epsilon}\mathbb{Z}$, and all coefficients of each $K_i^{(j)}$ lie in $p^\epsilon\mathbb{Z}_p$. We now define, for $i = 1, \ldots, d$,

$$U_i(X_1, \ldots, X_d) = F_i \circ (\mathbf{K}^{(1)}(X_1), \ldots, \mathbf{K}^{(d)}(X_d)).$$

Then for $\lambda \in p\mathbb{Z}_p^d$ we have

$$\begin{aligned}\mathbf{U}(\lambda) &= \mathbf{F}(\psi(u_1^{\lambda_1}), \ldots, \psi(u_d^{\lambda_d})) \\ &= \psi(u_1^{\lambda_1} \ldots u_d^{\lambda_d}) \\ &= \psi(\phi^{-1}(\lambda)),\end{aligned}$$

by Lemma 8.24.

To find $\mathbf{V}$ we use the 'Inverse Function Theorem', Theorem 6.37. To justify this step, we have to examine the linear part of $\mathbf{U}$: from the above, this is

$$\sum_{j=1}^d (\psi(u_j) + \nu_j)X_j = \left(\sum b_{1j}X_j, \ldots, \sum b_{dj}X_j \right),$$

say. Now it follows from the proof of Theorem 8.31 that the cosets of $\psi(u_1), \ldots, \psi(u_d)$ form a basis for the $\mathbb{F}_p$-vector space $p^\epsilon\mathbb{Z}/p^{1+\epsilon}\mathbb{Z}$. Hence the matrix $(p^{-\epsilon}b_{ij})$ lies in $\mathrm{GL}_d(\mathbb{Z}_p)$. Also the coefficients of $\mathbf{U}$ all lie in $p^\epsilon\mathbb{Z}_p$. Thus the conditions of Theorem 6.37 are satisfied, and we obtain power series $V_1, \ldots, V_d \in \mathbb{Q}_p[[\mathbf{X}]]$ such that

$$\mathbf{V}(\lambda) \in p\mathbb{Z}_p^d, \quad \mathbf{U}(\mathbf{V}(\lambda)) = \lambda \quad \text{for all } \lambda \in p^{1+\epsilon}\mathbb{Z}_p^d.$$

Then for $\lambda \in p^{1+\epsilon}\mathbb{Z}_p^d$ we have

$$\psi(\phi^{-1}(\mathbf{V}(\lambda))) = \mathbf{U}(\mathbf{V}(\lambda)) = \lambda,$$

giving

$$\mathbf{V}(\lambda) = \phi(\psi^{-1}(\lambda))$$

as required.

The main result of this section now follows easily:

9.4 Theorem *Let G_1 and G_2 be p-adic analytic groups. Then every continuous homomorphism $G_1 \to G_2$ is analytic.*

Proof Let $f : G_1 \to G_2$ be a continuous homomorphism. By Theorem 8.29, G_2 has a standard open subgroup B, and $f^{-1}(B)$ has a standard open subgroup A, which is then open in G_1. Put $H = P_2(A)$ and $K = P_2(B)$.

We shall show that $f|_H$ is analytic: the theorem will then follow by Lemma 8.16.

Choose topological generating sets $\{a_1, \dots, a_r\}$ and $\{b_1, \dots, b_d\}$ for A and B respectively, where $r = \dim(A)$ and $d = \dim(B)$. Lemma 9.3 shows that we have charts $(H, \phi|_H, r)$ of H and $(K, \phi'|_K, d)$ of K given by

$$\phi(a_1^{\lambda_1} \dots a_r^{\lambda_r}) = (\lambda_1, \dots, \lambda_r)$$
$$\phi'(b_1^{\mu_1} \dots b_d^{\mu_d}) = (\mu_1, \dots, \mu_d)$$

for all $\lambda \in p\mathbb{Z}_p^r$ and $\mu \in p\mathbb{Z}_p^d$. Now if $v_i = f(a_i)$ for each i, we have $v_i \in B$ and

$$f(a_1^{\lambda_1} \dots a_r^{\lambda_r}) = v_1^{\lambda_1} \dots v_r^{\lambda_r}$$

for each λ, since f is a continuous homomorphism. Proposition 8.21 shows that

$$v_1^{\lambda_1} \dots v_r^{\lambda_r} = b_1^{g_1(\lambda)} \dots b_d^{g_d(\lambda)}$$

where $\mathbf{g} = (g_1, \dots, g_d) : \mathbb{Z}_p^r \to \mathbb{Z}_p^d$ is analytic. Since

$$\phi' \circ f \circ \phi^{-1}(\lambda) = \mathbf{g}(\lambda)$$

for all $\lambda \in p\mathbb{Z}_p^r$, this shows that $f|_H$ is analytic, as claimed (note that for $\lambda \in p\mathbb{Z}_p^r$, $f(\phi^{-1}(\lambda)) \in f(H) \le K$).

9.5 Corollary *Let G be a topological group. Then G has at most one structure as a p-adic analytic group; and unless G is discrete, the prime p is uniquely determined.*

Proof Let G_1 and G_2 be p-adic analytic groups (for the same prime p) whose underlying topological group is G. Theorem 9.4 shows that the identity map $G_1 \to G_2$ and its inverse are both analytic. Hence the atlases defining G_1 and G_2 are compatible, so G_1 and G_2 are the same analytic group.

Now suppose that G is both p-adic analytic and q-adic analytic where p and q are distinct primes. By Theorem 8.1, G has an open subgroup H which is a pro-p group; and H, which is again q-adic analytic, has an open subgroup K which is a pro-q group. Then K is both pro-p and pro-q, so every open subgroup of K has index in K which is simultaneously a power of p and a power of q. Hence the only open subgroup of K is K itself. Since the open subgroups of K intersect in $\{1\}$, it follows that $K = 1$. But K is open in G, so G is discrete.

9.3 Subgroups, quotients, extensions

The results of this section show that there is a plentiful supply of interesting p-adic analytic groups, to supplement the meagre list of Examples 8.17. In particular, it follows from Theorem 9.6(i) that every linear algebraic group $\mathcal{G}$ defined over $\mathbb{Q}_p$ gives rise to the p-adic analytic group $\mathcal{G}(\mathbb{Q}_p)$: this is the zero-set in $\mathrm{GL}_n(\mathbb{Q}_p)$ of some family of polynomial equations, and is therefore a closed subgroup. (In Exercise 13.11 we exhibit an open standard subgroup of $\mathcal{G}(\mathbb{Q}_p)$ when $\mathcal{G}$ is a Chevalley group.)

9.6 Theorem *Let G be a p-adic analytic group, H a closed subgroup of G, and N a closed normal subgroup of G. Then*

(i) *H is p-adic analytic, and the inclusion map $H \to G$ is an analytic homomorphism;*

(ii) *G/N, with the quotient topology, is p-adic analytic, and the natural projection $G \to G/N$ is an analytic homomorphism.*

Proof G has an open pro-p subgroup K of finite rank, by Corollary 8.33 Then $K \cap H$ is a closed subgroup of K, so $K \cap H$ is a pro-p group of finite rank. But $K \cap H$ is an open subgroup of H, so H is p-adic analytic by Corollary 8.33 (and Corollary 4.3). The second claim of (i) follows from Theorem 9.4. Part (ii) is proved similarly, using the fact that KN/N is an open subgroup of G/N.

9.7 Theorem *Let G be a Hausdorff topological group, and N a closed normal subgroup. If both N and G/N are p-adic analytic (with the induced and quotient topologies respectively), then G is p-adic analytic.*

Proof This depends on some elementary facts about topological groups; these are proved in Appendix B. Now every p-adic analytic group is locally compact and totally disconnected, since it is locally homeomorphic to $\mathbb{Z}_p^d$, which clearly has both properties. Thus N and G/N are locally compact and totally disconnected, and it follows that G has these properties: see Appendix B, B8.

Corollary 8.33 shows that N has an open subgroup M which is a pro-p group of finite rank. Then $M = N \cap U$ for some open set U in G; since G is locally compact and totally disconnected, G has an open compact subgroup H_1 with $H_1 \subseteq U$: see **B7**. Then $N \cap H_1 \leq_c M$, so $N \cap H_1$ is a pro-p group of finite rank.

Again by Corollary 8.33, G/N has an open subgroup H_2/N which is a pro-p group of finite rank. Put $H = H_1 \cap H_2$. We claim that H is pro-p of finite rank: a final appeal to Corollary 8.33 then completes the proof.

Now H is compact and totally disconnected, hence profinite (Exercise 1.2, or Appendix B, B5). Also $H/(H \cap N) \cong HN/N \leq_c H_2/N$, and $H \cap N = N \cap H_1$, so both $H/(H \cap N)$ and $H \cap N$ are pro-p groups of finite rank. Hence if $K \leq_o H$ then $|H : K| = |H : K(H \cap N)| \cdot |H \cap N : H \cap N \cap K|$ is a power of p, showing that H is a pro-p group; and H has finite rank, by Exercise 3.1.

9.4 Powerful Lie algebras

This section may be read independently of Chapter 8: here we complete the story begun in Sections 4.5 and 7.2, by showing how the correspondence which assigns a Lie algebra to every uniform pro-p group can be reversed.

We fix

$$\epsilon = 1 \quad \text{if } p \text{ is odd}, \quad \epsilon = 2 \quad \text{if } p = 2.$$

A Lie algebra L over $\mathbb{Z}_p$ will be called *powerful* if $L \cong \mathbb{Z}_p^d$ for some finite d and

$$(L, L) \subseteq p^\epsilon L.$$

Now recall the *Campbell–Hausdorff formula*, discussed in §6.5:

$$\Phi(X, Y) = \sum_{n=1}^{\infty} u_n(X, Y)$$

$$u_1(X, Y) = X + Y, \quad u_2(X, Y) = \frac{1}{2}(X, Y)$$

$$u_n(X, Y) = \sum_{\mathbf{e}} q_{\mathbf{e}}(X, Y)_{\mathbf{e}} \quad (n \geq 3) \tag{3}$$

where $(X,Y)_{\mathbf{e}} = (X,Y,\dots,Y,X,\dots,X,\dots)$ denotes a left-normed repeated Lie bracket of total length $\langle \mathbf{e} \rangle + 1$, and the summation in (3) is over all vectors $\mathbf{e}$ of positive integers satisfying $\langle \mathbf{e} \rangle = n - 1$. The coefficients $q_{\mathbf{e}}$ are rational numbers satisfying

$$p^{\epsilon\langle\mathbf{e}\rangle} q_{\mathbf{e}} \in p^{\epsilon}\mathbb{Z}_p, \quad \left|p^{\epsilon\langle\mathbf{e}\rangle} q_{\mathbf{e}}\right| \to 0 \ \text{ as } \ \langle\mathbf{e}\rangle \to \infty. \qquad (4)$$

Since each $u_n(X,Y)$ is a finite sum, it may be evaluated in any Lie algebra over $\mathbb{Q}_p$; in Corollary 6.38 we showed that if L is a *powerful* $\mathbb{Z}_p$-Lie algebra, then in fact $u_n(X,Y)$ can be evaluated in L, and that for $x,y \in L$ the series

$$\widetilde{\Phi}(x,y) = \sum_{n=1}^{\infty} u_n(x,y)$$

converges in L. We may therefore define a binary operation $* : L \times L \to L$ by setting

$$x * y = \widetilde{\Phi}(x,y).$$

9.8 Theorem *Let L be a powerful Lie algebra. Then the operation $*$ makes L into a uniform pro-p group. If $\{a_1,\dots,a_d\}$ is a basis for L over $\mathbb{Z}_p$ then $\{a_1,\dots,a_d\}$ is a topological generating set for the group $(L,*)$, which has dimension d.*

Proof It is clear from the definition that

$$x * 0 = 0 * x = x$$
$$x * (-x) = 0$$

for all $x \in L$. So $(L,*)$ will be a group provided that the operation $*$ is *associative*; we postpone the proof of this fact to the next lemma. Taking it as given, we proceed as follows; multiplicative notation applied to the elements of L will refer to the operation $*$.

If $(x,y) = 0$ then $u_n(x,y) = 0$ for each $n \geq 2$, so $x * y = x + y$. It follows that

$$x^m = mx$$

for all $m \in \mathbb{N}$, and since $x^{-1} = -x$ this holds for all $m \in \mathbb{Z}$. In particular, for each $t \in \mathbb{N}$ we have

$$\{x^{p^t} \mid x \in L\} = p^t L,$$

which is a subgroup of $(L, *)$ since $p^t L$ is again a powerful Lie algebra. It follows that

$$L^{p^t} = \overline{L^{p^t}} = p^t L$$

for all t.

Next, we show that the multiplicative cosets of $p^t L$ in L are the same as the additive cosets. Suppose that $x - y \in p^t L$. Then $x - y = p^t z$ for some z; if $\langle e \rangle \geq 1$ we have

$$\begin{aligned}
(x, -y)_e &= (x, x - y, -y, \dots, x, \dots) \\
&= p^t(x, z, -y, \dots, x, \dots) \\
&\in p^{t + \epsilon \langle e \rangle} L
\end{aligned}$$

since L is powerful. Together with (3) and (4) this shows that $u_n(x, -y) \in p^t L$ for each $n \geq 2$, which in turn implies that

$$\begin{aligned}
xy^{-1} &= x * (-y) \\
&= x - y + \sum_{n \geq 2} u_n(x, -y) \in p^t L.
\end{aligned}$$

A similar argument shows that if $xy^{-1} = v \in p^t L$ then

$$x - y = v * y - y \in p^t L.$$

It follows that for each $t \geq 1$ the index of the subgroup L^{p^t} in $(L, *)$ is equal to $|L : p^t L| = p^{td}$. Now the additive cosets $x + p^t L$ $(x \in L, t \in \mathbb{N})$ form a base for open sets in the p-adic topology on L; as these are the same as the multiplicative cosets $x L^{p^t}$, this implies that $(L, *)$ is a topological group, and indeed a pro-p group.

To show that this pro-p group is *powerful*, we refer to parts (iii) and (iv) of Corollary 6.38. The former asserts that

$$x * y - (x + y) \in pL \tag{5}$$

for all $x, y \in L$. As we have seen that multiplicative and additive congruences modulo $pL = L^p$ are equivalent, this shows at once that the multiplicative group L/L^p is abelian, so if p is odd we are done. If $p = 2$, we argue as follows. According to Corollary 6.38(iv),

$$x * y - (x + y) - \frac{1}{2}(x, y) \in 4L$$

for $x, y \in L$. This gives

$$x * y - y * x \equiv \frac{1}{2}(x,y) - \frac{1}{2}(y,x) \,(\mathrm{mod}\,4L)$$
$$= (x,y) \in 4L$$

because L is powerful. Thus in this case L/L^4 is abelian, as required.

That $(L, *)$ is uniform, of dimension d, now follows from the fact that $|L : L^{p^t}| = p^{td}$ for all t. Finally, (5) shows that the identity mapping on $L/pL = L/L^p$ is an isomorphism between the additive and multiplicative structures, and hence that $\{a_1, \dots, a_d\}$ generates L modulo L^p; as L^p is the Frattini subgroup of $(L, *)$ it follows that $\{a_1, \dots, a_d\}$ is a topological generating set for the group $(L, *)$. This completes the proof, modulo the following lemma:

9.9 Lemma *The operation $*$ on a powerful Lie algebra is associative.*

Proof Let M be a positive integer. From (4), there exists an integer N_0 such that

$$p^{\epsilon\langle\mathbf{e}\rangle} q_{\mathbf{e}} \in p^M \mathbb{Z}_p \tag{6}$$

whenever $\langle\mathbf{e}\rangle \geq N_0$.

Put $N = (N_0 + 1)^2$ and let

$$\Phi_N(X,Y) = \sum_{n=1}^{N} u_n(X,Y);$$

this is a polynomial in the non-commuting variables X and Y. It follows from Proposition 6.29 (the 'formal associativity' of the power series Φ) that the polynomial

$$R_N(X,Y,Z) = \Phi_N(\Phi_N(X,Y),Z) - \Phi_N(X,\Phi_N(Y,Z))$$

in three non-commuting variables has no terms of degree $\leq N$. Using (3), we can expand $R_N(X,Y,Z)$ as a linear combination of 'Lie monomials' (that is, compound Lie brackets in X, Y and Z). Since each Lie monomial of length n is a homogeneous polynomial of degree n, it follows that the terms in our expansion involving Lie monomials of degree at most N all cancel out.

Now fix $n > N$. The homogeneous component of degree n in the expansion of $R_N(X,Y,Z)$ is a sum of terms

$$q_{\mathbf{e}} q_{\mathbf{f}_1} \dots q_{\mathbf{f}_k}((X,Y)_{\mathbf{f}_1}, Z, \dots, Z, (X,Y)_{\mathbf{f}_2}, \dots, (X,Y)_{\mathbf{f}_{1+e_2}}, Z, \dots) \tag{7}$$

or

$$-q_{\mathbf{e}}q_{\mathbf{f}_1}\cdots q_{\mathbf{f}_k}(X,(Y,Z)_{\mathbf{f}_1},\ldots,(Y,Z)_{\mathbf{f}_{e_1}},X,\ldots,X,(Y,Z)_{\mathbf{f}_{1+e_1}},\ldots),\quad (8)$$

where

$$\langle\mathbf{e}\rangle+\langle\mathbf{f}_1\rangle+\cdots+\langle\mathbf{f}_k\rangle=n-1,\quad k\le\max\{\langle\mathbf{e}\rangle,1\},$$

and just for now we allow $\mathbf{e}$, $\mathbf{f}_i$ to be (0) and interpret $q_{(0)}(X,Y)_{(0)}$ as either X or Y. Let $\mathbf{f}_i=\mathbf{g}$ be such that $\langle\mathbf{f}_i\rangle=\max\{\langle\mathbf{f}_1\rangle,\ldots,\langle\mathbf{f}_k\rangle\}$. Then either $\mathbf{e}=(0)$ and $\langle\mathbf{g}\rangle=n-1\ge N$ or $\langle\mathbf{e}\rangle\ge1$ and $\langle\mathbf{e}\rangle(1+\langle\mathbf{g}\rangle)\ge n-1\ge N$. It follows that either $\langle\mathbf{e}\rangle\ge\sqrt{N}>N_0$ or $\langle\mathbf{g}\rangle\ge\sqrt{N}-1\ge N_0$; with (6) and (4) this implies that

$$p^{\epsilon(n-1)}q_{\mathbf{e}}q_{\mathbf{f}_1}\cdots q_{\mathbf{f}_k}\in p^M\mathbb{Z}_p. \quad (9)$$

Now let L be a powerful Lie algebra, and let $x,y,z\in L$. Each compound Lie bracket of length n in x,y and z lies in $p^{\epsilon(n-1)}L$. Since $R_N(x,y,z)$ is a sum of terms like (7) and (8), with x,y,z replacing X,Y,Z and $n>N$, it follows from (9) that

$$R_N(x,y,z)\in p^M L.$$

On the other hand,

$$(x*y)*z-\Phi_N(\Phi_N(x,y),z)$$

is a sum of terms like (7), with x,y,z replacing X,Y,Z and $n>N$; the same argument therefore shows that

$$(x*y)*z-\Phi_N(\Phi_N(x,y),z)\in p^M L;$$

similarly we see that

$$x*(y*z)-\Phi_N(x,\Phi_N(y,z))\in p^M L.$$

Therefore

$$(x*y)*z-x*(y*z)\equiv R_N(x,y,z)$$
$$\equiv 0\pmod{p^M L}.$$

As M was an arbitrary positive integer this shows that $(x*y)*z=x*(y*z)$, and so concludes the proof. (Some variants of this proof are indicated in Exercise 4.)

Now let G be a uniform pro-p group. According to Theorem 4.30, or Corollary 7.14 and Exercise 7.9, the group G becomes a Lie algebra over $\mathbb{Z}_p$ when endowed with the operations $+_G,(\,,\,)_G$. We shall denote this Lie algebra by L_G. Since $[G, G]$ consists of p^ϵth powers in G, it follows from the definition in §4.5 that the Lie algebra L_G is powerful (see Exercise 4.2(ii)).

9.10 Theorem *The assignments*

$$G \mapsto L_G, \quad L \mapsto (L, *)$$

are mutually inverse isomorphisms between the category of uniform pro-p groups and the category of powerful Lie algebras over $\mathbb{Z}_p$.

Since an isomorphism of categories is supposed to be a functor, we have also to specify what happens to the morphisms: each morphism is sent to itself, as a map of the underlying sets.

Proof Suppose that $f : G \to H$ is a homomorphism of uniform pro-p groups. Since the Lie algebra operations in L_G and L_H are defined in terms of the group operations by taking limits, and because f is continuous (Corollary 1.22), it follows that $f : L_G \to L_H$ is a Lie algebra homomorphism. Thus we have a functor that sends each uniform pro-p group G to L_G and each morphism to itself, considered as a map of the underlying sets.

Now suppose that $f : L \to M$ is a homomorphism of powerful Lie algebras. Then f is continuous (think of L and M as additive pro-p groups). For $x, y \in L$ and each $n \geq 1$ we have

$$f(u_n(x, y)) = u_n(f(x), f(y)),$$

since $p^{2n} u_n(X, Y)$ is a $\mathbb{Z}_p$-linear combination of Lie monomials in X and Y. It follows by continuity that

$$f(x * y) = f(x) * f(y);$$

thus f is a group homomorphism from $(L, *)$ to $(M, *)$. Thus we have a functor that sends each powerful Lie algebra L to $(L, *)$ and each morphism to itself, considered as a map of the underlying sets.

To complete the proof we have to establish

(a) if L is a powerful Lie algebra then $L_{(L,*)} = L$;

(b) if G is a uniform pro-p group then $(L_G, *) = G$.

Now each of our functors preserves not only the underlying set but also the topology. We have seen in §4.3 and in the proof of Theorem 9.8, above, that for each positive integer m, the operations

$$x \mapsto x^m, \quad x \mapsto mx$$

correspond to each other. It follows in case (b) that the operation $x \mapsto$

x^m is the same in $(L_G, *)$ as it is in G; and in case (a) it follows by continuity that the operation of $\mathbb{Z}_p$ on $L_{(L,*)}$ is the same as it is on L.

To prove (a), therefore, it suffices to establish the following for all $a, b \in L$:

$$\lim_{n\to\infty} p^{-n}((p^n a) * (p^n b)) = a + b, \qquad (10)$$

$$\lim_{n\to\infty} p^{-2n}((-((p^n b) * (p^n a))) * (p^n a) * (p^n b)) = (a, b). \qquad (11)$$

Now

$$p^{-n}((p^n a) * (p^n b)) = p^{-n} \left(p^n a + p^n b + \sum_{j \geq 2} p^{jn} u_j(a, b) \right)$$

$$= a + b + p^n c_n$$

where $c_n = \sum_{j \geq 2} p^{(j-1)n} u_j(a, b) \in L$. This implies (10). Similarly, we have

$$- ((p^n b) * (p^n a)) = -p^n b - p^n a - \frac{p^{2n}}{2}(b, a) - p^{2n} r_n$$

$$= -(p^n b + p^n a) + \frac{p^{2n}}{2}(a, b) - p^{2n} r_n,$$

$$(p^n a) * (p^n b) = p^n a + p^n b + \frac{p^{2n}}{2}(a, b) + p^{2n} s_n,$$

where $r_n, s_n \in L$, giving

$$(-((p^n b) * (p^n a))) * ((p^n a) * (p^n b)) = p^{2n}(a, b) + p^{3n} t_n$$

for some $t_n \in L$. This implies (11).

To prove (b) we must show that $x * y = xy$ for all $x, y \in G$, where $x * y$ is defined using the Lie algebra operations in L_G. Now recall Corollary 7.14, which establishes a Lie algebra isomorphism

$$\log : L_G \to \Lambda = \log(G) \subseteq \widehat{A}_0,$$

where $\widehat{A}$ is the completed group algebra of G. Put $u = \log x, v = \log y$. Since $\log$ is continuous, we have

$$\log(x * y) = \log \widetilde{\Phi}(x, y)$$
$$= \widetilde{\Phi}(u, v) = \Phi(u, v)$$
$$= \log(\exp u \cdot \exp v) \quad \text{by Proposition 6.27}$$
$$= \log(xy).$$

Thus $x * y = xy$ as required.

Combining Propositions 4.31 and 7.15, we can add the

Scholium to Theorem 9.10 *Let S be a closed subset of the uniform pro-p group G. Then*

- S *is an isolated subgroup of G if and only if S is an isolated Lie subalgebra of L_G;*
- S *is an isolated normal subgroup of G if and only if S is an isolated ideal of L_G.*

Here, to say that S is *isolated* means that if $x \in G$ and $x^n \in S$ (or equivalently, if $x \in L_G$ and $nx \in S$) for some non-zero integer n, then $x \in S$. This implies that S is powerful, as a group or Lie algebra respectively.

Under certain conditions, the 'Scholium' can be extended to subgroups and Lie subalgebras in a uniform group that are not necessarily isolated: see Ilani (1995).

9.5 Analytic groups and their Lie algebras

Let us begin by defining the Lie algebra of a p-adic analytic group G. According to Theorem 8.32, G has an open subgroup which is a uniform pro-p group. If H_1 and H_2 are both open uniform subgroups of G, then $H = H_1 \cap H_2$ has finite index in both H_1 and H_2, so L_H has finite index in L_{H_i} for $i = 1, 2$. Hence

$$\mathbb{Q}_p \otimes_{\mathbb{Z}_p} L_{H_1} = \mathbb{Q}_p \otimes_{\mathbb{Z}_p} L_H = \mathbb{Q}_p \otimes_{\mathbb{Z}_p} L_{H_2}.$$

We may therefore unambiguously define

$$\mathcal{L}(G) = \mathbb{Q}_p \otimes_{\mathbb{Z}_p} L_H$$

where H is any uniform open subgroup of G. Thus $\mathcal{L}(G)$ is a Lie algebra over $\mathbb{Q}_p$, of dimension equal to $\dim(H) = \dim(G)$.

Now suppose that $f : G_1 \to G_2$ is a morphism of analytic groups. Choose a uniform open subgroup H_2 in G_2; as f is continuous, the subgroup $f^{-1}(H_2)$ is open in G_1, hence contains a uniform open subgroup H_1. The group homomorphism $f_0 = f|_{H_1} : H_1 \to H_2$ is at the same time a Lie algebra homomorphism from L_{H_1} to L_{H_2}, as observed in Theorem 9.10; it therefore induces a Lie algebra homomorphism

$$f^* = 1 \otimes f_0 : \mathcal{L}(G_1) \to \mathcal{L}(G_2);$$

it is clear that f^* does not depend on the choices of H_2 and H_1. It is also clear that if $f : G_1 \to G_2$ and $g : G_2 \to G_3$ are morphisms, then

$$(g \circ f)^* = g^* \circ f^*,$$

and that $(\mathrm{Id}_G)^* = \mathrm{Id}_{\mathcal{L}(G)}$. Thus we have the first part of

9.11 Theorem (i) *The assignment* $G \mapsto \mathcal{L}(G)$, $f \mapsto f^*$ *is a functor from the category of p-adic analytic groups (of dimension d) to the category of Lie algebras over* $\mathbb{Q}_p$ *(of dimension d).*

(ii) *Let* $f_1, f_2 : A \to B$ *be morphisms of p-adic analytic groups. Then* $f_1^* = f_2^* : \mathcal{L}(A) \to \mathcal{L}(B)$ *if and only if* $f_1|_U = f_2|_U$ *for some open subgroup* U *of* A.

(iii) *Let* G *be a p-adic analytic group and identify* $\mathcal{L}(G)$ *with* $\mathbb{Q}_p^d$ *by choosing a basis. Then* G *has a uniform open subgroup* H *such that the composition*

$$\phi : H \xrightarrow{\mathrm{Id}} L_H \xrightarrow{1 \otimes -} \mathcal{L}(G) = \mathbb{Q}_p^d$$

gives a chart (H, ϕ, d) *of* G.

To prove (ii), choose a uniform open subgroup H in A. Then $f_1^* = f_2^*$ if and only if $f_1|_H = f_2|_H$, so the 'only if' statement follows on putting $U = H$. On the other hand, if $f_1|_U = f_2|_U$ where U is open in A, we may choose H to be contained in U and then have $f_1|_H = f_2|_H$, whereupon $f_1^* = f_2^*$.

Part (iii) shows that $\mathcal{L}(G)$ may be identified with the 'tangent space at 1' of an analytic group G, as in the classical theory of Lie groups. Before proving it, we must take another look at the material of the preceding section, this time through analytic spectacles. Recall that the function $\widetilde{\Phi}$ was defined at the beginning of the previous section:

9.12 Lemma *Let* L *be a powerful Lie algebra. Then the function* $\widetilde{\Phi} : L \times L \to L$ *is analytic.*

Proof Choose a $\mathbb{Z}_p$-basis $\{a_1, \dots, a_d\}$ for L. Write

$$\lambda \cdot \mathbf{a} = \lambda_1 a_1 + \cdots + \lambda_d a_d$$

for $\lambda = (\lambda_1, \dots, \lambda_d) \in \mathbb{Z}_p^d$, and write

$$\widetilde{\Phi}(x, y) = \widetilde{\Phi_1}(x, y) a_1 + \cdots + \widetilde{\Phi_d}(x, y) a_d,$$

with each $\widetilde{\Phi}_1(x,y) \in \mathbb{Z}_p$. We shall show that for each i the function

$$(\lambda,\mu) \mapsto \widetilde{\Phi}_i(\lambda \cdot \mathbf{a}, \mu \cdot \mathbf{a})$$

is strictly analytic on $\mathbb{Z}_p^d \times \mathbb{Z}_p^d$.

For each $\mathbf{e}$, there exist homogeneous polynomials $P_{\mathbf{e}}^1, \ldots, P_{\mathbf{e}}^d$ in $2d$ variables over $\mathbb{Z}_p$, of degree $1 + \langle \mathbf{e} \rangle$, such that

$$(\lambda \cdot \mathbf{a}, \mu \cdot \mathbf{a})_{\mathbf{e}} = P_{\mathbf{e}}^1(\lambda,\mu)a_1 + \cdots + P_{\mathbf{e}}^d(\lambda,\mu)a_d.$$

Since $(L,L) \subseteq p^{\epsilon}L$, each term $(a_i,a_j)_{\mathbf{e}}$ lies in $p^{\epsilon \langle \mathbf{e} \rangle}L$; it follows that the coefficients of each $P_{\mathbf{e}}^k$ lie in $p^{(n-1)\epsilon}\mathbb{Z}_p$. Thus

$$u_n(\lambda \cdot \mathbf{a}, \mu \cdot \mathbf{a}) = \sum_{i=1}^d \sum_{\langle \mathbf{e} \rangle = n-1} q_{\mathbf{e}} P_{\mathbf{e}}^i(\lambda,\mu)a_i$$

$$= \sum_{i=1}^d \left(\sum_{\langle \mathbf{e} \rangle = n-1} (p^{\epsilon \langle \mathbf{e} \rangle} q_{\mathbf{e}})(p^{-\epsilon \langle \mathbf{e} \rangle} P_{\mathbf{e}}^i(\lambda,\mu)) \right) a_i$$

$$= W_n^1(\lambda,\mu)a_1 + \cdots + W_n^d(\lambda,\mu)a_d,$$

say. Here each W_n^i is a homogeneous polynomial of degree n in $2d$ variables, and it follows from (4) that the coefficients of each W_n^i lie in $p^{f(n)}\mathbb{Z}_p$ where $f(n) \geq 0$ for all n and $f(n) \to \infty$ as $n \to \infty$. Hence for each i, the power series

$$W^i(\mathbf{X},\mathbf{Y}) = \sum_{n=1}^{\infty} W_n^i(\mathbf{X},\mathbf{Y})$$

converges on $\mathbb{Z}_p^d \times \mathbb{Z}_p^d$, and the result follows since

$$\widetilde{\Phi}(\lambda \cdot \mathbf{a}, \mu \cdot \mathbf{a}) = W^1(\lambda,\mu)a_1 + \cdots + W^d(\lambda,\mu)a_d.$$

9.13 Corollary *Let H be a uniform pro-p group of dimension d, let $\{a_1, \ldots, a_d\}$ be a topological generating set for H, and define $\psi : H \to \mathbb{Z}_p^d$ by*

$$\psi(\lambda_1 a_1 + \cdots + \lambda_d a_d) = (\lambda_1, \ldots, \lambda_d).$$

Then $H' = P_{1+\epsilon}(H)$ is a standard group with respect to the global chart $(H', \psi|_{H'}, d)$.

Proof The results of §4.3 ensure that $\psi|_{H'}$ is a homeomorphism from H' onto $p^{\epsilon}\mathbb{Z}_p^d$. Theorem 9.10 shows that $xy = x * y$ for all $x,y \in H' = L_{H'}$; and the proof of Lemma 9.12 shows that for $i = 1, \ldots, d$ and $x,y \in L_{H'}$,

$$\psi_i(x * y^{-1}) = W^i(\psi(x), -\psi(y))$$

where each W^i is a power series with coefficients in $\mathbb{Z}_p$. All the requirements of Definition 8.22 are therefore satisfied.

This can be used to provide an alternative proof of Theorem 8.18, the analytic structure on a uniform group being given by the additive 'co-ordinates of the first kind' instead of the multiplicative 'co-ordinates of the second kind' used in §8.3.

Now let K be a uniform pro-p group of dimension d, put $H = P_\epsilon(K)$, and define $\psi : H \to \mathbb{Z}_p^d$ as in the last corollary. Thus ψ is an isomorphism (of $\mathbb{Z}_p$-modules) from L_H onto $\mathbb{Z}_p^d$, and the argument used to prove Corollary 9.13 shows equally that H is an analytic group with respect to the global chart (H, ψ, d).

We can now complete the proof of Theorem 9.11(iii). Let K be a uniform open subgroup of the p-adic analytic group G, and let $H = P_\epsilon(K)$ be as above. Then H has two structures as an analytic group: the one given by the global chart (H, ψ, d), and its induced structure as an open subgroup of G. Now Corollary 9.5 shows that these structures are the same; in other words, (H, ψ, d) is also a chart of the p-adic analytic group G. The map ϕ given in Theorem 9.11(iii) is just the composition of ψ with a linear automorphism of $\mathbb{Q}_p^d$ corresponding to a change of basis; hence (H, ϕ, d) is compatible with (H, ψ, d) and the result follows.

Remark The construction of $\mathcal{L}(G)$ given above used only the (topological) group structure of the analytic group G. There is another quite different way to associate a Lie algebra to an analytic group, in which the Lie algebra structure is 'read off' directly from the quadratic part of the power series representing multiplication in the group: this gives rise to the Lie algebra of the *formal group law* associated to an analytic group, and is the procedure adopted in Chapter 13 where we consider groups analytic over rings more general than $\mathbb{Z}_p$. Exercise 13 shows that for p-adic analytic groups, the Lie algebras obtained by the two constructions are nevertheless isomorphic.

In the preceding section, we saw that the functor $H \mapsto L_H$ on uniform pro-p groups is invertible, giving an isomorphism of categories. The functor $G \mapsto \mathcal{L}(G)$ on p-adic analytic groups is of course not invertible, since $\mathcal{L}(G)$ only depends on a 'small' open subgroup of G. However, Theorem 9.10 will enable us to set up an equivalence between a suitably defined category of p-adic analytic groups and the category of finite-dimensional Lie algebras over $\mathbb{Q}_p$.

For p-adic analytic groups A and B, let $\mathcal{H}(A, B)$ denote the set of all analytic homomorphisms whose domain is an open subgroup of A and whose codomain is B. Now define an equivalence relation $\sim$ on $\mathcal{H}(A, B)$: if A_1 and A_2 are open subgroups of A, $f_1 : A_1 \to B$ and $f_2 : A_2 \to B$, then $f_1 \sim f_2$ if there exists $A_3 \leq_o A$, with $A_3 \leq A_1 \cap A_2$, such that $f_1|_{A_3} = f_2|_{A_3}$. Let us call the equivalence classes *germs*. It is easy to see that composition of germs is well defined, and that the germ of an identity mapping acts as an identity when composed on either side with any germ (see Exercise 5). Thus we may define a category

$$\mathfrak{G}_p,$$

of 'local p-adic analytic groups', whose objects are the p-adic analytic groups, in which the morphisms from A to B are the germs of $\mathcal{H}(A, B)$.

Writing

$$\mathfrak{L}_p$$

to denote the category of finite-dimensional Lie algebras over $\mathbb{Q}_p$, we can state our final theorem:

9.14 Theorem *The functor $\mathcal{L}$ induces a category equivalence between $\mathfrak{G}_p$ and $\mathfrak{L}_p$.*

Proof The main difficulty is to remember what this means! First of all, if A and B are objects of $\mathfrak{G}_p$, a morphism $\mathbf{f} : A \dashrightarrow B$ is represented by a homomorphism $f : A_1 \to B$, for some $A_1 \leq_o A$. (We denote morphisms of $\mathfrak{G}_p$ by dotted arrows $\dashrightarrow$, to distinguish them from group homomorphisms.) Then f induces a Lie algebra homomorphism $f^* : \mathcal{L}(A_1) \to \mathcal{L}(B)$. Now $\mathcal{L}(A_1) = \mathcal{L}(A)$, from the definition; and Theorem 9.11(ii) shows that f^* depends only on the germ $\mathbf{f}$ of f. Thus we may define $\mathcal{L}(\mathbf{f}) = f^* : \mathcal{L}(A) \to \mathcal{L}(B)$, and so consider $\mathcal{L}$ as a functor from $\mathfrak{G}_p$ to $\mathfrak{L}_p$.

Now we have to construct a functor in the reverse direction. Let V be a finite-dimensional Lie algebra over $\mathbb{Q}_p$, with basis $\{v_1, \ldots, v_d\}$. Choose h so large that

$$(p^h v_i, p^h v_j) \in p^\epsilon \sum_k \mathbb{Z}_p \cdot p^h v_k,$$

and put $L = \sum_k \mathbb{Z}_p \cdot p^h v_k$. Evidently L is then a powerful Lie subalgebra of V, and we obtain the corresponding group $(L, *)$. Having made this (arbitrary) choice of basis, we now define

$$\mathcal{F}(V) = (L, *).$$

Given a Lie algebra homomorphism $f : V \to V'$, let f_0 be the restriction of f to $L_1 = L \cap f^{-1}(L')$, where L' is the chosen powerful Lie subalgebra of V'. Then the mapping f_0 is also an analytic group homomorphism from the open subgroup $(L_1, *)$ of $(L, *) = \mathcal{F}(V)$ into $(L', *) = \mathcal{F}(V')$. We now define $\mathcal{F}(f)$ to be the germ of f_0. It is routine to verify that this makes $\mathcal{F}$ into a functor from $\mathfrak{L}_p$ to $\mathfrak{G}_p$.

To prove the theorem, it will suffice to verify that $\mathcal{F} \circ \mathcal{L}$ is naturally equivalent to the identity functor on $\mathfrak{G}_p$, and that $\mathcal{L} \circ \mathcal{F}$ is naturally equivalent to the identity functor on $\mathfrak{L}_p$.

A natural equivalence from $\mathcal{L} \circ \mathcal{F}$ to the identity functor on $\mathfrak{L}_p$ is a family of isomorphisms $\theta_V : \mathcal{L}(\mathcal{F}(V)) \to V$, one for each object V of $\mathfrak{L}_p$, such that the diagram

$$\mathcal{L}(\mathcal{F}(V)) \xrightarrow{\theta_V} V$$

$$\downarrow_{\mathcal{L}(\mathcal{F}(f))} \qquad\qquad \downarrow_f$$

$$\mathcal{L}(\mathcal{F}(V')) \xrightarrow{\theta_{V'}} V'$$

commutes for every morphism $f : V \to V'$ in $\mathfrak{L}_p$. Now

$$\mathcal{L}(\mathcal{F}(V)) = \mathbb{Q}_p \otimes_{\mathbb{Z}_p} L_{(L, *)}$$
$$= \mathbb{Q}_p \otimes_{\mathbb{Z}_p} L \quad \text{by Theorem 9.10,}$$

where L is a powerful Lie subalgebra of V that spans V as a vector space; so we may define θ_V to be the natural isomorphism from $\mathbb{Q}_p \otimes L$ to $V = \mathbb{Q}_p L$. The commutativity of the given diagram is then more or less obvious: we leave it to the reader to check the details.

We also need to construct a natural equivalence from $\mathcal{F} \circ \mathcal{L}$ to the identity functor on $\mathfrak{G}_p$. Let G be an object of $\mathfrak{G}_p$. Then $\mathcal{L}(G) = \mathbb{Q}_p \otimes L_H$ for a suitable open subgroup H of G, and $\mathcal{F}(\mathcal{L}(G)) = (L, *)$ where L is a suitable $\mathbb{Z}_p$-Lie subalgebra of $\mathcal{L}(G)$. Identify L_H with $1 \otimes L(H) \subseteq \mathcal{L}(G)$, and put $L_1 = L \cap L_H$. Then $K = (L_1, *)$ is an open subgroup both of $(L, *) = \mathcal{F}(\mathcal{L}(G))$ and of $(L_H, *)$, which by Theorem 9.10 is equal to the open subgroup H of G. Thus the inclusion map $\iota : K \to G$ belongs to $\mathcal{H}(\mathcal{F}(\mathcal{L}(G)), G)$, and we define ϕ_G to be the germ of ι. Then ϕ_G is an isomorphism from $\mathcal{F}(\mathcal{L}(G))$ to G in the category $\mathfrak{G}_p$. Again, following through the definitions, it is easy to see (though tedious to prove in writing) that each morphism $\mathbf{f} : G \dashrightarrow G'$ in $\mathfrak{G}_p$ gives rise to a commutative diagram

$$
\begin{array}{ccc}
\mathcal{F}(\mathcal{L}(G)) & \xrightarrow{\ \phi_G\ } & G \\[2em]
\Big\downarrow{\scriptstyle\mathcal{F}(\mathcal{L}(\mathbf{f}))} & & \Big\downarrow{\scriptstyle\mathbf{f}} \\[2em]
\mathcal{F}(\mathcal{L}(G')) & \xrightarrow{\ \phi_{G'}\ } & G'
\end{array}
$$

This completes the proof.

Notes

Again, this is mostly due to Lazard [L]; alternative accounts of most of the material are given in Serre (1965), Part II and Bourbaki (1989b), Chapter III, §§3,4; our construction of the Lie algebra differs from theirs, but leads to the same object (Exercise 13). The category equivalence of §9.5, though expressed in different language, is essentially given in both Serre and Bourbaki.

Theorem 3.2.6 of [L], Chapter IV, establishes an isomorphism between the category of p-saturable groups and a certain category of Lie algebras over $\mathbb{Z}_p$; Theorem 9.10 is our version of that result.

Exercises

1. Let $c : \mathbb{N} \to \mathbb{Q}_p$ be a function and $f \in \mathbb{Q}_p[T]$ a polynomial of degree n. Suppose that $f(0) = c(1) = 0$ and that $c(m+1) - c(m) = f(m)$ for all positive integers m. Prove that there exists a polynomial $g \in \mathbb{Q}_p[T]$, of degree at most $n+1$, such that $g(0) = 0$ and $g(m) = c(m)$ for all positive integers m.

2. Let $g \in \mathbb{Q}_p[T]$ have degree d. Show that if $g(m) \in \mathbb{Z}_p$ for $m = 0, 1, \ldots, d$, then there exist $\xi_j \in \mathbb{Z}_p$ such that

$$g(m) = \sum_{j=0}^{d} \xi_j \binom{m}{j} \quad \text{for all } m \in \mathbb{N}.$$

3. Let G be a topological group containing an open subgroup which is a uniform pro-p group. Use Corollary 9.13 and Proposition 8.15 to show that G is p-adic analytic.

4. Let $L \cong \mathbb{Z}_p^d$ be a Lie algebra over $\mathbb{Z}_p$. This exercise shows, without the nasty power series calculations of Lemma 9.9 and Proposition 6.29, that if the integer m is sufficiently large then $(p^m L, *)$ is a group.

(i) Using Ado's Theorem, show that if the integer m is sufficiently large then $p^m L = L_0$, say, may be identified with a Lie subalgebra of $p^\epsilon M_n(\mathbb{Z}_p)$, for some n (the Lie bracket on $p^\epsilon M_n(\mathbb{Z}_p)$ being the usual commutator).

(ii) Having made this identification, show that $\exp(x * y) = (\exp x)(\exp y)$ (matrix product) for all $x, y \in L_0$. Deduce that $(L_0, *)$ is a group. [Use Proposition 6.27.].

(iii) Show that $(L_0, *)$ is both a torsion-free pro-p group and a p-adic analytic group. [See Theorem 5.2 and Example 8.17(iv); use Theorem 9.6.]

5. Let $A \leq_c B$ be p-adic analytic groups, K a uniform open subgroup of B and $H \leq A \cap K$ a uniform open subgroup of A. Then L_H is a Lie subalgebra of L_K; we identify L_K with $1 \otimes L_K \subseteq \mathbb{Q}_p \otimes L_K = \mathcal{L}(B)$, and we identify $\mathcal{L}(A) = \mathbb{Q}_p \otimes L_H$ with the Lie subalgebra $\mathbb{Q}_p \cdot L_H$ of $\mathbb{Q}_p \cdot L_K = \mathcal{L}(B)$.

Show that if A_1 and A_2 are closed subgroups of B then

$$\mathcal{L}(A_1 \cap A_2) = \mathcal{L}(A_1) \cap \mathcal{L}(A_2).$$

[*Hint*: Let H_3 be a uniform open subgroup of $A_1 \cap A_2$ contained in $H_1 \cap H_2$, where $H_i \leq A_i \cap K$ is a uniform open subgroup of A_i for $i = 1, 2$. Show that H_3 has finite index in $H_1 \cap H_2$ and deduce that $\mathbb{Q}_p \cdot (L_{H_1} \cap L_{H_2}) = \mathbb{Q}_p \cdot L_{H_3}$.]

6. Let $f : A \to B$ be a morphism of p-adic analytic groups. Prove that $\ker f^* = \mathcal{L}(N)$ where $N = \ker f$.

[*Hint*: Let K be a uniform open subgroup of B and $H \leq f^{-1}(K)$ a uniform open subgroup of A. Show that $N \cap H$ is a uniform open subgroup of N (use Proposition 4.31(ii) and Theorem 4.5). Then observe that $L_{N \cap H}$ is the kernel of $f^*|_{L_H}$.]

7. Let $f : A \to B$ be a surjective morphism of p-adic analytic groups. Prove that $f^* : \mathcal{L}(A) \to \mathcal{L}(B)$ is surjective.

[*Hint:* Compare the dimensions of Im f^* and $\mathcal{L}(B)$, using Exercise 6 and Exercise 8.6.]

8. Let $A_1 \leq_o A$, $B_1 \leq_o B$ and C be p-adic analytic groups, and let $f : A_1 \to B$, $g : B_1 \to C$ be morphisms.

(i) Show that the equivalence class of $g \circ f$ in $\mathcal{H}(A, C)$ depends only on the equivalence classes of f in $\mathcal{H}(A, B)$ and of g in $\mathcal{H}(B, C)$.

(ii) Show that if $A = B$ and f is the inclusion map of A_1 in A, then $g \circ f \sim g$, while if $B = C$ and g is the inclusion $B_1 \to C$ then $g \circ f \sim f$.

(iii) Show that if f is injective and $f(A_1)$ is open in B then the germ of f is an isomorphism in the category $\mathfrak{G}_p$.

9. [For an alternative approach to this exercise, see Exercise 13 below.]

(i) Let G be the p-adic analytic group $\mathrm{GL}_n(\mathbb{Q}_p)$. Show that $\mathcal{L}(G) \cong \mathfrak{gl}_n(\mathbb{Q}_p)$, where $\mathfrak{gl}_n(\mathbb{Q}_p)$ is the algebra $\mathrm{M}_n(\mathbb{Q}_p)$ with the usual commutator as Lie bracket.

[*Hint*: recall Exercise 7.8(iii).]

(ii) Let det : $G \to \mathbb{Q}_p^* = \mathrm{GL}_1(\mathbb{Q}_p)$ and tr : $\mathrm{M}_n(\mathbb{Q}_p) \to \mathbb{Q}_p$ denote the determinant and trace functions, respectively. Let e_{ij} be the matrix with 1 in the (i, j) place and 0 elsewhere. Show that if $\mu \in p^\epsilon \mathbb{Z}_p$ then

$$\log(1 + \mu e_{ij}) = \log(1 + \mu) \cdot \delta_{ij} e_{ij} \quad \text{(Kronecker } \delta\text{)};$$

deduce that for $g = 1 + \mu e_{ij}$,

$$\mathrm{tr}(\log(g)) = \log(\det(g)).$$

Hence show that $(\det)^* = \mathrm{tr}$.

[*Hint*: for the final statement, see Exercise 7.8(i), and use Exercise 5.6.]

(iii) Deduce that $\mathcal{L}(\mathrm{SL}_n(\mathbb{Q}_p)) \cong \mathfrak{sl}_n(\mathbb{Q}_p)$, the Lie subalgebra of $\mathfrak{gl}_n(\mathbb{Q}_p)$ consisting of matrices with trace zero.

[*Hint*: Use Exercise 6. A different method is given in Exercise 13.10]

10. The adjoint representation. Let G be a uniform pro-p group and let Λ be the Lie algebra $\log(G)$. Write $\mathrm{GL}(\Lambda)$ for the group of all $\mathbb{Z}_p$-module automorphisms of Λ and $\mathfrak{gl}(\Lambda)$ for the Lie algebra of all $\mathbb{Z}_p$-module endomorphisms of Λ, and put

$$\Delta = \{\gamma \in \mathrm{GL}(\Lambda) \mid \Lambda(\gamma - 1) \subseteq p^\epsilon \Lambda\}\,;$$

recall (Exercise 7.8) that

$$\log(\Delta) = p^\epsilon \mathfrak{gl}(\Lambda).$$

Define $\mathrm{Ad}_\Lambda : G \to \mathrm{GL}(\Lambda)$ by $u \cdot \mathrm{Ad}_\Lambda(g) = g^{-1}ug$ $(u \in \Lambda, g \in G)$. Define $\mathrm{ad}_\Lambda : \Lambda \to \mathfrak{gl}(\Lambda)$ by $u \cdot \mathrm{ad}_\Lambda(v) = (u, v)$ $(u, v \in \Lambda)$.

(i) Verify that Ad_Λ is a homomorphism from G into Δ, and that ad_Λ is a Lie algebra homomorphism from Λ into $p^\epsilon \mathfrak{gl}(\Lambda)$.

(ii) Let $\mathrm{Ad}^\ddagger : \Lambda \to \log(\Delta)$ be the Lie algebra homomorphism induced by $\mathrm{Ad}_\Lambda : G \to \Delta$; so $\mathrm{Ad}^\ddagger = \log \circ \mathrm{Ad}_\Lambda \circ \exp$. Prove that

$$\mathrm{Ad}^\ddagger = \mathrm{ad}_\Lambda.$$

[*Hint*: use Exercise 6.12 to show that if $g \in G$ and $v = \log g$ then $\log(\mathrm{Ad}_\Lambda(g)) = \mathrm{ad}_\Lambda(v)$.]

(iii) Let

$$\Gamma = \{\gamma \in \mathrm{GL}(L_G) \mid L_G(\gamma - 1) \subseteq p^\epsilon L_G\}\,,$$

and define $\mathrm{Ad} : G \to \mathrm{GL}(L_G)$ by $u \cdot \mathrm{Ad}(g) = g^{-1}ug$ $(u \in L_G, g \in G)$. Show that Ad maps G into Γ. Let $\mathrm{Ad}^* : L_G \to p^\epsilon \mathfrak{gl}(L_G) = \log(\Gamma)$ be the Lie algebra homomorphism induced by $\mathrm{Ad} : G \to \Gamma$ (so $\mathrm{Ad}^* = \log \circ \mathrm{Ad}$). Show that

$$a \cdot \mathrm{Ad}^*(v) = \exp((\log a) \cdot \log(\mathrm{Ad}_\Lambda(v)))$$

for all $a, v \in L_G$. Hence justify the following steps, where $u = \log a$:

$$\begin{aligned}
a \cdot \mathrm{ad}(v) &= \exp(u \cdot \mathrm{ad}_\Lambda(\log v)) \\
&= \exp(u \cdot \mathrm{Ad}^\ddagger(\log v)) \\
&= \exp(u \cdot \log(\mathrm{Ad}_\Lambda(v))) = a \cdot \mathrm{Ad}^*(v).
\end{aligned}$$

Conclude that

$$\mathrm{Ad}^* = \mathrm{ad}.$$

[*Hint*: Use Lemma 7.12.]

11. The adjoint representation, again. Let G be a p-adic analytic group. For $g \in G$ define $\widehat{g} : G \to G$ by $\widehat{g}(h) = g^{-1}hg$ $(h \in G)$.

(i) Show that $\widehat{g}$ is a morphism of analytic groups. Deduce that $\mathcal{L}(\widehat{g})$ is an automorphism of the Lie algebra $\mathcal{L}(G) = L$. Denoting this automorphism by $\mathrm{Ad}(g)$ show that

$$\mathrm{Ad} : G \to \mathrm{GL}(L)$$

is a morphism of analytic groups.

(ii) According to Exercise 9(i), $\mathcal{L}(\mathrm{GL}(L))$ is isomorphic to $\mathfrak{gl}(L)$. Choosing a suitable isomorphism to identify $\mathcal{L}(\mathrm{GL}(L))$ with $\mathfrak{gl}(L)$, show that

$$\mathcal{L}(\mathrm{Ad}) = \mathrm{ad} : \mathcal{L}(G) \to \mathfrak{gl}(L).$$

[*Hint*: Let H be a uniform open subgroup of G, consider the restriction of $\mathcal{L}(\mathrm{Ad})$ to $L_H \subseteq \mathcal{L}(G)$, and use Exercise 10(iii).]

12. Let G be a p-adic analytic group. Show that G has closed normal subgroups $Z \leq K$ such that Z is abelian, K/Z is discrete, and G/K is isomorphic to a subgroup of $\mathrm{GL}_d(\mathbb{Q}_p)$ where $d = \dim(G)$.

[*Hint*: Let K be the kernel of $\mathrm{Ad} : G \to \mathrm{GL}(L)$ where $L = \mathcal{L}(G)$. Let G^+ be the subgroup of G generated by all uniform open subgroups of G. Show that $K = C_G(G^+)$, and that G/G^+ is discrete. Then put $Z = K \cap G^+$.]

13. The Lie algebra of a formal group. This exercise shows how to recognise the Lie algebra of an analytic group directly from the power series that define the group law; in Chapter 13 this method will be used to *define* the Lie algebra.

(i) Let H be a uniform pro-p group. Let $\psi : H \to \mathbb{Z}_p^d$ be the homeomorphism given in Corollary 9.13, and let $W^1, \ldots, W^d$ be the power series defined in Lemma 9.12 for the powerful Lie algebra L_H. Recall that W_n^i denotes the homogeneous component of degree n in W^i. Show that

$$W_1^i(\mathbf{X}, \mathbf{Y}) = X_i + Y_i,$$
$$W_2^i(\mathbf{X}, \mathbf{Y}) \text{ is bilinear in } \mathbf{X} \text{ and } \mathbf{Y},$$

and that for $x, y \in H$,

$$\psi_i((x,y)_H) = W_2^i(\psi_i(x), \psi_i(y)) - W_2^i(\psi_i(y), \psi_i(x)).$$

Hence show that the binary operation $\langle \cdot, \cdot \rangle$ on $\mathbb{Z}_p^d$ given by

$$\langle \lambda, \mu \rangle_i = W_2^i(\lambda, \mu) - W_2^i(\mu, \lambda) \quad (i = 1, \ldots, d)$$

makes $\mathbb{Z}_p^d$ into a Lie algebra, so that $\psi : L_H \to (\mathbb{Z}_p^d, \langle \cdot, \cdot \rangle)$ is a Lie algebra isomorphism.

[*Hint*: recall that $u_1(a, b) = a + b$ and $u_2(a, b) = \frac{1}{2}(a, b)$.]

(ii) Let H be a standard group with respect to the global chart (H, ϕ, d), and let $M_i(\mathbf{X}, \mathbf{Y}) \in \mathbb{Z}_p[[\mathbf{X}, \mathbf{Y}]]$ $(i = 1, \ldots, d)$ be the power series representing group multiplication in H (see (ii$'$) in §8.4). According to Corollary 8.28, the homogeneous component of degree 2 in $M_i(\mathbf{X}, \mathbf{Y})$ takes the form $B_i(\mathbf{X}, \mathbf{Y}) = \sum b_{i,rs} X_r Y_s$. Define a binary operation $\langle \cdot, \cdot \rangle^{\mathbf{M}}$ on $\mathbb{Q}_p^d$ by

$$\langle \lambda, \mu \rangle_i^{\mathbf{M}} = B_i(\lambda, \mu) - B_i(\mu, \lambda) \quad (i = 1, \ldots, d).$$

We shall see in Section 13.3 that this operation makes $\mathbb{Q}_p^d$ into a Lie algebra, the 'Lie algebra of the formal group law $\mathbf{M}$'; moreover, this Lie algebra is independent of $\mathbf{M}$ in the sense that

$$(\mathbb{Q}_p^d, \langle \cdot, \cdot \rangle^{\mathbf{M}}) \cong (\mathbb{Q}_p^d, \langle \cdot, \cdot \rangle^{\mathbf{M}'})$$

if $\mathbf{M}'$ is the formal group law arising from another global chart on H, compatible with (H, ϕ, d) and with respect to which H is again a standard group.

Now let G be a p-adic analytic group. Prove that

$$\mathcal{L}(G) \cong (\mathbb{Q}_p^d, \langle \cdot, \cdot \rangle^{\mathbf{M}})$$

if M is the formal group law representing multiplication in any standard open subgroup of G.

[*Hint*: Use (i) and Theorem 8.31.]

(iii) The description of $\mathcal{L}(G)$ provided by (ii) has the great advantage over both previous constructions (via L_H and via $\log(H)$) that it involves no limiting processes. Use it to give a simple proof that $\mathcal{L}(\mathrm{GL}_n(\mathbb{Q}_p)) \cong \mathfrak{gl}_n(\mathbb{Q}_p)$.

Part III

Further topics

10

Pro-p groups of finite coclass

Until relatively recently, it was received wisdom that there was no point in trying to classify finite p-groups up to isomorphism: there are just too many of them. Received wisdom was flouted in 1980 by Charles Leedham-Green and Mike Newman, who formulated a series of conjectures that amount to no less than a programme for the classification of all finite groups of prime-power order.

Quite remarkably, not only did the conjectures turn out to be exactly correct, but they were proved, over a period of about ten years, by following the guidelines proposed in the original paper of Leedham-Green and Newman.

We shall not repeat the conjectures, nor relate the interesting history and prehistory of their ultimate success; some pointers to the relevant literature are provided in the Notes at the end of the chapter. Our aim here is to explain how pro-p groups come into the picture, and to give a self-contained account of that part of the theory that relates to pro-p groups. Apart from its original motivation in the classification of finite groups, this makes a beautiful chapter in the theory of pro-p groups of finite rank.

The primary invariant used in the classification is the *coclass* of a finite p-group: the coclass of a group G of order p^n is $n - c$ where c is the nilpotency class of G. Thus G is abelian if the coclass is as large as possible, namely $n - 1$, and the groups of smallest coclass, namely coclass 1, are the furthest from being abelian (these are the so-called 'groups of maximal class'). It is one of the more surprising consequences of the programme that, when viewed in the right light, the groups of small coclass also turn out to be rather close to abelian.

To classify the p-groups of a fixed coclass $r \geq 1$, we form a directed graph $\Gamma = \Gamma(p, r)$ as follows. The vertices of Γ correspond to p-groups of coclass r, each isomorphism type being represented exactly once. A directed edge joins P to Q if and only if there exists an epimorphism from P onto Q with kernel of order p; it is easy to see that this holds just when $P/\gamma_c(P) \cong Q$, where c is the nilpotency class of P. The classification programme now falls into two parts. The first part consists in describing all the *infinite chains* in Γ; the second, in describing an explicit procedure whereby any p-group of coclass r can be obtained by suitably modifying one of the groups occurring in one of the infinite chains. We are going to concentrate on the first part.

An infinite chain in Γ has the form

$$\mathcal{C}: \quad \cdots \to P_n \to P_{n-1} \to \cdots \to P_2 \to P_1.$$

Let us assume that $\mathcal{C}$ is a *maximal* chain; in other words, it cannot be continued to the right. Choosing a specific epimorphism $P_n \to P_{n-1}$ for each n, we make $\mathcal{C}$ into an inverse system of finite p-groups, in the obvious manner; let G be its inverse limit. Elementary considerations, spelt out in the exercises, show (a) that the isomorphism type of G is independent of the chosen epimorphisms, (b) that, conversely, G determines $\mathcal{C}$, and (c) that G has *coclass* r, in the following sense:

Definition An infinite pro-p group G has *coclass r* if

$$|G : \overline{\gamma_{c+1}(G)}| = p^{c+r} \quad \text{for all sufficiently large } c.$$

Thus to determine the infinite chains in Γ is equivalent to finding all the infinite pro-p groups of coclass r. The feasibility of this task rests on the following theorem, the main result of this chapter:

10.1 Theorem *Let G be an infinite pro-p group of coclass r. Then G has an open normal subgroup $A \cong \mathbb{Z}_p^d$, where $d = (p-1)p^s$ for some $s < r$ if p is odd, $d = 2^s$ for some $s \leq r+1$ if $p = 2$. Moreover, G/A has coclass r and $|G : A| = p^{r+p^r}$ if p is odd, $|G : A| = 2^{r+(r+1)2^{r+1}}$ if $p = 2$.*

With Theorem 5.8, this implies

10.2 Theorem *For given p and r, there are only finitely many isomorphism types of infinite pro-p groups of coclass r.*

As regards the classification of finite p-groups, Theorem 10.2 has the following interpretation: *the graph $\Gamma(p,r)$ contains only a finite number of maximal infinite chains.*

The rest of the chapter is devoted to the proof of Theorem 10.1.

10.1 Coclass and rank

Our aim in this section is to establish

10.3 Proposition *Let G be an infinite pro-p group of finite coclass r. Then*

- $\gamma_{p^r}(G)$ *is open in G and powerful, of rank at most p^r, if $p \neq 2$;*
- $\gamma_{2^{r+2}}(G)$ *is open in G and powerful, of rank at most 2^{r+1}, if $p = 2$.*

This already shows that groups of finite coclass have finite rank, and that the rank is bounded in terms of the coclass.

Throughout the section, G will denote an infinite pro-p group of finite coclass r. For each $i \geq 1$ we write

$$G_i = \gamma_i(G).$$

Recall that, according to the definition of coclass, there exists c_0 such that

$$|G : \overline{G_{n+1}}| = p^{n+r} \quad \text{for all } n \geq c_0.$$

This is equivalent to saying that $|G : \overline{G_{c_0+1}}| = p^{c_0+r}$ and $|\overline{G_i} : \overline{G_{i+1}}| = p$ for all $i \geq c_0$, which makes it clear that every quotient of G has coclass at most r. We shall use this fact without special mention.

Since G is infinite, we have $\overline{G_{n+1}} < \overline{G_n}$ for every n; in particular G cannot be procyclic, so $|G : \overline{G_2}| \geq p^2$. It follows that

$$p^{n+1} \leq |G : \overline{G_{n+1}}| \leq p^{n+r} \quad \text{for all } n,$$

and as $\overline{G_2} \geq \Phi(G)$ this implies in turn that $d(G) \leq r + 1$. We saw in Exercise 1.17 that if G is a finitely generated pro-p group then in fact $\gamma_i(G)$ is closed for each i, so we have

10.4 Lemma $G_i = \overline{G_i}$ *is open in G for each i.*

The following definition introduces a concept which is crucial for understanding groups of finite coclass:

Definition Suppose that G acts on a finite p group U. The action is *uniserial* if $|N : [N, G]| = p$ for every non-trivial G-invariant normal subgroup N of U.

We shall also simply say 'U is uniserial', the reference to G being understood. When this holds, U has for each n with $p^n \leq |U|$ a unique

G-invariant normal subgroup of index p^n, namely $[U,_n G]$, so the G-invariant normal subgroups of U are linearly ordered by inclusion. To see this we argue by induction, starting with $U = [U,_0 G]$. Suppose that $K = K^G \lhd U$ has index p^{n+1}, and let $N/K = Z(U/K) \cap C_{U/K}(G)$. Then $|U : N| = p^m$ for some $m \leq n$. As $|N : [N,G]| = p$ and $[N,G] \leq K < N$ it follows that $[N,G] = K$ and that $m = n$. Assuming inductively that $N = [U,_n G]$ we infer that $K = [[U,_n G],G] = [U,_{n+1} G]$.

The next lemma shows that uniserial actions are going to be ubiquitous (when U is of the form A/B for open normal subgroups $B < A$ of G, we shall assume that G is acting on U by conjugation):

10.5 Lemma *Let $G = Q_1 \geq Q_2 \geq \cdots \geq Q_n \geq Q_{n+1}$ be a chain of open normal subgroups in G, and put $|Q_i : Q_{i+1}| = p^{f_i}$ for each i.*

(i) If $[Q_i, G] \leq Q_{i+1}$ for each i then

$$\sum_{i=1}^{n} (f_i - 1) \leq r.$$

(ii) If $n > r$ then at least $n - r$ of the factors Q_i/Q_{i+1} are uniserial.

Proof (i) is clear if we rewrite it in the form $|G : Q_{n+1}| = p^{\sum f_i} \leq p^{n+r}$, since $Q_{n+1} \geq G_{n+1}$. For (ii), we may assume that $Q_i > Q_{i+1}$ for each i (factors of order 1 are trivially uniserial). Let us suppose that k of the given factors are not uniserial. We can then refine the series to a central series of G in which at least k factors have order at least p^2. Applying part (i) to this new series we infer that $k \leq r$, and (ii) follows.

Now we bring in the 'power structure' of G, and state a lemma that lies at the heart of the whole theory. We write $\Pi_0 = G$ and for $i \geq 1$ define

$$\Pi_i = \overline{\Pi_{i-1}^p}.$$

10.6 Lemma *Let V be a finite $\mathbb{F}_p[G]$-module and k a non-negative integer.*

(i) If $p^k \geq \dim_{\mathbb{F}_p}(V)$ then $[V, \Pi_k] = 0$.

(ii) If V is uniserial and $p^k < \dim_{\mathbb{F}_p}(V)$ then

$$\dim_{\mathbb{F}_p}(V/[V, \Pi_k]) = p^k,$$

in particular $[V, \Pi_k] \neq 0$.

Proof Part (i) is an easy exercise, left to the reader. Let us prove (ii). Replacing G by $G/C_G(V)$, we may assume that G is finite. Write $V_i = [V,_i G]$ for each i. Since $[V, \Pi_k] = [V, G]$ the result is clear when $k = 0$; let us suppose that $k \geq 1$ and that the lemma is true for smaller values of k. Consider the series

$$V = A_0 > B_0 > A_1 > B_1 > \cdots > A_p > B_p \geq 0,$$

where $A_j = V_{jp^{k-1}}$ and $B_j = [A_j, G] = V_{1+jp^{k-1}}$ for $0 \leq j \leq p$. For each $j \geq 1$, the factor A_{j-1}/B_j is a uniserial G-module of $\mathbb{F}_p$-dimension $1+ p^{k-1}$, so by the inductive hypothesis it is acted on non-trivially by Π_{k-1}. Hence $C_{\Pi_{k-1}}(A_{j-1}/B_j) = H_j$, say, is a proper subgroup of Π_{k-1}. Now a finite p-group cannot be the union of p proper subgroups (Exercise 3), so we may choose an element $x \in \Pi_{k-1} \setminus \bigcup_{j=1}^{p} H_j$.

Let ξ denote the endomorphism of V induced by $x - 1$. Since $\dim_{\mathbb{F}_p}(A_{j-1}/A_j) = p^{k-1}$, part (i) shows that $A_{j-1}\xi \leq A_j$, while the choice of x ensures that $A_{j-1}\xi \not\leq B_j$, for each j. As A_j/B_j is one-dimensional it follows that

$$A_{j-1}\xi + B_j = A_j$$

for $j = 1, \ldots, p$. Again, by (i) we have $B_j\xi \leq B_{j+1}$ for each j, so applying ξ p times to $A_0 = V$ we get

$$\begin{aligned} A_0\xi^p + B_p &= (A_0\xi + B_1)\xi^{p-1} + B_p \\ &= A_1\xi^{p-1} + B_p = \cdots \\ &= A_p + B_p = A_p. \end{aligned}$$

Since ξ acts like $x^p - 1$ on V, and $x^p \in \Pi_k$, it follows that $[V, \Pi_k] + B_p \geq A_p$. On the other hand, (i) also shows that $[V, \Pi_k] \leq A_p$; as $[V, \Pi_k]$ is a G-submodule of the uniserial module V this implies that $[V, \Pi_k] = A_p$, which has codimension p^k in V as required.

For each $i \geq 0$ put

$$X_i = \Pi_i G_{p^i}.$$

Then each X_i is open in G, by 10.4, and we have

$$X_{i+1} = \Pi_i^p G_{p^{i+1}} \leq X_i^p G_{p^{i+1}} \leq X_i^p G_{2p^i} \leq X_i^p [X_{i,_{p^i}} G]. \tag{1}$$

10.7 Lemma *Assume that p is odd. For some t with $1 \leq t \leq r$, the following hold:*

- $X_t/\Phi(X_t)$ *is uniserial as a G-module;*

- $d(X_t) \leq p^t$;
- *if $N \lhd_o G$ and $N \leq X_t$ then N p.e. X_t.*

Proof It follows from Lemma 10.5(ii) that X_t/X_{t+1} is uniserial for some $t \leq r$; and $t \neq 0$ because $X_1 \leq \Phi(G)$, while $G/\Phi(G)$ is a central factor of G of order at least p^2. Put $X = X_t$, and choose a G-submodule $Y/\Phi(X)$ of $X/\Phi(X)$ minimal subject to X/Y being uniserial. Since $[X, \Pi_t] \leq [X, X] \leq \Phi(X) \leq Y$ and X/Y is uniserial, Lemma 10.6(ii) shows that $\dim_{\mathbb{F}_p}(X/Y) \leq p^t$.

Suppose now that $Y > \Phi(X)$. We may choose a G-submodule $Z/\Phi(X)$ such that $Z < Y$ and $|Y : Z| = p$, and then $\dim_{\mathbb{F}_p}(X/Z) \leq p^t + 1$. Since X/Z is not uniserial, there is a central series of G running from X to Z in which at least one factor has order at least p^2; it follows that

$$[X,_{p^t} G] \leq Z.$$

As $X^p \leq \Phi(X) \leq Z$, this shows with (1) that $X_{t+1} \leq Z$. But X/X_{t+1} is uniserial by hypothesis, forcing X/Z to be uniserial also, in contradiction to our minimal choice of Y.

Thus $Y = \Phi(X)$, so $X_t/\Phi(X_t)$ is uniserial as required. The second claim also follows since $d(X) = \dim_{\mathbb{F}_p}(X/\Phi(X))$. (Note that the argument so far works as well when $p = 2$.)

To establish the third claim, recall from §3.1 that $V(G, n)$ denotes the intersection of the kernels of all homomorphisms from G into $GL_n(\mathbb{F}_p)$. It follows from Lemma 10.6(i), **0.8** and **0.7** that $X \leq V(G, p^t)$. Proposition 3.9 now shows that X is powerful; so if $N \lhd_o G$ and $N \leq X$ then $d(N) \leq d(X) \leq p^t$, and by Proposition 3.9 again it follows that Np.e.X.

Proposition 10.3, when p is an odd prime, follows on taking $N = G_{p^r}$ in Lemma 10.7. The proof for $p = 2$ is a little more complicated and we defer it to the next section.

10.2 The case $p = 2$

This section is included for completeness; it can safely be omitted by the reader who is prepared to take this case on trust. Throughout this section we assume that $p = 2$.

When $p = 2$, the argument of Lemma 10.7 breaks down at the point where we apply Proposition 3.9: having established that $X = X_t \leq V(G, 2^t)$ and that $d(X) \leq 2^t$ we cannot infer that X is powerful; for that, we need to show that in fact $X \leq V(G, 2^t)^2$. We shall see that this is the case for a suitable choice of t.

10.8 Lemma *Let W be a uniserial $\mathbb{F}_2[G]$-module. Then for each $i \geq 0$,*

$$[W, \Pi_i] = [W,_{2^i} G],$$
$$[W, [\Pi_i, G]X_i^2] \leq [W,_{2^i+1} G].$$

Proof Put $W_n = [W,_n G]$ for each n. That $[W, \Pi_i] = W_{2^i}$ follows from Lemma 10.6. Applying this also with W_1 in place of W and using the Three-subgroup Lemma, we infer that $[W, [\Pi_i, G]] \leq W_{2^i+1}$. Next, note that $[W, G_{2^i}] \leq W_{2^i}$ (see **0.7**(i)); hence $[W, X_i] = [W, \Pi_i][W, G_{2^i}] \leq W_{2^i}$, and similarly $[W_{2^i}, X_i] \leq W_{2^i+2^i} \leq W_{2^i+1}$. This now implies (see **0.7**(ii)) that $[W, X_i^2] \leq W_{2^i+1}$, and the lemma follows.

10.9 Lemma *There exist $k > j > i \geq 1$ with $k \leq r + 2$ such that*

- X_ℓ/X_ℓ^2 *is uniserial as a G-module for $\ell = i, j, k$;*
- $X_\ell^2 \geq X_{\ell+1}$ *for $\ell = i, j, k$;*
- *either* $\mathrm{d}(X_k) \leq 2^j$ *or* $\mathrm{d}(X_j) \leq 2^i$.

Proof Applying Lemma 10.5(ii) to the chain $G = X_0 \geq X_1 \geq \cdots \geq X_{r+3}$, we find three values of ℓ in the range $1 \leq \ell \leq r + 2$ such that $X_\ell/X_{\ell+1}$ is uniserial; these values we call i, j, k. For each such ℓ, the proof of Lemma 10.7 shows that X_ℓ/X_ℓ^2 is uniserial as a G-module and that $d(X_\ell) \leq 2^\ell$.

It follows that for $\ell \in \{i, j, k\}$ we have $X_{\ell+1} = \Pi_\ell^2[G_{2^\ell,2^\ell} G] \leq X_\ell^2$.

Now write $U = X_j/X_j^2$ and $V = X_k/X_k^2$, and let $U = U_0 > U_1 > \cdots > U_a = 0$, $V = V_0 > V_1 > \cdots > V_b = 0$ be the (unique) G-composition series in U and V respectively. Thus $|U_{n-1} : U_n| = 2 = |V_{n-1} : V_n|$ for each n, and $a = \dim_{\mathbb{F}_2}(U) = \mathrm{d}(X_j)$, $b = \dim_{\mathbb{F}_2}(V) = \mathrm{d}(X_k)$. Assuming that $b > 2^j$, we shall deduce that then $a \leq 2^i$; this will suffice to complete the proof.

Since $\Pi_i X_i^2/X_i^2$ is contained in the uniserial G-module X_i/X_i^2, the quotient $\Pi_i/(\Pi_i \cap X_i^2)[\Pi_i, G]$ has order at most 2; so there exists an element $g \in \Pi_i$ such that $\Pi_i = (\Pi_i \cap X_i^2)[\Pi_i, G] \langle g \rangle$. Then Lemma 10.8 shows that, in additive notation,

$$V_{n+2^i} = V_{n+2^i+1} + V_n(g - 1)$$

for each $n \geq 0$. It follows that

$$V(g^{2^{j-i}} - 1) = V(g - 1)^{2^{j-1}} \equiv V_{2^j-i2^i} = V_{2^j} \not\equiv 0 \pmod{V_{2^j+1}},$$

since we have assumed that $2^j < b$. Hence, by Lemma 10.8 again,

$g^{2^{j-i}} \notin [\Pi_j, G]X_j^2$. But $g^{2^{j-i}} \in \Pi_j$ and $\Pi_j/(\Pi_j \cap X_j^2)[\Pi_j, G]$ has order at most 2, as above; so we conclude that $\Pi_j = (\Pi_j \cap X_j^2)[\Pi_j, G] \left\langle g^{2^{j-i}} \right\rangle$. It follows that $U = U_1 + \langle u \rangle$ where u is the image of $g^{2^{j-i}}$ in $U = X_j/X_j^2$.

As g commutes with $g^{2^{j-i}}$ we have $u(g-1) = 0$. Applying Lemma 10.8 to U and to U_1 we therefore have

$$U_{2^i} = U_{2^i+1} + U(g-1) = U_{2^i+1} + U_1(g-1) = U_{2^i+1}.$$

It follows that $a = \dim_{\mathbb{F}_2}(U) \leq 2^i$ as required.

Now put $t = k$ if $d(X_k) \leq 2^j$, $t = j$ if $d(X_j) \leq 2^i$. In either case we have $d(X_t) \leq 2^{t-1}$ and $X_t \leq X_{t-1}^2$; arguing as in the last part of Lemma 10.7 (with $t - 1$ in place of t) we may therefore infer

10.10 Corollary *For some t with $1 \leq t \leq r + 2$, the following hold:*

- $X_t/\Phi(X_t)$ *is uniserial as a G-module;*
- $d(X_t) \leq 2^{t-1}$;
- *if $N \lhd_o G$ and $N \leq X_t$ then $Np.e.X_t$.*

The proof of Proposition 10.3 for $p = 2$ now follows as before.

10.3 The dimension

We now start to pin down some finer details of the structure of a pro-p group of finite coclass.

An open normal subgroup N of G will be called *uniserial in G* if G acts uniserially on N/K for every $K \lhd_o G$ with $K < N$. The main results of this section are

10.11 Proposition *Let G be an infinite pro-p group of finite coclass r. Let $m \geq rp^r$ if p is odd, $m \geq (r+1)2^{r+1}$ if $p = 2$. Then the group $\gamma_m(G)$ is both uniform and uniserial in G.*

10.12 Proposition *Let G be an infinite pro-p group of finite coclass r. Then $\dim(G) = (p-1)p^s$ for some $s < r$ if p is odd, $\dim(G) = 2^s$ for some $s \leq r + 1$ if $p = 2$.*

Note that Proposition 10.12 is a very strong structural restriction: it shows that the dimension of a pro-p group of coclass r can take at most r values ($r + 2$ when $p = 2$). Exercise 6 shows that in fact all of these values do occur (at least for odd p). The fact that they take the precise form they do turns out, surprisingly, to be a crucial ingredient in the solubility proof given later in the chapter.

Henceforth, G denotes an infinite pro-p group of coclass r. We know from §1 that G has finite rank, and write

$$d = \dim(G)$$

for its dimension. As before we write $G_n = \gamma_n(G)$.

We shall need the following simple observation:

10.13 Lemma *Suppose that G acts on a finite p-group A and that $C < B$ are G-invariant normal subgroups of A. If both B and A/C are uniserial, then A is uniserial.*

Proof Let $M < N$ be G-invariant normal subgroups of A with $[N, G] \leq M$. If $N \leq B$ then $|N : M| = p$. If not, then $NC > B$, which implies that $MC \geq B$. Then $B = (M \cap B)C$, which implies that $B = M \cap B$, so $C < M < N < A$ and again it follows that $|N : M| = p$.

Now let $X = X_t = \Pi_t G_{p^t}$ be the subgroup of G introduced in Lemma 10.7 or Corollary 10.10; thus $1 \leq t \leq r$, $d(X) \leq p^t$ if $p \neq 2$, $1 \leq t \leq r+2$ and $d(X) \leq p^{t-1}$ if $p = 2$; moreover, $X/\Phi(X)$ is uniserial in G and every open normal subgroup of G contained in X is powerfully embedded in X.

Let $T = \tau(X)$ be the torsion subgroup of X. According to Theorem 4.20, T is finite and X/T is torsion-free. Let p^e be the exact exponent of T. It follows from Theorem 3.6(iii) that $X^{p^e} \cap T = 1$. Now consider the series

$$G \geq X^{p^e} \cdot T \geq X^{p^{e+1}} \cdot T^p \geq \cdots \geq X^{p^{2e-1}} \cdot T^{e-1} \geq X^{p^{2e}} \cdot T^{p^e} = X^{p^{2e}}.$$

None of the factors in this series is uniserial: for the first one has $G/\Phi(G)$ as an image, while each of the others has the form

$$X^{p^{e+i-1}} T^{p^{i-1}} / X^{p^{e+i}} T^{p^i} \cong (X^{p^{e+i-1}}/X^{p^{e+i}}) \times (T^{p^{i-1}}/T^{p^i}).$$

It follows by Lemma 10.5(ii) that $e + 1 \leq r$, so $e \leq r - 1$.

Next, we show that X/T is uniserial in G/T. Replacing G by G/T, we may as well assume for this part of the argument that $T = 1$. Then X is uniform, of dimension d. For $i \geq 1$ put $U_i = X^{p^{i-1}} = P_i(X)$. From Theorem 3.6 we know that for each i, the mapping $x \mapsto x^{p^{i-1}}$ induces an isomorphism of G-modules $U_1/U_2 = X/\Phi(X) \rightarrow U_i/U_{i+1}$.

Hence U_i/U_{i+1} is a uniserial $\mathbb{F}_p[G]$-module of dimension d, for each i. In particular, putting $V_i = [U_i, G]U_{i+1}$ we have $|U_i : V_i| = p$.

Now since V_i is open and normal in G, it is powerfully embedded in X; it follows that V_i/V_i^p is central in U_i/V_i^p, and hence that U_i/V_i^p is abelian (since U_i/V_i is cyclic). As V_i is also uniform (as it is torsion-free) and of dimension d, we have $|V_i : V_i^p| = p^d$, so V_i^p/U_{i+1} is an $\mathbb{F}_p[G]$-submodule of index p in U_{i+1}/U_{i+2}; it follows that $V_i^p = V_{i+1}$. Thus U_i/V_{i+1} is abelian. If $p = 2$, a similar argument shows that U_i/V_{i+2} is abelian.

We claim that the mapping $x \mapsto x^p$ induces a G-module isomorphism from U_i/V_{i+1} onto U_{i+1}/V_{i+2}, for each i. This will be clear provided we have

$$(xy)^p \equiv x^p y^p \pmod{V_{i+2}}$$

for $x, y \in U_i$. If $p = 2$ this is immediate; if $p \neq 2$, it holds (by **0.2** (iii)) because $[x, y] \in V_{i+1}$ and $V_{i+1}^p[V_{i+1}, U_i] \leq V_{i+2}$. Now according to Lemma 10.5(ii), at least one of the factors in the chain $U_1 > U_3 > \cdots > U_{2r+3}$ must be uniserial; but if U_j/U_{j+2} is uniserial then so is its quotient U_j/V_{j+1}, and as we have just seen that these G-modules are all isomorphic we may conclude that U_i/V_{i+1} is uniserial for every $i \geq 1$. Repeated applications of Lemma 10.13 then show that X/U_n is uniserial for every n, and as the subgroups U_n form a base for the neighbourhoods of 1 in X, it follows that X is uniserial in G, as required.

It is now easy to complete the

Proof of Proposition 10.11 Suppose that $p \neq 2$, and let $m \geq rp^r$. Since $X^{p^e} \cap T = 1$, we see that $X^{p^e} \cong X^{p^e} T/T$ is both torsion-free and uniserial in G. Since $d(X) \leq p^t$ and X is powerful, we have $|X/X^{p^e}| \leq p^{ep^t} \leq p^{(r-1)p^r}$. As $X \geq G_{p^r}$ it follows that

$$G_m \leq G_{rp^r} \leq [X,_{(r-1)p^r} G] \leq X^{p^e}.$$

Hence G_m is uniserial in G, and being both torsion-free and powerful it is uniform, by Theorem 4.5. The same argument with slight notational changes deals with the case $p = 2$.

In order to prove *Proposition 10.12*, we may as well replace G by G/T, so we assume henceforth that X is torsion-free. Now fix an integer $m \geq 3p^t$ and put

$$A = G_m \Pi_t.$$

Then $A \leq X$, so A is *uniform* and *uniserial in* G. For $i \geq 1$ write

$$A_i = \overline{[A_{i-1}, G]},$$

so $A = A_1 > A_2 > \ldots$ is the unique maximal descending chain of G-invariant normal subgroups of A, and $|A_1 : A_{i+1}| = p^i$ for each i. Moreover, since every open normal subgroup of G contained in X is powerfully embedded in X, each A_i is uniform, of the same dimension $d = \dim X \leq p^t$. It follows that for each i we have

$$[A_i, A] \leq A_i^p = \Phi(A_i) = A_{i+d}.$$

Proposition 10.12 asserts that $d = (p-1)p^s$ (the bounds on s then follow from Proposition 10.3). To prove this, we have to examine the structure of A as a G-operator group; this is more transparent if we think of A as an *additive* group, with the structure introduced in Section 4.3. According to Corollary 4.15, for each n the identity map on $A^{p^n}/A^{p^{n+1}} = A_{1+nd}/A_{1+(n+1)d}$ is an isomorphism between the multiplicative and additive structures on this group; hence the additive group $A_{1+nd}/A_{1+(n+1)d}$ is a uniserial G-module, having the unique composition series $(A_j/A_{1+(n+1)d})_{nd<j\leq 1+(n+1)d}$. For exactly the same reason, the additive group $A_{2+nd}/A_{2+(n+1)d}$ is a uniserial G-module. Provided that $d > 1$, it follows by Lemma 10.13 that the $\mathbb{Z}_p$-module $(A, +)$ is uniserial for G, in the sense that A/B is uniserial for every open G-submodule B of A; moreover, the series $(A_i)_{i \geq 1}$ is the unique descending composition series in this $\mathbb{Z}_p[G]$-module. If $d = 1$, the same holds because in that case the additive and multiplicative structures are the same.

Now a function f is defined as follows:

$$f(0) = 1;$$
$$f(i) = \min\{pf(i-1), d + f(i-1)\} \text{ for } i \geq 1.$$

Evidently $f(i) \leq p^i$ for each i.

10.14 Lemma *Unless* $d = (p-1)p^s$ *for some* s, *we have*

$$A_i(\Pi_j - 1) = A_{i+f(j)}$$

for all i and j.

This refers to the additive group $(A, +)$, considered as a G-module. The analogous statement for the multiplicative group A is also true, but less simple to prove, and we can manage without it; however we do need the (relatively easy)

10.15 Lemma $[A_i, \Pi_j] \leq A_{i+f(j)}$ *for all* i *and* j.

Given 10.14 and 10.15, the proof of Proposition 10.12 is completed as follows. Let us assume that d is not of the form $(p-1)p^s$.

We show first that $f(t) \geq d$. Recall that $A_{1+d} = A^p = \Phi(A)$. It follows by Corollary 4.15 that the identity map on A/A_{1+d} is a G-module isomorphism between the additive and multiplicative structures on this set. Since $[A, \Pi_t] \leq [A, A] \leq A_{1+d}$ it follows that $A(\Pi_t - 1) \leq A_{1+d}$. With Lemma 10.14 this implies that $1 + f(t) \geq 1 + d$.

A glance at the definition of f then shows that $f(t+n) = f(t) + nd$ for each $n \geq 0$.

Now let $x \in A \setminus A_2$ and let $y \in A$. Since $|A : A_2| = p$, we have $y = zx^h$ for some $z \in A_2$ and some integer h, so for each n the following holds:

$$\begin{aligned}
[x^{p^n}, y] = [x^{p^n}, z] &\in [A_2, A^{p^n}] \\
&\leq [A_2, G_m \Pi_{t+n}] \\
&\leq A_{2+f(t+n)} [A_2, G_m]
\end{aligned}$$

by Lemma 10.15. Choosing n to be the least integer such that $nd \geq 1 + f(t)$, we have

$$f(t+n) = f(t) + nd \leq 2f(t) + d \leq 3p^t \leq m$$

since $f(t) \leq p^t$ and $d \leq p^t$; it follows by stability group theory (**0.7**) that $[A_2, G_m] \leq A_{2+f(t+n)}$, so we now have

$$[x^{p^n}, y] \in A_{2+f(t+n)} = A_{2+f(t)+nd} = A^{p^n}_{2+f(t)}.$$

Note now that $A^{p^n} = A_{1+nd} \leq A_{2+f(t)}$. According to the remark following Definition 4.12, it follows that

$$\begin{aligned}
x(y-1) &= x^{-1} + x^y \\
&\equiv x^{-1} +_n x^y \pmod{A_{2+f(t)}} \\
&= (x^{-p^n} x^{yp^n})^{p^{-n}} \\
&= [x^{p^n}, y]^{p^{-n}} \in A_{2+f(t)}.
\end{aligned}$$

Thus $x(y-1) \in A_{2+f(t)}$ whenever $x \in A \setminus A_2$ and $y \in \Pi_t$. The same holds if $x \in A_2$, by Lemma 10.14, and we conclude that $A_1(\Pi_t - 1) \leq A_{2+f(t)}$. But this contradicts Lemma 10.14; hence the initial hypothesis must be false, and so $d = (p-1)p^s$ for some s. Thus the proof of Proposition 10.12 is complete, modulo the two lemmas stated above.

Proof of Lemma 10.15 This is by induction on j, the case $j = 1$ being clear. Fix $j \geq 1$, put $k = f(j)$ and assume that $[A_n, \Pi_j] \leq A_{n+k}$ for all n. Now the following commutator identity is given in Lemma 11.9 (in the next chapter): if a and x are elements of any group then

$$[x^p, a] \equiv [x, a]^p \pmod{\gamma_2(K)^p \gamma_p(K)},$$

where $K = \langle x, [x, a] \rangle$. We apply this with $x \in \Pi_j$ and $a \in A_i$. In that case, we have $K \leq \Pi_j$ and $[x, a] \in A_{i+k}$, so $\gamma_2(K) \leq A_{i+2k}$ and $\gamma_{2+n}(K) \leq A_{i+2k+nk}$ for each n. In particular, $\gamma_p(K) \leq A_{i+pk}$.

As $A_{i+k}^p = A_{i+k+d}$, we see that $[x, a]^p \gamma_2(K)^p \subseteq A_{i+k+d}$. It follows that $[a, x^p] \in A_{i+k+d} A_{i+pk} = A_{i+f(j+1)}$; this shows that $[A_i, \Pi_{j+1}] \leq A_{i+f(j+1)}$, since Π_{j+1} is generated modulo any open normal subgroup of G by elements like x^p.

Proof of Lemma 10.14 We now consider A as an additive group. In place of the commutator identity above, we use the trivial polynomial identity

$$x^p - 1 \equiv (x - 1)^p + p(x - 1) \pmod{p(x - 1)^2 \mathbb{Z}[x]}.$$

This implies that if $a \in A_i$ and $x \in \Pi_j$ then

$$a(x^p - 1) \equiv a(x - 1)^p + pa(x - 1) \pmod{pA_i(x - 1)^2}. \qquad (2)$$

As in the preceding proof, it follows easily that if $[A_i, \Pi_j] \leq A_{i+f(j)}$ for all i, then also $[A_i, \Pi_{j+1}] \leq A_{i+f(j+1)}$ for all i; it follows by induction that this holds for all j.

It remains to establish the reverse inclusion, under the assumption that d is not of the form $(p-1)p^s$ for any s. Let us fix $j \geq 0$ and assume, inductively, that $[A_n, \Pi_j] = A_{n+f(j)}$ for every n. Let $i \geq 1$. We shall show that there exist $a \in A_i$ and $x \in \Pi_j$ such that $a(x^p - 1) \notin A_{i+f(j+1)+1}$. From this it follows that $[A_i, \Pi_{j+1}] \not\leq A_{i+f(j+1)+1}$; since $[A_i, \Pi_{j+1}]$ is a closed G-submodule of $A_{i+f(j+1)}$ it must therefore be equal to $A_{i+f(j+1)}$, and the result follows by induction.

Let us take a closer look at (2), where $a \in A_i$ and $x \in \Pi_j$. The first point is that

$$pA_i(x - 1)^2 \leq pA_{i+2f(j)} = A_{i+2f(j)+d} \leq A_{i+f(j+1)+1}.$$

Next, note that

$$a(x - 1)^p \in A_{i+pf(j)},$$
$$pa(x - 1) \in pA_{i+f(j)} = A_{i+f(j)+d}.$$

The significance of our condition on d now becomes apparent: if $pf(j) = f(j) + d$ then $d = (p-1)f(j)$, and $f(j)$ must be equal to p^j since for each $n < j$ we have $(p-1)f(n) < (p-1)f(j)$ (giving $pf(n) < f(n) + d$). Assuming that $d \neq (p-1)p^j$, we infer that $pf(j) \neq f(j) + d$, and consequently that *either* $f(j+1) = pf(j) < f(j) + d$ *or* $f(j+1) = f(j) + d < pf(j)$. In the first case it follows that $a(x^p - 1) \equiv a(x-1)^p \bmod A_{i+f(j+1)+1}$, while in the second case $a(x^p-1) \equiv pa(x-1) \bmod A_{i+f(j+1)+1}$. We consider the cases separately, writing $k = f(j)$.

Case 1 where $f(j+1) = pk$. For $n = 1, \dots, p$ let $C_n = C_{\Pi_j}(A_{i+(n-1)k}/A_{i+nk+1})$. The inductive hypothesis asserts that $C_n < \Pi_j$ for each n, so (by Exercise 3) there exists $x \in \Pi_j \setminus \bigcup_{n=1}^p C_n$. Now let $a \in A_i \setminus A_{i+1}$. Then $A_i = A_{i+1} + \langle a \rangle$, and as $A_{i+1}(x-1) \le A_{i+k}$ it follows that $a(x-1) \notin A_{i+1+k}$, and hence that $A_{i+k} = A_{i+k+1} + \langle a(x-1) \rangle$. Repeating this argument we see that $a(x-1)^p \notin A_{i+1+pk} = A_{i+f(j+1)+1}$. (When $p = 2$, it follows that this case cannot occur.)

Case 2 where $f(j+1) = k + d$. Let $a \in A_i \setminus A_{i+1}$ and $x \in \Pi_j \setminus C_1$. As above we have $A_{i+k} = A_{i+k+1} + \langle a(x-1) \rangle$. It follows that

$$A_{i+k+d} = pA_{i+k} = pA_{i+k+1} + \langle pa(x-1) \rangle$$
$$= A_{i+k+d+1} + \langle pa(x-1) \rangle,$$

giving in this case $pa(x-1) \notin A_{i+k+d+1} = A_{i+f(j+1)+1}$.

In each case, it follows that $a(x^p - 1) \notin A_{i+f(j+1)+1}$, as required.

10.4 Solubility

In this section we complete the proof of Theorem 10.1, modulo two results about Lie algebras to be proved in the next section. The main task is to show that pro-p groups of finite coclass are *soluble*; once this is established the main result will follow quickly. We begin with this reduction.

As before, G denotes an infinite pro-p group of finite coclass r, with lower central series $G_i = \gamma_i(G)$. Recall from §1 that G has finite rank, and that each G_i is open in G, so $(G_i)_{i \ge 1}$ forms a base for the neighbourhoods of 1 in G.

10.16 Lemma *If G is soluble then G_n is abelian for some finite n.*

Proof As G is infinite and soluble, there is a (unique) term N of the derived series of G such that G/N is finite and $N/[N, N]$ is infinite. Put

$K = [N, N]$. It follows from Proposition 1.19 that N is open and K is closed in G, so $K = \bigcap_{i \geq 1} KG_i$. As $\gamma_i(G/K) = KG_i/K$ has finite index in G/K for each i, while G/K is infinite, we see that $KG_i > KG_{i+1}$ for each i. Now let n be so large that $G_n \leq N$ and $|G_i : G_{i+1}| = p$ for all $i \geq n$. It follows from the modular law that $K \cap G_i = K \cap G_{i+1}$ for all such i; hence

$$[G_n, G_n] \leq [N, N] \cap G_n = K \cap G_n = \bigcap_{i=n}^{\infty} K \cap G_i = 1.$$

We can now give the

Proof of Theorem 10.1 (soluble case) Let us assume that G is soluble. Let $m = rp^r$ if $p \neq 2$, $m = (r + 1)2^{r+1}$ if $p = 2$, and put $A = G_m$. Proposition 10.11 shows that A is uniserial in G and that A is uniform. Lemma 10.16 shows that A is virtually abelian, and it follows by Exercise 4.9 that $A \cong \mathbb{Z}_p^d$ for some d. It then follows by Proposition 10.12 that $d = \dim(G) = (p - 1)p^s$ for some $s < r$ if $p \neq 2$, for some $s \leq r + 2$ if $p = 2$.

There exists c_0 such that $|G : G_{i+1}| = p^{i+r}$ for all $i \geq c_0$. Taking $i \geq \max\{c_0, m\}$ we have $|A : G_{i+1}| = p^{i-m}$, since A is uniserial in G, whence $|G : A| = p^{r+m}$. Thus the coclass of G/A is exactly r, and the proof is complete.

The rest of the chapter is devoted to the proof of

10.17 Proposition *Every pro-p group of finite coclass is soluble.*

This depends on the following result, to be proved in the next section:

10.18 Kreknin's Theorem *Let L be a finite-dimensional Lie algebra over a field of characteristic zero. If L has an automorphism γ of finite order such that $C_L(\gamma) = 0$ then L is soluble.*

We know from §1 that the group G has a uniform open normal subgroup U. Suppose we can find an element $g \in G$ such that $C_G(g)$ is finite. Then $C_U(g) = 1$ since U is torsion-free; and it follows that g has finite order, since some positive power of g lies in $U \cap C_G(g) = C_U(g)$. Conjugation by g therefore induces an automorphism γ of finite order on the Lie algebra $L_U = (U, +, (\,,\,))$ discussed in §4.6, and $C_{L_U}(\gamma) = 0$. Now let $L = L_U \otimes_{\mathbb{Z}_p} \mathbb{Q}_p$. This is a finite-dimensional Lie algebra over $\mathbb{Q}_p$, and γ extends by linearity to an automorphism of L, with $C_L(\gamma) = 0$.

Pro-p groups of finite coclass

It follows by Kreknin's Theorem that L is soluble; therefore L_U is soluble, and therefore so is U, by Corollary 7.16. Hence G is soluble. To establish Proposition 10.17, therefore, it now suffices to prove

10.19 Proposition *There exists* $g \in G$ *such that* $\mathrm{C}_G(g)$ *is finite.*

We begin by constructing a second Lie algebra, this time over $\mathbb{F}_p$. Fix a multiple m of $d = \dim(G)$ with $m \geq \max\{(r+1)p^{r+1}, 2d\}$. According to Proposition 10.11, the group G_m is uniform and uniserial in G, so we have

$$G_n^p = \Phi(G_n) = G_{n+d} \text{ for all } n \geq m.$$

For each $i \geq 0$ put $A_i = G_{m+i}/G_{m+i+1}$; thus A_i is cyclic of order p. Writing A_i additively we now put

$$A = \bigoplus_{i=0}^{\infty} A_i.$$

An element of A is called *homogeneous* if it lies in one of the summands A_i.

Because $[G_i, G_j] \leq G_{i+j}$ for all i and j, the formula

$$(xG_{i+1}, yG_{j+1}) = [x, y]G_{i+j+1} \text{ for } x \in G_i, \, y \in G_j \qquad (3)$$

gives a well defined, skew-symmetric bilinear map $(\,,\,) : G_i/G_{i+1} \times G_j/G_{j+1} \to G_{i+j}/G_{i+j+1}$, for each i and j (to see this, use the elementary commutator identity **0.1**).

10.20 Lemma (i) *A is a Lie algebra over* $\mathbb{F}_p$, *with Lie bracket given on homogeneous elements by the formula (3). For each* $i, j \geq 0$,

$$(A_i, A_j) \subseteq A_{i+j+m}. \qquad (4)$$

(ii) *For* $y \in G_j \setminus G_{j+1}$ *the linear mapping* $\delta_y : A \to A$ *given on homogeneous elements by*

$$a\delta_y = (a, \overline{y}) \in A_{i+j} \text{ for } a \in A_i, \, \overline{y} = yG_{j+1} \in G_j/G_{j+1}$$

is a derivation of the Lie algebra A.

(A *derivation* of A is a linear mapping $\delta : A \to A$ such that $(a, b)\delta = (a\delta, b) + (a, b\delta)$ for all $a, b \in A$.)

Proof To prove that A is a Lie algebra it suffices to verify the Jacobi identity on homogeneous elements; this will follow from (ii) with $j \geq m$. The relation (4) simply translates the fact that $[G_{m+i}, G_{m+j}] \leq G_{2m+i+j}$. The proof of (ii) is an easy application of the Hall–Witt identity (see **0.3**). (In fact the lemma amounts to saying that the formula (3) defines a Lie ring structure on the whole of $\bigoplus_{i=1}^{\infty} G_i/G_{i+1}$.)

Now by Lemma 4.10, the mapping $x \mapsto x^p$ induces for each n an isomorphism $G_n/G_{n+d} = G_n/G_n^p \to G_n^p/G_n^{p^2} = G_{n+d}/G_{n+2d}$; as this is clearly a G-module isomorphism it induces in turn an isomorphism $\tau_n : G_n/G_{n+1} \to G_{n+d}/G_{n+d+1}$. Thus τ_{m+i} is an isomorphism $A_i \to A_{i+d}$, and we may define a linear mapping $\tau : A \to A$ by putting $a\tau = a\tau_{m+i}$ for $a \in A_i$. Thus A becomes a module for the polynomial ring $\mathbb{F}_p[t]$, with t acting like τ; then $A_i t^n = A_{i+nd}$ for $i = 0, 1, \ldots, n$, so

$$A = \bigoplus_{i=0}^{d-1} A_i \mathbb{F}_p[t].$$

10.21 Lemma A *is a Lie algebra over* $\mathbb{F}_p[t]$, *and for* $y \in G$ *the derivation* δ_y *is* $\mathbb{F}_p[t]$-*linear.*

Proof Let $x \in G_i$, where $i \geq m$, let $y \in G_j \setminus G_{j+1}$ and put $K = \langle x, [x, y] \rangle$. As $[x, y] \in G_{i+j}$ we have $[K, K] \leq G_{2i+j} \leq G_{i+j+d+1}$, since $i \geq m > d$. It follows (see **0.2**) that $[x^p, y] \equiv [x, y]^p \mod G_{i+j+d+1}$. Thus if $a = xG_{i+1} \in A_{i-m}$ we have

$$(at)\delta_y = [x^p, y]G_{i+d+j+1}$$
$$= [x, y]^p G_{i+d+j+1} = (a\delta_y)t.$$

This shows that δ_y is linear over $\mathbb{F}_p[t]$. Applying this for $j \geq m$ shows that the Lie bracket on A is $\mathbb{F}_p[t]$-linear in the second argument, hence bilinear by skew-symmetry; this is the claim of the lemma.

It follows from Lemma 10.21 that $A(t-1)$ is an ideal in the Lie algebra A. We now define the $\mathbb{F}_p$-Lie algebra

$$B = A/A(t-1).$$

Writing $B_i = (A_i + A(t-1))/A(t-1)$ for $i = 0, 1, \ldots, d-1$ we have $B = B_0 \oplus \cdots \oplus B_{d-1}$, each B_i being a 1-dimensional $\mathbb{F}_p$-subspace of B. Let n^* denote the least non-negative residue of an integer n modulo d. Since $A_k t = A_{k+d}$ for each k, it follows from (4) that

$$(A_i, A_j) \subseteq A_{(i+j+m)^*} t^e \subseteq A_{(i+j+m)^*} + A(t-1)$$

for some $e \geq 0$. Now we have chosen m to be a multiple of d: so $(i+j+m)^* = (i+j)^*$, and it follows that

$$(B_i, B_j) \subseteq B_{(i+j)^*} \quad \text{for } 0 \leq i, j < d. \tag{5}$$

We have done enough scene-setting; it is time to prove something. We say that a derivation δ of B has *degree* 1 if $B_i \delta \subseteq B_{(i+1)^*}$ for $i = 0, \ldots, d-1$.

10.22 Lemma *For each non-zero element $b \in B$ there exists a derivation δ of B of degree 1 such that $b\delta \neq 0$.*

Proof Suppose first that b is homogeneous; that is, $b = a + A(t-1)$ where $0 \neq a \in A_i$ for some $i < d$. Say $a = xG_{m+i+1}$, where $x \in G_{m+i} \backslash G_{m+i+1}$. Then $G_{m+i} = \langle x \rangle G_{m+i+1}$, and it follows that $[\langle x \rangle, G]$ is not contained in G_{m+i+2}; thus there exists $y \in G$ such that $[x, y] \notin G_{m+i+2}$. Note that $y \notin G_2$, since $[G_{m+i}, G_2] \leq G_{m+i+2}$. Therefore $a\delta_y = [x, y]G_{m+i+2} \in A_{i+1} \backslash 0$. It follows similarly that $A_j \delta_y \subseteq A_{j+1}$ for every j. As δ_y is $\mathbb{F}_p[t]$-linear, it induces a derivation δ of B, which has degree 1 by the preceding remark.

We claim that $b\delta \neq 0$. For $b\delta = a\delta_y + A(t-1)$, so $b\delta = 0$ only if $a\delta_y \in A(t-1)$; but it is easy to see that $A(t-1)$ contains no non-zero homogeneous element of A (think about leading coefficients), so the claim follows from the fact that $0 \neq a\delta_y \in A_{i+1}$.

Finally, suppose $b = b_0 + \cdots + b_{d-1}$, where $b_j \in B_j$ for each j, and $b_i \neq 0$ for some i. We can find a derivation δ of degree 1 such that $b_i \delta \neq 0$; the component of $b\delta$ in $B_{(i+1)^*}$ is just $b_i \delta$, so in fact $b\delta \neq 0$. This completes the proof.

The proof of the next step is deferred to the next section:

10.23 Proposition *Let $B = B_0 \oplus \cdots \oplus B_{d-1}$ be a Lie algebra over $\mathbb{F}_p$, where each B_i is a 1-dimensional subspace of B and (5) holds. If*

- $d = ap^s$ *for some $s \geq 0$ and some $a \in \{1, \ldots, p-1\}$, and*
- *the assertion of **10.22** holds,*

then B is abelian.

It is at this stage that Proposition 10.12 is invoked. With Lemma 10.22 and Proposition 10.23 it now shows that B *is abelian*; we are ready for the

Proof of Proposition 10.19 We have shown that the Lie algebra $B = A/A(t-1)$ is abelian, so $(A, A) \subseteq A(t-1)$. It is clear from the definition that (A, A) is the direct sum of its homogeneous components; on the other hand, as observed above, $A(t-1)$ contains no non-zero homogeneous elements. It follows that $(A, A) = 0$, in other words that *A is abelian.*

We can now forget about Lie algebras and return to our group G. To say that A is abelian means that $[G_i, G_j] \leq G_{i+j+1}$ for all $i, j \geq m$. Let n be the least integer such that

$$[G_n, G_j] \leq G_{n+j+1} \text{ for all sufficiently large } j;$$

thus $n \leq m$ and $n > 1$ (since $[G_1, G_j] = G_{j+1}$ for all j).

Let $x \in G_{n-1}$, and suppose that $[G_i, x] \leq G_{n+i}$ for some large value of i. Then $H = G_n \langle x \rangle \lhd G$, and applying the Three-subgroup Lemma gives

$$\begin{aligned}
[G_{i+i}, H] &= [G_i, G, H] \\
&\leq [H, G_i, G][G, H, G_i] \\
&\leq [G_{n+i}, G][G_n, G_i] \leq G_{n+i+1},
\end{aligned}$$

so $[G_{i+1}, x] \leq G_{n+i+1}$. It follows by induction that $[G_j, x] \leq G_{n+j}$ for all large j.

As $[G_j, G_n] \leq G_{n+j}$ for all j and G_{n-1}/G_n is finite, one of the following must hold:

(a) there exists $x \in G_{n-1}$ such that $[G_i, x] \leq G_{n+i}$ for no large value of i; or

(b) $[G_j, G_{n-1}] \leq G_{n+j}$ for all sufficiently large j.

The second possibility is excluded by the minimal choice of n; so we have (a). We claim that now $C_G(x)$ is finite: this will complete the proof.

Put $C = C_G(x)$, and let k be maximal such that $[G_k, x] \leq G_{n+k}$. Then for $i > k$ we have $G_i \not\leq CG_{i+1}$, since $[G_i, x] \not\leq G_{n+i}$ while $[G_{i+1}, x] \leq [G_{i+1}, G_{n-1}] \leq G_{n+i}$. If also $i \geq m$, this implies by the modular law that $C \cap G_i = C \cap G_{i+1}$, since then $|G_i : G_{i+1}| = p$. Hence putting $q = \max\{k+1, m\}$ we have

$$C \cap G_q = \bigcap_{i=q}^{\infty} C \cap G_i = 1.$$

It follows that C is finite as claimed.

Let us recap the rather extraordinary strategy of the whole proof, which proceeds via three distinct Lie algebras. In logical order it goes like this. *Step 1*: prove that the finite Lie algebra B is abelian; this depends on the uniserial embedding of G_m in G, and on the apparent arithmetical accident that $\dim(G)$ has the form $(p-1)p^s$. *Step 2*: deduce that the graded Lie algebra A is abelian. This translates into the rather weak statement that $(G_i, G_j) \leq G_{i+j+1}$ for large i and j. *Step 3*: deduce that G contains an element with finite centraliser, again using the uniserial embedding of G_m in G. *Step 4*: Use this element to give an automorphism of finite order of the intrinsic Lie algebra L_U. *Step 5*: deduce via Kreknin's Theorem that L_U is soluble. *Step 6*: Deduce that U is soluble, hence that G is soluble. *Step 7*: Deduce that G is virtually abelian, since it has finite coclass.

Schematically,

$$B \text{ abelian} \Rightarrow A \text{ abelian} \Rightarrow L_U \text{ soluble} \Rightarrow G \text{ soluble}$$

$$\Rightarrow G \text{ virtually abelian} \Rightarrow U \text{ abelian} \Rightarrow L_U \text{ abelian};$$

we mention the last two implications to indicate the roundabout route that leads from properties of the graded $\mathbb{F}_p$-Lie algebra B to those of the 'intrinsic' $\mathbb{Z}_p$-Lie algebra L_U.

10.5 Two theorems about Lie algebras

Here we prove the two results quoted in the preceding section.

10.18 Kreknin's Theorem *Let L be a finite-dimensional Lie algebra over a field of characteristic 0, having an automorphism γ of finite order such that $C_L(\gamma) = 0$. Then L is soluble.*

Proof This is by induction on $\dim(L)$. If $(L, L) \neq L$, we may suppose inductively that (L, L) is soluble: then L is soluble and we are done. So let us assume that $L = (L, L)$, and aim to deduce that then $L = 0$.

Say γ has order q. We may assume without loss of generality that the field contains a primitive qth root of unity ζ. Then L decomposes as a direct sum

$$L = L_0 \oplus \cdots \oplus L_{q-1},$$

where $L_i = \ker(\gamma - \zeta^i 1_L)$ is the ζ^i-eigenspace of γ, for each i. Note that $L_0 = C_L(\gamma) = 0$. If $a \in L_i$ and $b \in L_j$ then

$$(a, b)^\gamma = (a^\gamma, b^\gamma) = (\zeta^i a, \zeta^j b) = \zeta^{i+j}(a, b),$$

so we have

$$(L_i, L_j) \subseteq L_{(i+j)^*} \tag{6}$$

for each i and j, where n^* denotes the least non-negative residue of an integer n modulo q. We shall write '$\equiv$' to mean 'congruent modulo q'.

Put $H_{q-1} = 0$ and for $n < q - 1$ let H_n be the Lie subalgebra of L generated by H_{n+1} and L_{n+1}; thus $L = H_0 \supseteq H_1 \supseteq \cdots \supseteq H_{q-1} = 0$. Now fix $n \geq 1$ and suppose we have shown that $L = H_{n-1}$. Then L is spanned by Lie products of the form

$$w = (u_1, \ldots, u_s), \text{ with } u_j \in L_{k(j)} \text{ and } n \leq k(j) < q \text{ for each } j \tag{7}$$

(Exercise 4); and $w \in L_i$ where $\sum_{j=1}^r k(j) \equiv i$. It follows that (L, L) is spanned by elements of the form (v, w) with w as in (7) and $v \in L_k$, for $1 \leq k < q$; and $(v, w) \in L_{(k+i)^*}$ if $w \in L_i$. With (6) this implies that $L_n = (L, L) \cap L_n$ is spanned by such elements for which

$$k + \sum_{j=1}^r k(j) \equiv n.$$

By repeated applications of the Jacobi identity (see Exercise 4), we can re-write $(v, w) = (v, (u_1, \ldots, u_r))$ as a linear combination of terms of the form $(v, u_{1\sigma}, \ldots, u_{r\sigma})$, where σ runs over certain permutations of $\{1, \ldots, r\}$. We claim now that each such term lies in H_n. Given this claim, it follows that $L_n \subseteq H_n$, and hence that $H_{n-1} = H_n$. As $L = H_0$ and $H_{q-1} = 0$ we see by induction that $L = 0$, as required.

It only remains to establish the claim. Put $t = k(r\sigma)$ and $s = k + \sum_{j=1}^{r-1} k(j\sigma)$. Then $(v, u_{1\sigma}, \ldots, u_{r\sigma}) = (a, u)$ where $a \in L_{s^*}$ and $u = u_{r\sigma} \in L_t$; and we have

$$n \leq t < q, \quad s + t \equiv n.$$

If $s \equiv 0$ then $a \in L_0 = 0$ so $(a, u) = 0 \in H_n$. Suppose that $s \not\equiv 0$. Then $t \not\equiv n$ so $n < t < q$. It follows that $s^* = n + q - t$ also lies strictly between n and q, showing that both a and u are in H_n. Thus again $(a, u) \in H_n$, and the proof is complete. (It is not in fact necessary to assume that the characteristic is zero; the argument is valid provided only that the characteristic of the field does not divide the order of γ.)

The second result concerns a *cyclically-graded* $\mathbb{F}_p$-Lie algebra $B = B_0 \oplus \cdots \oplus B_{d-1}$: that is, $(B_i, B_j) \subseteq B_{i+j}$ for all i and j, subscripts being interpreted in $\mathbb{Z}/d\mathbb{Z}$ (this convention remains in force for the rest of this section). A derivation δ of B is said to have *degree* f if $B_i\delta \subseteq B_{i+f}$

for each i; the set of all such derivations is denoted $\mathrm{Der}_f(B)$; it is clearly a vector subspace of $\mathrm{Hom}_{\mathbb{F}_p}(B,B)$. Finally, we say that $\mathrm{Der}_f(B)$ *acts without constants* if for each $b \in B \setminus 0$ there exists $\delta \in \mathrm{Der}_f(B)$ such that $b\delta \neq 0$.

10.24 Lemma *Assume that* $\dim_{\mathbb{F}_p}(B_i) = 1$ *for each* i. *If* $\mathrm{Der}_1(B)$ *acts without constants then so does* $\mathrm{Der}_{p^k}(B)$ *for every* $k \geq 0$.

Proof Let $B_i = \mathbb{F}_p b_i$ for each i. Suppose that for some f we know that $\mathrm{Der}_f(B)$ acts without constants. Then for each i, the annihilator D_i of b_i in $\mathrm{Der}_f(B)$ is a proper subspace of $\mathrm{Der}_f(B)$. It follows that for each j, the set $D_j \cup D_{j+f} \cup \ldots \cup D_{j+(p-1)f}$ is properly contained in $\mathrm{Der}_f(B)$ (see Exercise 3), so we can find $\delta_j \in \mathrm{Der}_f(B) \setminus (D_j \cup D_{j+f} \cup \ldots \cup D_{j+(p-1)f})$. Then $b_j \delta_j^p \neq 0$. If $0 \neq b \in B$ then $b = \sum_{j=0}^{d-1} \alpha_j b_j$ with $\alpha_j \neq 0$ for some j; the B_{j+pf}-component of $b\delta_j^p$ being $\alpha_j b_j \delta_j^p \neq 0$ we see that $b\delta_j^p \neq 0$. An elementary calculation (Exercise 5) shows that δ_j^p is a derivation. Thus $\delta_j^p \in \mathrm{Der}_{pf}(B)$ and we conclude that $\mathrm{Der}_{pf}(B)$ acts without constants.

The lemma follows by induction.

10.23 Proposition *Let* $B = B_0 \oplus \cdots \oplus B_{d-1}$ *be a cyclically-graded Lie algebra over* $\mathbb{F}_p$, *where each* B_i *is a 1-dimensional subspace of* B. *If* $\mathrm{Der}_1(B)$ *acts without constants and* $d = ap^s$, *where* $1 \leq a < p$ *and* $s \geq 0$, *then* B *is abelian.*

Proof As above, let $B_i = \mathbb{F}_p b_i$ for each i, and define $\lambda_i \in \mathbb{F}_p$ by $(b_0, b_i) = \lambda_i b_i$. The Jacobi identity applied to $(b_0, (b_i, b_j))$ gives $\lambda_{i+j}(b_i, b_j) = (\lambda_i + \lambda_j)(b_i, b_j)$ for each i and j, whence

$$(b_i, b_j) \neq 0 \Rightarrow \lambda_{i+j} = \lambda_i + \lambda_j. \tag{8}$$

Suppose now that $j = p^k$ for some $k \geq 0$, and that $(b_i, b_j) = 0$. By Lemma 10.24, we can find a derivation $\delta \in \mathrm{Der}_{p^k}(B)$ such that $\delta b_i \neq 0$, so $\delta b_i = \mu b_{i+j}$ where $\mu \neq 0$. Then

$$\begin{aligned}
\lambda_i \mu b_{i+j} = (b_0, b_i)\delta &= (b_0 \delta, b_i) + (b_0, b_i \delta) \\
&= (\nu b_j, b_i) + (b_0, \mu b_{i+j}) \quad \text{for some } \nu \in \mathbb{F}_p \\
&= 0 + \lambda_{i+j} \mu b_{i+j};
\end{aligned}$$

thus we have

$$(b_i, b_{p^k}) = 0 \Rightarrow \lambda_{i+p^k} = \lambda_i. \tag{9}$$

Let $t \geq 2$ be an integer. Repeated applications of (9) and/or (8) show that

$$\begin{aligned}
\lambda_{tp^k} &= \lambda_{2p^k + (t-2)p^k} \\
&= \lambda_{2p^k} + n'\lambda_{p^k} \text{ where } 0 \leq n' \leq t - 2 \qquad (10) \\
&= \lambda_{p^k} + n'\lambda_{p^k} = n\lambda_{p^k} \text{ where } 1 \leq n < t;
\end{aligned}$$

note that if $t \leq p$ then n is invertible in $\mathbb{F}_p$. Taking $t = a$ if $a \geq 2$, it follows that

$$\lambda_{p^s} = n^{-1}\lambda_{ap^s} = n^{-1}\lambda_d = n^{-1}\lambda_0 = 0.$$

Repeated applications of (10) with $t = p$ now show that $\lambda_1 = 0$, and it follows by repeated applications of (9) and/or (8) that $\lambda_i = 0$ for all i.

Thus $(b_0, B) = 0$. Now let $j \geq 0$ and suppose inductively that $(b_j, B) = 0$. There exists $\delta \in \mathrm{Der}_1(B)$ such that $b_j\delta = \mu b_{j+1}$ for some $\mu \neq 0$, and then for $b \in B$ we have

$$(b_{j+1}, b) = \mu^{-1}(b_j\delta, b) = \mu^{-1}((b_j, b)\delta - (b_j, b\delta)) = 0.$$

It follows by induction that $(b_j, B) = 0$ for all j, and the proof is complete.

Notes

Prehistory The theory of p-groups of maximal class (i.e. coclass 1) was developed by Blackburn (1958). It was advanced by R.T. Shepherd in his 1970 Chicago Ph.D., and completed by Leedham-Green and McKay (1976, 1978); in these papers they showed that the p-groups of maximal class can be classified by relating them to certain *p-adic space groups*.

History Leeham-Green and Newman (1980) defined coclass, and proposed a classification of p-groups of fixed coclass in terms of p-adic space groups. Their programme was expressed in a series of five conjectures, of which we state

Conjecture A *Given p and r, there is a positive integer f such that every finite p-group of coclass r has a normal subgroup which is nilpotent of class at most 2 and has index at most p^f.*

Conjecture C *Every pro-p group of finite coclass is soluble.*

Conjecture D   *Given p and r, there are only finitely many infinite pro-p groups of coclass r (up to isomorphism).*

It is the case (though not obviously) that Conjecture A implies all the others. A large number of papers contributed to the ultimate proof of these conjectures; we only mention a few highlights:

(a) Leedham-Green (1994a) proved that every pro-p group of finite coclass is p-adic analytic, of dimension $(p-1)p^s$ for some s.

(b) Donkin (1987) proved that if $p \geq 5$, every pro-p group of finite coclass satisfying this conclusion is soluble; this proof depends on the classification of semisimple algebraic groups over $\mathbb{Q}_p$. Together with (a) this establishes Conjecture C for primes $p \geq 5$.

(c) Leedham-Green (1994b) completes the proof of Conjecture A (for $p \geq 5$), and shows explicitly how every finite p-group of a given coclass may be constructed by suitably 'twisting' a finite image of some infinite pro-p group of finite coclass.

(d) Shalev and Zel'manov (1992) give an elementary self-contained proof of Donkin's theorem (b), valid for all primes p.

(e) Shalev (1994) gives a 'constructive' self-contained proof of Conjecture A, valid for all primes; this is independent of all the previous work (logically though not conceptually), and gives explicit bounds. It does not involve pro-p groups at all, but makes intensive use of the theory of powerful p-groups.

This chapter Theorems 10.1 and 10.2 imply Conjectures C and D. Sections 1–3 are based on arguments from Shalev (1994); some of these were modelled on the earlier work of Leedham-Green (1994a) (which was in circulation for some years before the publication date). Section 4, with Proposition 10.23, is essentially an exposition of Shalev and Zel'manov (1992); Theorem 10.18 is due to Kreknin (1963).

Exercises

In the following exercises, r denotes a fixed positive integer.

1. (i) Let P and Q be finite p-groups of coclass r, with $p|P| = |Q| = p^{c+r} > p^r$. Show that if $\psi : Q \to P$ is an epimorphism then $\ker \psi = \gamma_c(Q)$. Deduce that if $\varphi, \psi : Q \to P$ are surjective and $\alpha \in \operatorname{Aut}(Q)$ then there exists $\beta \in \operatorname{Aut}(P)$ such that $\alpha\phi = \psi\beta$.

(ii) Let $(P_n)_{n \in \mathbb{N}}$ be a family of p-groups of coclass r with $|P_{n+1}| = p|P_n|$ for each n, and let $\psi_n : P_{n+1} \to P_n$ be an epimorphism for each n. Put

$$G_\psi = \varprojlim_{n \to \infty} (P_n, \psi_n).$$

Prove that if (ϕ_n) is another such sequence of epimorphisms then $G_\phi \cong G_\psi$.

[*Hint:* Use (i) to construct a suitable inverse system $\cdots \to \operatorname{Aut}(P_{n+1}) \to \operatorname{Aut}(P_n) \to \cdots$.]

(iii) Let $G = G_\psi$ as in (ii), and write $\pi_n : G \to P_n$ for the natural homomorphism, $K_n = \ker \pi_n$. Show that π_n is surjective for each n. Deduce that for each $i \geq 1$,

$$|G : K_n\gamma_{i+1}(G)| = p^{i+r} \quad \text{for all sufficiently large } n.$$

Hence show that the pro-p group G has coclass r.

2. Let G be a pro-p group of coclass r, and let c_0 be the least positive integer such that $|G : \gamma_{i+1}(G)| = p^{i+r}$ for all $i \geq c_0$. Put $Q_i = G/\gamma_{c_0+i}(G)$ for each $i \geq 1$.

(i) Show that $G = \varprojlim_{n \to \infty} (Q_n)$.

(ii) Suppose that G is the group G_ψ defined in Exercise 1. Show that $P_i \cong Q_{i+k}$ for all $i \geq 1$, for a certain integer $k \geq 0$. Deduce that if P_1 has no proper image of coclass r then $P_i \cong Q_i$ for all $i \geq 1$.

3. Show that a finite p-group cannot be the union of p proper subgroups. [*Hint:* count elements!]

Deduce that a finitely generated pro-p group cannot be the union of p proper closed subgroups.

In Exercises 4 and 5, L denotes a Lie algebra over a field F, and a, a_i, b, b_i, etc. elements of L. 'Left-normed brackets' are defined by

$$(a_1, a_2, \dots, a_n) = ((a_1, a_2, \dots, a_{n-1}), a_n).$$

4. (i) Prove that

$$(a, (b_1, b_2, \ldots, b_n)) = \sum_{\sigma \in \Sigma} \pm (a, b_{1\sigma}, b_{2\sigma}, \ldots, b_{n\sigma})$$

where Σ is a certain subset of $\mathrm{Sym}(n)$.

[*Hint:* Put $c = (b_1, b_2, \ldots, b_{n-1})$, expand $(a, (c, b_n))$ by the Jacobi identity, and apply an induction hypothesis to expand (a, c) and $((a, b_n), c)$.]

(ii) Let S be a subset of L, and let V be the subspace of L spanned by S and all left-normed Lie products $(a_1, a_2, \ldots, a_n)$ with $a_i \in S$ for all i and $n \geq 2$. Deduce that $(V, V) \subseteq V$, hence that V is the Lie subalgebra of L generated by S.

5. Let δ be a derivation of L. Prove *Leibniz's formula*:

$$(a, b)\delta^n = \sum_{i=0}^{n} \binom{n}{i} (a\delta^i, b\delta^{n-i}).$$

Deduce that if $\mathrm{char}(F) = p \neq 0$ then δ^p is a derivation.

We recall some facts from algebraic number theory (see Fröhlich and Taylor (1991), Chapter VI). Let ζ be a complex primitive p^tth root of unity, where $t \geq 1$, and put $d = \varphi(p^t) = p^{t-1}(p-1)$. Then

$$\mathfrak{o} = \mathbb{Z}[\zeta] = \bigoplus_{j=0}^{d-1} \zeta^j \mathbb{Z}$$

is the ring of integers in the algebraic number field $\mathbb{Q}(\zeta)$, and $\mathfrak{p} = (\zeta - 1)\mathfrak{o}$ is the unique maximal ideal of $\mathfrak{o}$ containing p. It satisfies $\mathfrak{o}/\mathfrak{p} \cong \mathbb{F}_p$. We write $\mathfrak{o}_p = \mathbb{Z}_p[\zeta] = \bigoplus_{j=0}^{d-1} \zeta^j \mathbb{Z}_p$. Multiplication by ζ gives an automorphism z of the additive group of $\mathfrak{o}$, and of the $\mathbb{Z}_p$-module $\mathfrak{o}_p$.

6. (i) Let $\Gamma = \mathfrak{o} \rtimes \langle z \rangle$. Show that $\gamma_i(\Gamma) = \mathfrak{p}^i$ for each $i \geq 2$. Deduce that each $\Gamma/\gamma_i(\Gamma)$ is a finite p-group of order p^{t+i-1} and coclass t.

(ii) Let $G = \mathfrak{o}_p \rtimes \langle z \rangle$. Show that $G \cong \widehat{\Gamma}_p$, and that G is a pro-p group of coclass t and dimension d.

(iii) For each s in the range $0 \leq s < r$ construct a pro-p group of coclass r and dimension $p^s(p-1)$.

[*Hint:* consider $G \times C_{p^n}$, where t and n are suitably chosen.]

(A more interesting exercise is to construct *p-adic space groups* of given coclass and dimension; a p-adic space group is an extension of $A \cong \mathbb{Z}_p^d$

by a finite p-group in which A is equal to its own centraliser. See Section 3 of Leedham-Green and Newman (1980).)

7. Let us say that a pro-p group G has *p-coclass* k if $|G : P_i(G)| = p^{k+i}$ for all sufficiently large i. Prove that a pro-p group G has finite p-coclass if and only if *either* G has finite coclass *or* G is infinite and virtually pro-cyclic.

[*Hint for* 'only if': Put $G_i = P_i(G)$. There exists m such that $|G_i : G_{i+1}| = p$ for all $i \geq m$. *Case 1*: there exists $i \geq m$ such that $G_{i+1} = \overline{G_i^p}$. Show that then G_i is pro-cyclic. *Case 2*: for all $i \geq m$, $G_{i+1} > G_i^p$. Show that then $G_{i+1} = \overline{[G_i, G]}$ for each $i \geq m$. Show that for some f and some $r \geq m$, $\gamma_f(G) = G_r$, and deduce that G has finite coclass.]

11

Dimension subgroup methods

The main result of Chapter 3 was that a finitely generated pro-p group G has finite rank if and only if G is virtually powerful (Theorem 3.13). We also saw that this is equivalent to 'polynomial growth' (relative to p^n) of either of the functions

$$\sigma_n(G) = |\{H \leq_o G \mid |G : H| \leq p^n\}|$$
$$q_n(G) = |G : \overline{G^{p^n}}|.$$

As an application of the main result of this chapter, we shall give refinements of these results; the characterisations obtained this way are best possible. The main result is a characterisation of a different sort: namely in terms of the behaviour of the *modular dimension series*. It is explained in §11.1, and proved in §11.3; the main work is done in §11.2, where we establish various properties of the modular dimension subgroups (valid in all groups, not just pro-p groups). The applications are given in Sections 11.1 and 11.4.

11.1 Modular dimension subgroups

In this section and the next, G denotes an arbitrary group (not necessarily a pro-p group); p, as usual, denotes a fixed prime.

We start by defining a series of characteristic subgroups $D_n = D_n(G)$ by

11.1 Definition $D_1 = G$, and for $n > 1$

$$D_n = D_{n^*}^p \cdot \prod_{i+j=n} [D_i, D_j]$$

where $n^* = \lceil n/p \rceil$ is the least integer r such that $pr \geq n$.

270

Thus $[D_i, D_j] \leq D_{i+j}$ and $D_i^p \leq D_{pi}$ for all i and j, and the series (D_n) is the fastest descending series with these properties starting at $D_1 = G$. It is clear that $D_i \geq \gamma_i(G)$ for each i. The subgroups D_i are the *dimension subgroups* of G in characteristic p: that is, D_n is the kernel of the natural homomorphism of G into the unit group of $k[G]/I^n$, where k is any field of characteristic p and I is the augmentation ideal of $k[G]$. However, we shall not use this fact in this chapter (except to justify the title); it will be proved in Chapter 12, along with further properties of the 'dimension series' (D_n). However, in the next section we establish the following useful 'closed formula', discovered by Lazard:

11.2 Theorem *For each* n,

$$D_n(G) = \prod_{ip^j \geq n} \gamma_i(G)^{p^j}.$$

Suppose now that G is a pro-p group of finite rank. Then, by Theorem 3.16, there exist constants c and s such that $|G : G^{p^k}| \leq cp^{ks}$ for all k. It follows from the definition that $D_{p^k} \geq G^{p^k}$ for each k, so we have

$$|G : D_{p^k}| \leq cp^{ks} \qquad (1)$$

for all k. If there exists an m such that $D_n > D_{n+1}$ for every $n \geq p^m$ then

$$|G : D_{p^k}| \geq |G : D_{p^m}| \cdot p^{p^k - p^m}$$

for every $k \geq m$; this contradicts (1) for large values of k, so we have established

11.3 Proposition *If* G *is a pro-p group of finite rank then* $D_n(G) = D_{n+1}(G)$ *for infinitely many values of* n.

The main result of this chapter is the remarkable fact that Proposition 11.3 has a converse, indeed a very strong one:

11.4 Theorem *Let* G *be a finitely generated pro-p group. Then* G *has finite rank if and only if* $D_n(G) = D_{n+1}(G)$ *for some* n.

We shall see (Lemma 11.16 below) that if G is a finitely generated pro-p group then each of the subgroups D_n is *open* in G. So the 'if' part of Theorem 11.4 will follow by the results of Chapter 3 from the next result (due to David Riley):

11.5 Theorem *Let G be a finitely generated pro-p group and n a positive integer. If $D_n(G) = D_{n+1}(G)$ then $D_i(G)$ is powerful for all $i \geq n$.*

This will be proved in §11.3, along with a number of further characterisations of pro-p groups of finite rank, essentially due to Lazard.

We conclude this section with two applications of Theorem 11.4. The first concerns the function $q_n(G) = |G : \overline{G^{p^n}}|$; it follows by an argument like the proof of Proposition 11.3, above, which we leave as an exercise:

11.6 Corollary *Let G be a finitely generated pro-p group. If $q_n(G) < p^{p^n}$ for some $n \geq 1$ then G has finite rank.*

This result can be improved; the best-possible version will be established in §11.4. The second application, due to Aner Shalev, concerns subgroup growth :

11.7 Theorem *Let G be a finitely generated pro-p group and let $c < \frac{1}{8}$ be a positive constant. If*

$$\sigma_k(G) \leq p^{ck^2} \tag{2}$$

for all sufficiently large k, then G has finite rank.

We shall see in Exercise 13.13 that this result is best possible, in the sense that there exist finitely generated pro-p groups of infinite rank which satisfy (2) for all k, with some constant c. The proof depends on the following lemma; here, G denotes a finitely generated pro-p group,

$$d_n = \mathrm{d}(D_{2^n}) \text{ and } i_n = \log_p |G : D_{2^n}|.$$

11.8 Lemma *Suppose that there exist $\varepsilon > 0$ and $m \in \mathbb{N}$ such that*

$$d_n < (1 - \varepsilon)i_n \quad \text{for all } n \geq m.$$

Then $D_i = D_{i+1}$ for some i.

Proof We quote the fact that each D_i is open in G (Lemma 11.16, below). This implies that $D_{2^{n+1}} \leq \Phi(D_{2^n})$ for each n, and hence that

$$p^{d_n} = |D_{2^n} : \Phi(D_{2^n})| \geq |D_{2^n} : D_{2^{n+1}}| = p^{i_{n+1} - i_n}.$$

It follows that

$$i_{n+1} - i_n \leq d_n < (1 - \varepsilon)i_n \text{ for } n \geq m.$$

Thus $i_{n+1} \leq (2 - \varepsilon)i_n$ for $n \geq m$, and so

$$i_{m+k} \leq (2 - \varepsilon)^k i_m < 2^{m+k} - 1$$

for large enough values of k. But if $D_i > D_{i+1}$ for all $i < 2^{m+k}$, then $i_{m+k} \geq 2^{m+k} - 1$; hence it must be the case that $D_i = D_{i+1}$ for some $i < 2^{m+k}$.

Proof of Theorem 11.7 Suppose that G has infinite rank. Choose a small positive constant ε. Then (keeping the above notation) Theorem 11.4 and Lemma 11.8 show that

$$d_n \geq (1 - \varepsilon)i_n \qquad (3)$$

for infinitely many values of n. Now choose such a value of n and put

$$H = D_{2^n}.$$

The $\mathbb{F}_p$-vector space $H/\Phi(H)$ has dimension d_n; so if the positive integer r satisfies

$$3r \leq (1 - \varepsilon)i_n < 3(r + 1)$$

then $H/\Phi(H)$ will have at least p^{2r^2} subspaces of codimension r. Each of these corresponds to a subgroup of index p^{i_n+r} in G, so putting $k = i_n + r$ we have $\sigma_k(G) \geq p^{2r^2}$.

Now

$$2r^2 > 2 \left(\frac{(1 - \varepsilon)k - 3}{4 - \varepsilon} \right)^2 > ck^2,$$

provided ε is small and k is large relative to the positive constant $\frac{1}{8} - c$. Thus (2) fails for this value of k. Since $k \geq i_n + 1 \geq 2^n$, and we may choose n arbitrarily large, we see that (2) fails for arbitrarily large k. The result follows.

Of course, the converse of Theorem 11.7 follows from Theorem 3.19.

11.2 Commutator identities

Throughout this section, G denotes an arbitrary group. We write

$$D_n = D_n(G), \quad \Gamma_n = \gamma_n(G)$$

for each n; we shall constantly be using the fundamental properties

$$[D_i, D_j] \leq D_{i+j}, \quad D_i^p \leq D_{pi},$$

without special mention. Recall that for a positive integer n,

$$n^* = \lceil n/p \rceil$$

denotes the least integer r such that $pr \geq n$.

The basis of all our calculations is

11.9 Lemma *Let x and y be elements of a group H and let $K = \langle x, [x,y] \rangle$. Then for all $j \geq 1$,*
   (i)

$$(xy)^{p^j} \equiv x^{p^j} y^{p^j} \mod \gamma_2(H)^{p^j} \prod_{\ell=1}^{j} \gamma_{p^\ell}(H)^{p^{j-\ell}};$$

   (ii)

$$[x^{p^j}, y] \equiv [x,y]^{p^j} \mod \gamma_2(K)^{p^j} \prod_{\ell=1}^{j} \gamma_{p^\ell}(K)^{p^{j-\ell}}.$$

Proof We quote the well-known identity of *Hall and Petrescu*, proved in Appendix A: for each $n \geq 1$,

$$x^n y^n = (xy)^n c_2(x,y)^{\binom{n}{2}} \dots c_n(x,y)^{\binom{n}{n}}, \qquad (4)$$

where, for each r, $c_r(x,y) \in \gamma_r\langle x,y \rangle$.

Now take $n = p^j$. If $r = sp^\ell$, where $1 \leq s < p$ and $\ell < j$, then $\binom{n}{r}$ is divisible by $p^{j-\ell}$ (Exercise 1). Hence $c_r(x,y)^{\binom{n}{r}} \in \gamma_r(H)^{p^{j-\ell}} \subseteq \gamma_{p^\ell}(H)^{p^{j-\ell}}$. Thus part (i) follows from (4).

To deduce (ii), we replace y by $[x,y]$ and H by K. Then (i) becomes

$$x^{p^j}[x,y]^{p^j} \equiv (x[x,y])^{p^j} \mod \gamma_2(K)^{p^j} \prod_{\ell=1}^{j} \gamma_{p^\ell}(K)^{p^{j-\ell}}. \qquad (5)$$

Since

$$(x[x,y])^{p^j} = (x^y)^{p^j} = x^{p^j}[x^{p^j}, y],$$

we can cancel x^{p^j} on each side of (5) to obtain (ii).

11.10 Corollary *Let $x \in D_r$ and $y \in D_s$.*
   (i) *If $s \geq r$ then $(xy)^p \equiv x^p y^p \mod D_{2pr}\gamma_p(D_r)$;*
   (ii) *if $s > r$ then $(xy)^p \equiv x^p \mod D_{pr+1}$;*
   (iii) *$[x^p, y] \in \langle w \rangle D_{pr+s+t}$ where $w = [y,_p x]$ and $t = \min\{r,s\}$.*

Proof Parts (i) and (ii) are direct applications of 11.9(i), with $j = 1$:

note that if $s \geq r$ and $H = \langle x, y \rangle$ then $H \leq D_r$, $\gamma_2(H) \leq D_{r+s}$ and $\gamma_p(H) \leq D_{(p-1)r+s}$. Part (iii) follows similarly from Lemma 11.9(ii); for if $K = \langle x, [x, y] \rangle$ then $\gamma_2(K) \leq D_{2r+s}$, and $\gamma_p(K) \leq \langle w \rangle D_{pr+s+t}$: to see this, note that any commutator of length $\geq p$ in x and $[x, y]$ either involves at least two occurrences of $[x, y]$, in which case it lies in D_{pr+2s}, or it involves at least p occurrences of x, in which case it lies in $D_{(p+1)r+s}$, or it is one of $[[x, y],_{p-1} x]$, $[x, [x, y],_{p-2} x]$; each of these is conjugate in K to w or w^{-1}, and hence lies in $\langle w \rangle D_{(p+1)r+2s}$.

The next lemma is a stepping-stone to the first main result of this section; it will subsequently be strengthened (see 11.14):

11.11 Lemma *Let i and j be positive integers. Then*

(i)

$$[D_i, D_j] \leq \Gamma_{i+j} D_{i+j+1};$$

(ii)

$$\gamma_p(D_i) \leq \Gamma_{pi} D_{pi+1}.$$

Proof We start by proving that

$$[D_i, \Gamma_j] \leq \Gamma_{i+j} D_{i+j+1}. \tag{6}$$

This is clear if $i = 1$; now let $i > 1$ and argue by induction on i. Putting $r = i^*$, what we have to show is that

$$h + k = i \Rightarrow [D_h, D_k, \Gamma_j] \leq \Gamma_{i+j} D_{i+j+1} \tag{7}$$
$$[D_r^p, \Gamma_j] \leq \Gamma_{i+j} D_{i+j+1}.$$

The inductive hypothesis gives

$$[D_h, \Gamma_j] \leq \Gamma_{h+j} D_{h+j+1}$$
$$[D_k, \Gamma_{h+j}] \leq \Gamma_{k+h+j} D_{k+h+j+1},$$

whence

$$[D_h, \Gamma_j, D_k] \leq [\Gamma_{h+j}, D_k][D_{h+j+1}, D_k]$$
$$\leq \Gamma_{h+j+k} D_{h+j+k+1} = \Gamma_{i+j} D_{i+j+1}.$$

By symmetry the same holds with h and k interchanged, and the first line of (7) follows by the Three-subgroup Lemma (0.3).

To prove the second line of (7), let $x \in D_r$ and $y \in \Gamma_j$. Applying 11.10(iii), we see that

$$[x^p, y] \in \langle w \rangle D_{pr+j+1},$$

where

$$w = [y,_p x] \in [[D_r, \Gamma_j],_{p-1} D_r].$$

Now $[D_r, \Gamma_j] \leq \Gamma_{r+j} D_{r+j+1}$ by the inductive hypothesis, so

$$[D_r, \Gamma_j, D_r] \leq [\Gamma_{r+j}, D_r][D_{r+j+1}, D_r] \leq \Gamma_{2r+j} D_{2r+j+1}$$

by the inductive hypothesis again, and repeating this argument we obtain

$$w \in \Gamma_{pr+j} D_{pr+j+1}.$$

Thus $[x^p, y] \in \Gamma_{i+j} D_{i+j+1}$, since $pr \geq i$, and the second line of (7) follows.

We have now established (6), and are ready to prove (i). Again, the proof is by induction on i. Since $D_1 = G = \Gamma_1$, the case $i = 1$ follows from (6) (taking $j = 1$ and $i = j$). The proof of the inductive step now exactly follows the argument above, with Γ_j replaced by D_j (but keeping Γ_{h+j}, Γ_{r+j}, etc.), and quoting (6) at appropriate points instead of the inductive hypothesis.

To deduce (ii), take $j = (p-1)i$ in (i): this gives

$$\gamma_p(D_i) = [\gamma_{p-1}(D_i), D_i] \leq [D_{(p-1)i}, D_i] \leq \Gamma_{pi} D_{pi+1}.$$

The first main result follows quickly:

11.12 Proposition

$$D_n = D_{n^*}^p \Gamma_n D_{n+t}$$

for all positive integers n and t.

Proof Lemma 11.11 shows that if $h + k = n$ then $[D_h, D_k] \leq \Gamma_n D_{n+1}$. Hence $D_n = D_{n^*}^p \Gamma_n D_{n+1}$; the proposition follows on replacing n by $n+1, \ldots, n+t-1$.

Next, we need a simple observation:

11.13 Lemma *If G is a finite p-group then*

$$D_{p^{n-1}} \leq P_n(G)$$

for all $n \geq 1$.

Proof This is trivial if $n = 1$. Let $n \geq 1$ and suppose that $D_{p^{n-1}} \leq P_n(G)$. Then

$$
\begin{aligned}
D_{p^n} &= \prod_{h+k=p^n} [D_h, D_k] \cdot D_{p^{n-1}}^p \\
&\leq [D_{p^{n-1}}, G] D_{p^{n-1}}^p \\
&\leq [P_n(G), G] P_n(G)^p = P_{n+1}(G),
\end{aligned}
$$

since $h + k = p^n$ implies $h \geq p^{n-1}$ or $k \geq p^{n-1}$. The result follows by induction.

We are now ready to give the

Proof of Theorem 11.2 We have to show that $D_n = X_n$ for each n, where $X_n = X_n(G)$ is defined by

$$
X_n = \prod_{ip^j \geq n} \Gamma_i^{p^j}.
$$

Note that this is really equal to a finite product, since whenever $i > n$ or $p^j > n$, the factor $\Gamma_i^{p^j}$ is contained in $\Gamma_n \Gamma_1^{p^f}$ where f is the least integer with $p^f \geq n$.

Since $\Gamma_i^{p^j} \leq D_i^{p^j} \leq D_{ip^j}$, it is clear that $X_n \leq D_n$ for each n. The theorem is therefore equivalent to the validity of

$$
[X_i, X_j] \leq X_{i+j}, \quad X_i^p \leq X_{pi} \tag{8}
$$

for all i and j.

Suppose our result is known whenever G is finitely generated. To deduce it in general, we only have to verify (8); to this end, let $x \in X_i$, $y \in X_j$ and put $H = \langle x, y \rangle$. It is clear that $X_n(H) \leq X_n(G)$ for each n, so applying (8) for the finitely generated group H we deduce that

$$
[x, y] \in X_{i+j}(H) \leq X_{i+j}(G), \quad x^p \in X_{pi}(H) \leq X_{pi}(G).
$$

Hence (8) holds for G, as required.

Thus we may assume that G is finitely generated. Now to verify (8) for a given pair i, j, we may replace G by G/X_n provided $n \geq \max\{i+j, pi\}$. If G is finitely generated, then G/X_n is a finite p-group (see Exercise 2); so it will suffice to prove (8) when G itself is a finite p-group. In this case, Lemma 11.13 shows that $D_k = 1$ for large k. We claim now that, for each $n \geq 1$,

$$
D_n = X_n D_{n+1}. \tag{9}
$$

Repeated applications of (9) give $D_n = X_n X_{n+1} \ldots X_{k-1} D_k = X_n$; so the theorem will be proved once (9) is established.

Certainly (9) is true when $n = 1$. Now let $m > 1$ and suppose that (9) holds for all $n < m$. Since $m^* < m$, Proposition 11.12 gives

$$D_m = D_{m^*}^p \cdot \Gamma_m D_{m+1} = \left(\prod_{ip^j \geq m^*} \Gamma_i^{p^j} \right)^p \Gamma_m D_{m+1}. \qquad (10)$$

Let $x \in D_{m^*}$ and $y \in \Gamma_i$ where $ip^j \geq m^*$; then $y^{p^j} \in D_{m^*}$ also. Since $pm^* \geq m$, we have $D_{2pm^*} \leq D_{m+1}$. On the other hand, Lemma 11.11(ii) shows that $\gamma_p(D_{m^*}) \leq \Gamma_{pm^*} D_{pm^*+1} \leq \Gamma_m D_{m+1}$. Applying 11.10(i) we deduce that

$$(xy^{p^j})^p \equiv x^p y^{p^{j+1}} \bmod \Gamma_m D_{m+1}. \qquad (11)$$

Now each element of the bracketed factor in (10) is a finite product of the form

$$w = x_1^{p^{j_1}} \ldots x_s^{p^{j_s}},$$

where $x_\ell \in \Gamma_{i(\ell)}$ and $i(\ell)p^{j_\ell} \geq m^*$ for each $\ell = 1, \ldots, s$. Arguing by induction on s and applying (11), we infer that

$$w^p \equiv \prod x_\ell^{p^{j_\ell+1}} \bmod \Gamma_m D_{m+1}.$$

Since $i(\ell)p^{j_\ell+1} \geq pm^* \geq m$ for each ℓ, it follows from (10) that $D_m \leq \prod_{ip^j \geq m} \Gamma_i^{p^j} \cdot \Gamma_m D_{m+1} = X_m D_{m+1}$. The reverse inclusion is clear, so (9) holds for $n = m$, and the result follows by induction.

The following improved version of Lemma 11.11 will be important in the next section:

11.14 Lemma

 (i) *If* $m \geq n$ *then*

$$[D_m, D_n] \leq \Gamma_{m+n} D_{m+pn};$$

 (ii) *for any positive integers* m *and* k,

$$[D_m, _k G] \leq \Gamma_{m+k} D_{m+pk}.$$

Proof We begin with the case where $n = k = 1$; here (i) and (ii) say the same thing. In view of Theorem 11.2, it will suffice to show that if $y \in G$ and $x \in \Gamma_i$, where $ip^j \geq m$, then

$$[x^{p^j}, y] \in \Gamma_{m+1} D_{m+p}. \tag{12}$$

This is clear if $j = 0$, so let us assume that $j \geq 1$. Now Lemma 11.9(ii) gives

$$[x^{p^j}, y] \equiv [x, y]^{p^j} \bmod \gamma_2(K)^{p^j} \prod_{\ell=1}^{j} \gamma_{p^\ell}(K)^{p^{j-\ell}}, \tag{13}$$

where $K = \langle x, [x, y] \rangle$. We examine the factors on the right-hand side in turn:

$$[x, y]^{p^j} \in \Gamma_{i+1}^{p^j} \leq D_{(i+1)p^j} \leq D_{m+p};$$

$$\gamma_2(K)^{p^j} \leq \Gamma_{2i+1}^{p^j} \leq D_{(2i+1)p^j} \leq D_{m+p};$$

$$\gamma_{p^\ell}(K)^{p^{j-\ell}} \leq \Gamma_{p^\ell i+1}^{p^{j-\ell}} \leq D_{p^{j-\ell}(p^\ell i+1)} \leq D_{m+p} \text{ if } 1 \leq \ell < j;$$

$$\gamma_{p^j}(K) \leq \Gamma_{p^j i+1} \leq \Gamma_{m+1}.$$

Thus (12) follows, and we have established that

$$[D_m, G] \leq \Gamma_{m+1} D_{m+p},$$

for every m. Repeated applications of this now give (ii).

Before proceeding, we note the following consequence:

$$[D_m, \Gamma_k] \leq \Gamma_{m+k} D_{m+pk}; \tag{14}$$

this follows from (ii) by stability group theory (see **0.7**).

We turn to the general case of (i). As above, it will suffice to show that if $y \in D_n$ and $x \in \Gamma_i$, where $ip^j \geq m$, then

$$[x^{p^j}, y] \in \Gamma_{m+n} D_{m+pn}. \tag{15}$$

If $j = 0$ then $x \in \Gamma_m$, and (14) shows that

$$[x, y] \in [\Gamma_m, D_n] \leq \Gamma_{m+n} D_{n+pm} \leq \Gamma_{m+n} D_{m+pn}$$

since $m \geq n$. Suppose $j \geq 1$, and put $K = \langle x, [x, y] \rangle$. Using (14), we see that $[x, y] \in \Gamma_{i+n} D_{n+pi} \leq D_{i+n}$, and that for each $h \geq 1$ we have

$$\gamma_h(K) \leq [D_{n,h} \Gamma_i] \leq \Gamma_{n+hi} D_{n+hpi} \leq D_{n+hi}.$$

Bearing this in mind we now again examine the factors on the right-hand side of (13):

$$[x, y]^{p^j} \in D_{i+n}^{p^j} \leq D_{(i+n)p^j} \leq D_{m+pn};$$

$$\gamma_2(K)^{p^j} \leq D_{2i+n}^{p^j} \leq D_{(2i+n)p^j} \leq D_{m+pn};$$

$$\gamma_{p^\ell}(K)^{p^{j-\ell}} \leq D_{p^\ell i+n}^{p^{j-\ell}} \leq D_{p^{j-\ell}(p^\ell i+n)} \leq D_{m+pn} \text{ if } 1 \leq \ell < j;$$

$$\gamma_{p^j}(K) \leq \Gamma_{p^j i+n} D_{n+p^{j+1} i} \leq \Gamma_{m+n} D_{m+pn}.$$

Thus (15) follows from (13), and the proof is complete.

To conclude this section, we translate some of the results into the language of *Lie Algebras*. This material is not needed for the applications to be given in the next section; the topic will be dealt with in earnest in Chapter 12.

For each n, we put

$$L_n = D_n/D_{n+1}.$$

Since $D_n^p[D_n, D_n] \leq D_{n+1}$, each L_n is an elementary abelian p-group, and we treat it as a vector space over $\mathbb{F}_p$, writing the group operation additively. We make

$$L = \bigoplus_{n=1}^{\infty} L_n$$

into a graded Lie algebra over $\mathbb{F}_p$ in the following manner: if $\overline{x} = xD_{i+1} \in L_i$ and $\overline{y} = yD_{j+1} \in L_j$, we set

$$(\overline{x}, \overline{y}) = [x, y]D_{i+j+1} \in L_{i+j}.$$

This gives a well-defined bilinear mapping $L_i \times L_j \rightarrow L_{i+j}$, because $[D_i, D_{j+1}]$ and $[D_{i+1}, D_j]$ are both contained in D_{i+j+1}; and it extends uniquely to a bilinear product $(\cdot, \cdot)$ on L. It is a routine matter (left to the exercises) to verify that this product is *skew-symmetric* and satisfies the *Jacobi identity*.

A second operation can also be defined on L, that makes L into a *restricted* Lie algebra. We are not ready to give the full definition at this stage: this will be done in Chapter 12; but we can give a partial definition, which suffices for the present purpose. Recall (11.10(ii)) that if $x \in D_i$ and $y \in D_{i+1}$ then $(xy)^p \equiv x^p \bmod D_{pi+1}$; hence we may define a mapping

$$[p] : L_i \rightarrow L_{pi}$$

by setting

$$\bar{x}^{[p]} = x^p D_{pi+1}$$

when $\bar{x} = xD_{i+1}$ and $x \in D_i$. The extension of $[p]$ to an operation defined on the whole of L is not completely straightforward (see §12.2); but we have gone far enough to be able to state

11.15 Proposition *Let L^* be the Lie subalgebra of L generated by L_1, and for each n put*

$$L_n^* = \Gamma_n D_{n+1}/D_{n+1} \leq L_n.$$

Then

(i)

$$L^* = \bigoplus_{n=1}^{\infty} L_n^* \quad and \quad L_n^* = L^* \cap L_n \quad for \ each \ n.$$

(ii)

$$L_{n+1}^* = (L_n^*, L_1) \quad for \ each \ n.$$

(iii)

$$L_n = L_n^* \qquad if \quad p \nmid n$$
$$L_n = L_m^{[p]} + L_n^* \quad if \quad n = pm.$$

Proof Parts (i) and (ii) follow easily from the definition of the bracket operation $(\cdot, \cdot)$. Part (iii) is a translation of Proposition 11.12. To see this, suppose first that $p \nmid n$. Then $pn^* \geq n + 1$, so $D_{n^*}^p \leq D_{n+1}$; in this case, Proposition 11.12 gives $D_n = \Gamma_n D_{n+1}$, which is equivalent to $L_n = L_n^*$. Now suppose that $n = pm$. Let $x, y \in D_m$. Using Corollary 11.10(i) and Lemma 11.11(ii) we see that

$$x^p y^p \equiv (xy)^p \bmod \Gamma_n D_{n+1}.$$

It follows that

$$D_m^p \subseteq \{g^p \mid g \in D_m\} \cdot \Gamma_n D_{n+1}.$$

With Proposition 11.12 this shows that

$$D_n = \{g^p \mid g \in G_m\} \cdot \Gamma_n D_{n+1},$$

which is equivalent to $L_n = L_m^{[p]} + L_n^*$.

11.3 The main results

The hard work is over and we now reap the rewards. Throughout this section, G denotes a *finitely generated pro-p group*. We keep the notation of §11.2, so $D_n = D_n(G)$ and $\Gamma_n = \gamma_n(G)$ for each n.

11.16 Lemma *The family* $(D_n)_{n\geq 1}$ *forms a base for the neighbourhoods of* 1 *in* G.

Proof We show that for each n,

$$D_{p^{n-1}} \leq P_n(G) \leq D_n;$$

the lemma will follow by Proposition 1.16(iii).

Now $D_1 = G = P_1(G)$. Let $n \geq 1$ and suppose that $D_n \geq P_n(G)$. Since $(n + 1)^* \leq n$, we then have

$$D_{n+1} \geq [D_n, G]D_{(n+1)^*}^p \geq [P_n(G), G]P_n(G)^p = P_{n+1}(G),$$

by Corollary 1.20. Thus $D_n \geq P_n(G)$ for all n.

To establish the other inclusion, we may factor out $P_n(G)$, and so reduce to the case where G is a finite p-group; in that case, it follows from Lemma 11.13.

Next we give the

Proof of Theorem 11.5 Given that $D_n = D_{n+1}$, we have to show that D_i is powerful for all $i \geq n$. In view of Lemma 11.16, it will suffice to show that D_i/D_k is powerful whenever $k \geq i \geq n$; so, fixing some large k and replacing G by G/D_k, we may assume that in fact $D_k = 1$ (and hence that G is a finite p-group).

Let j be an integer with $j \geq n - \frac{n+1}{p}$. We claim that for each $m \geq 2$,

$$\Gamma_{mj} \leq D_{(m-1)j}^p. \tag{16}$$

Indeed,

$$\Gamma_{mj} \leq [\Gamma_n, {}_{mj-n}\, G] \leq [D_n, {}_{mj-n}\, G] = [D_{n+1}, {}_{mj-n}\, G]$$
$$\leq \Gamma_{mj+1}D_{n+1+p(mj-n)} \quad \text{by 11.14(ii)}$$
$$\leq \Gamma_{mj+1}D_{(m-1)j}^p \quad\quad \text{by 11.12;}$$

note that the condition on j implies that $(n + 1 + p(mj - n))^* \geq mj - j$. It follows that if $\widetilde{G} = G/D_{(m-1)j}^p$ then $\gamma_{mj}(\widetilde{G}) = \gamma_{mj+1}(\widetilde{G})$, and hence that $\gamma_{mj}(\widetilde{G}) = \gamma_k(\widetilde{G}) = 1$, and this is equivalent to (16).

Now using 11.14(i) we get

$$[D_j, D_j] \leq \Gamma_{2j} D_{pj+j}$$
$$\leq \Gamma_{2j}\Gamma_{pj+j} D_j^p \quad \text{by 11.12}$$
$$= D_j^p \quad \text{by (16)}.$$

Thus if p is *odd*, D_j is powerful, and we are done (indeed we have proved slightly more than was claimed).

Suppose now that $p = 2$. Let $r \geq 1$. Using Proposition 11.12 and (16), we find

$$D_{rj}^2 \leq D_{2rj} \leq \Gamma_{2rj} D_{rj}^2 \leq D_{(2r-1)j}^2 D_{rj}^2 = D_{rj}^2. \tag{17}$$

Now Lemma 11.14(i) and (16) give

$$[D_{2j}, D_j] \leq \Gamma_{3j} D_{4j} = D_{4j}, \tag{18}$$
$$[D_{4j}, D_{2j}] \leq \Gamma_{6j} D_{8j} = D_{8j}; \tag{19}$$

we deduce

$$[D_{2j}, D_{2j}] = [D_{2j}, D_j^2]$$
$$\leq [D_{2j}, D_j]^2 [D_{2j}, D_j, D_j] \quad \text{by 11.9(ii)}$$
$$\leq D_{4j}^2 [D_{2j}^2, D_j] \quad \text{by (18) and (17)}$$
$$\leq D_{4j}^2 [D_{2j}, D_j]^2 [D_{2j}, D_j, D_{2j}] \quad \text{by 11.9(ii)}$$
$$\leq D_{4j}^2 [D_{4j}, D_{2j}] \leq D_{4j}^2 D_{8j} \quad \text{by (18) and (19)}$$
$$= (D_{2j}^2)^2 \quad \text{by (17)}.$$

This implies that D_{2j} is powerful, by Exercise 2.4.

Finally, let $i \geq n$. If i is *even* then $i = 2j$ where j is as above, and we are done. If i is *odd*, we know from Proposition 11.12 that

$$D_i = \Gamma_i D_{\lceil i/2 \rceil}^2 \leq \Gamma_i D_{i+1};$$

but (as in the proof of (16)) we also have $\Gamma_i \leq [D_{n+1,i-n} G] \leq D_{i+1}$. Thus $D_i = D_{i+1}$ and the previous case applies. This completes the proof.

Next we deduce a strengthening of the main result of Chapter 3 (characterising pro-p groups of finite rank as those which have a powerful finitely generated open subgroup):

11.17 Corollary *The following are equivalent for a finitely generated pro-p group G:*

(a) *for some n, there exists h with $p^h > n$ such that*

$$\gamma_n(G) \leq \overline{G^{p^h}} \gamma_{n+1}(G);$$

(a') *the statement of (a) holds for infinitely many values of n;*

(b) *for some n, every commutator of length n in G is a p^hth power with $p^h > n$;*

(b') *the statement of (b) holds for infinitely many values of n;*

(c) *for some n, $\gamma_n(G)$ consists of p^hth powers with $p^h > n$;*

(c') *the statement of (c) holds for infinitely many values of n;*

(d) *G has finite rank.*

Proof We prove that (a) implies (d) and that (d) implies (c'); the other implications are then more or less obvious.

Suppose that (a) holds, and put $m = p^h - 1$. We claim that then $D_m = D_{m+1}$; this being so, Theorem 11.5 shows that D_m is powerful, and this in turn implies (d) by Theorem 3.13.

To establish the claim, write $\widetilde{G} = G/\overline{G^{p^h}}$. Then $\gamma_n(\widetilde{G}) = \gamma_{n+1}(\widetilde{G})$, which implies $\gamma_n(\widetilde{G}) = \gamma_{n+k}(\widetilde{G})$ for all k. It follows that

$$\Gamma_m \leq \Gamma_n \subseteq \bigcap_{k \geq 1} \overline{G^{p^h}} \Gamma_{n+k}$$

$$\subseteq \bigcap_{k \geq 1} \overline{G^{p^h}} D_{n+k} = \overline{G^{p^h}}$$

$$\leq D_{p^h} = D_{m+1}.$$

With Proposition 11.12 this now gives

$$D_m = \Gamma_m D_{m^*}^p \cdot D_{m+1} = D_{m^*}^p \cdot D_{m+1} = D_{m+1},$$

since $m^* = \lceil (p^h - 1)/p \rceil = p^{h-1}$. Thus (a) implies (d).

Now suppose that G has finite rank. By Corollary 4.3, G has a uniform open normal subgroup K, of rank k, say. There exists r_0 such that $\gamma_{r_0}(G) \leq K$. Let $r \geq r_0$, let h be large enough so that

$$p^h > r + hk,$$

and put $n = r + hk$. Since $|K : P_{h+1}(K)| = p^{hk}$, we have

$$\gamma_n(G) \leq [K,_{hk} G] \leq P_{h+1}(K).$$

However, Corollary 3.5 shows that $P_{h+1}(K)$ consist of p^hth powers in K; as r can be arbitrarily large, and $n > r$, we have established (c'). Thus (d) implies (c'), and the proof is complete.

11.4 Index growth

In this section we prove the following refinement of Corollary 11.6:

11.18 Theorem *Let G be a finitely generated pro-p group, where p is an odd prime, and let $k \geq 1$. If*

$$|G : D_{p^k}(G)| < p^{p^k + k - 1} \tag{20}$$

then $D_{p^k}(G)$ is powerful and has rank at most $(p-1)p^{k-1}$.

An immediate consequence, given Theorem 3.18, is

11.19 Corollary *Let G be as above. If, for some k,*

$$|G : \overline{G^{p^k}}| < p^{p^k + k - 1},$$

then G has finite rank and $\dim G \leq (p-1)p^{k-1}$.

Although only a little stronger than Corollary 11.6, this result is interesting because it is best possible: the simplest example of a finitely generated pro-p group of infinite rank, namely $G = C_p \wr \mathbb{Z}_p$, satisfies $|G : \overline{G^{p^k}}| = p^{p^k + k - 1}$ for every k (see Exercise 8). It is interesting also because, unexpectedly, the proof uses some of the theory of groups of finite coclass (and so, unlike most of the main results of this chapter, could not have been known to Lazard).

Henceforth, p denotes an *odd* prime, G is a finitely generated pro-p group, and $D_n = D_n(G)$ for each n. We start with an easy observation (for the notation $V(G, n)$ see Definition 2.10):

11.20 Lemma *For every $n \geq 1$,*

$$D_n \leq V(G, n).$$

Proof Let i and j be integers such that $ip^j \geq n$, and let M be an n-dimensional $\mathbb{F}_p[G]$-module. It follows from stability group theory (see **0.7**) that

$$M(\gamma_i(G) - 1)^{p^j} \subseteq M(G - 1)^{ip^j} = 0;$$

hence if $x \in \gamma_i(G)$ then $M(x^{p^j} - 1) = M(x - 1)^{p^j} = 0$ (alternatively, represent the action of G by unitriangular matrices, and note that the matrix of x has $i - 1$ rows of zeros below the main diagonal). Thus

$\gamma_i(G)^{p^j}$ acts trivially on M. It follows that $\gamma_i(G)^{p^j} \leq V(G, n)$, and this gives the result in view of Theorem 11.2.

With Proposition 3.9 and Theorem 3.8, the lemma gives

11.21 Corollary *Suppose that* $\mathrm{d}(D_n) = n_0 \leq n$. *If* $N \lhd_o G$ *and* $N \leq D_n$ *then* N *is powerful, of rank at most* n_0.

We now put

$$L_n = D_n/D_{n+1},$$

as at the end of §11.2, and write

$$d_n = \mathrm{d}(D_n), \quad \delta_n = \dim_{\mathbb{F}_p} L_n.$$

By results from Chapter 1, we have

$$d_n = \dim_{\mathbb{F}_p}(D_n/\Phi(D_n)) \geq \delta_n.$$

To prove Theorem 11.18, it will suffice to show that if (20) holds then there exists $i < k$ such that

$$d_{p^{i+1}} \leq (p-1)p^i; \tag{21}$$

the theorem will then follow on taking $n = p^{i+1}$, $n_0 = d_{p^{i+1}}$ and $N = D_{p^k}$ in Corollary 11.21. Two more lemmas are needed (the first does not require p to be odd).

11.22 Lemma *Suppose that* $\delta_s = 0$, *where* $s \geq 1$. *Then for each integer* $m \geq s/p$ *we have*
(i) $|D_m : D_{pm}| \geq |D_{pm} : D_{p^2 m}|$;
(ii) $D_{pm} = \overline{D_m^p}$.

Proof We use the notation of Proposition 11.15. The hypothesis is that $L_s = 0$. Hence $L_s^* = 0$, and it follows that $L_n^* = 0$ for all $n \geq s$ (see 11.15(ii)). Then 11.15(iii) shows that

$$L_n = 0 \quad \text{if } n \geq s \text{ and } p \nmid n,$$
$$L_n = L_r^{[p]} \quad \text{if } n = pr \geq s.$$

Thus $\delta_n = 0$ in the first case and $\delta_n \leq \delta_r$ in the second case. Therefore

$$|D_{pm} : D_{p^2m}| = \sum_{n=pm}^{p^2m-1} \delta_n = \sum_{n=pm}^{p^2m-p} \delta_n$$

$$\leq \sum_{n=m}^{pm-1} \delta_n = |D_m : D_{pm}|.$$

This proves (i).

Since $L_n^* = 0$ for all $n \geq s$, we have $\gamma_n(G) \leq D_{n+1}$ for $n \geq s$; as $pm \geq s$, Proposition 11.12 and Lemma 11.16 now give

$$D_{pm} = \bigcap_{\ell \geq 0} D_m^p D_{pm+\ell} = \overline{D_m^p},$$

so we have (ii).

11.23 Lemma *Suppose that $\delta_s = 1$ for each s in the range $p^j \leq s \leq 2p^j$, where $j \geq 1$. Then $\delta_{2p^j+1} = 0$.*

Proof Put $M = D_{p^j}/D_{2p^j+1}$. Since $p > 2$, we have

$$D_{p^j}^p \leq D_{p^{j+1}} \leq D_{2p^j+1},$$

so M has exponent p. Also M is abelian: for $[D_{p^j+1}, D_{p^j}] \leq D_{2p^j+1}$, while D_{p^j}/D_{p^j+1} is cyclic since $\delta_{p^j} = 1$; but any group which is cyclic modulo its centre is abelian. Thus we may consider M as a vector space over $\mathbb{F}_p$, with

$$\dim_{\mathbb{F}_p}(M) = \sum_{s=p^j}^{2p^j} \delta_s = p^j + 1.$$

Now recall Lemma 10.6: this shows that if M were *uniserial* as an $\mathbb{F}_p[G]$-module, then $[M, \pi_j(G)]$ would be non-zero. But $\pi_j(G) \leq D_{p^j}$, and $[M, D_{p^j}] = 0$ since M is abelian. Consequently M is *not* uniserial, and it follows that $[M, _{p^j}G] = 0$. Since $\gamma_{p^j}(G) \leq D_{p^j}$, we deduce that

$$\gamma_{2p^j+1}(G) \leq [D_{p^j}, _{p^j+1}G]$$

$$\leq [D_{2p^j+1}, G] \leq D_{2p^j+2}.$$

In the notation of Proposition 11.15, this says that $L_{2p^j+1}^* = 0$, and 11.15(iii) now shows that $L_{2p^j+1} = 0$; this is the claim of the lemma.

We can now complete the

Proof of Theorem 11.18 If $\delta_1 \leq 1$ then G is procyclic, since $D_1/D_2 = G/\Phi(G)$, and the theorem holds trivially. We assume henceforth that $\delta_1 \geq 2$. The hypothesis is that

$$\sum_{n=1}^{p^k-1} \delta_n < p^k + k - 1, \tag{22}$$

and we have to show that (21) holds for some $i < k$. Write

$$\Delta_j = \sum_{n=p^j}^{p^{j+1}-1} \delta_n$$

$$= \log_p |D_{p^j} : D_{p^{j+1}}|.$$

Since $p^k + k - 1 = \sum_{j=0}^{k-1}((p-1)p^j + 1)$, while the left-hand side of (22) is equal to $\sum_{j=0}^{k-1} \Delta_j$, there exists $i < k$ with $\Delta_i \leq (p-1)p^i$. This in turn implies that *either* $\delta_s = 0$ for some s between p^i and $p^{i+1} - 1$, *or* $\delta_s = 1$ for *every* s in this range. If $i \geq 1$ the latter possibility is excluded by Lemma 11.23, since for $i \geq 1$ we have $p^i < 2p^i + 1 \leq p^{i+1} - 1$, while if $i = 0$ it is excluded by our assumption that $\delta_1 \geq 2$.

Thus there exists $s < p^{i+1}$ such that $\delta_s = 0$. Taking $m = p^i$ in Lemma 11.22, we deduce that $\Delta_{i+1} \leq \Delta_i$, and that $D_{p^{i+2}} = \overline{D_{p^{i+1}}^p} \leq \Phi(D_{p^{i+1}})$. It follows that

$$d_{p^{i+1}} = \dim_{\mathbb{F}_p}(D_{p^{i+1}}/\Phi(D_{p^{i+1}})) \leq \log_p |D_{p^{i+1}} : D_{p^{i+2}}|$$

$$= \Delta_{i+1} \leq \Delta_i \leq (p-1)p^i.$$

Thus (21) holds, and the proof is complete.

Notes

Theorem 11.2 is essentially the same as [L], Appendice 3.14.5. Theorem 11.4, and most of its applications in Corollary 11.17, are established in Appendice, §1 of [L]. Lazard's proof is however quite different from ours, and depends on the construction of a 'mixed' Lie algebra ([L] Chapter II, §1.2); this combines the graded Lie algebra L that we introduce in §11.2 with a graded Lie algebra associated to the lower p-series.

Our approach, based on Theorem 11.5, is more direct, depending only on an application of the Hall–Petrescu formula. Theorem 11.5 is due to Riley (1993).

Theorems 11.7 and 11.18 are due to Shalev (1992b).

Exercises

1. Using the result of Exercise 6.5, calculate $v_p(\binom{n}{r})$.

2. Show that if G is a d-generator group and $G^{p^j} = \gamma_n(G) = 1$ then $|G| \mid p^{jd^n}$.

3. A series of subgroups $G = G_1 \geq G_2 \geq \ldots$ is called an *N-series* if $[G_i, G_j] \leq G_{i+j}$ for all i and j. Verify that if (G_i) is an N-series, then the direct sum $\bigoplus_{i=1}^{\infty} G_i/G_{i+1}$ has a Lie algebra structure given by

$$(xG_{i+1}, yG_{j+1}) = [x,y]G_{i+j+1} \qquad (x \in G_i, y \in G_j).$$

In the following exercises, G is a finitely generated pro-p group, and L is the Lie algebra defined at the end of §11.2.

4. Suppose that G is *powerful*. Prove that

$$D_i(G) = P_{n+1}(G) \quad \text{where } p^{n-1} < i \leq p^n.$$

[*Hint:* Examine the proof of 11.13, and recall Theorem 3.6.]

5. Prove that the following are equivalent: (*a*) G is powerful; (*b*) L is abelian; (*c*) $\gamma_2(G) \leq D_3(G)$.
[*Hint:* Use Exercise 4 to show that $(a) \Rightarrow (b)$. Use Exercise 2.4 to show that $(c) \Rightarrow (a)$ when $p = 2$.]

6. Show that if $D_n = D_{n+1}$ then $[D_i, D_j] \subseteq D_{i+j+1}$ whenever $i+j \geq n$. [Use 11.15 and 11.14.]

7. Prove that G *has finite rank if and only if L is nilpotent* (Shalev (1993b)).
[Use 11.4, 11.15 and Exercise 6.]

8. Let W_n be the wreath product $C_p \wr C_{p^n}$; this is the semi-direct product of the $\langle x \rangle$-module $\mathbb{F}_p[\langle x \rangle]$ by a cyclic group $\langle x \rangle$ of order p^n.
  (i) Fix $k \leq n$ and write $\phi(Y) = 1 + \cdots + Y^{p^k-1}$. Show that $\phi(x^i) \in \phi(x)\mathbb{F}_p[\langle x \rangle]$ for every i. Deduce that

$$W_n^{p^k} = (\phi(x)\mathbb{F}_p[\langle x \rangle]) \cdot \langle x^{p^k} \rangle,$$

and hence that $|W_n : W_n^{p^k}| = p^{p^k+k-1}$.
  (ii) For each n there is a natural epimorphism $W_{n+1} \to W_n$ (cf. Exercise 3.5). Using these we construct the inverse limit $G = \varprojlim W_n$; this is

the wreath product $C_p \wr \mathbb{Z}_p$ in the category of pro-p groups. Show that

$$|G : \overline{G^{p^k}}| = p^{p^k + k - 1}$$

for every positive integer k.

   Deduce that if p is odd, then $|D_i(G) : D_{i+1}(G)| = p$ for all $i \geq 2$.

   (iii) Show that G may be identified with the semi-direct product $\mathbb{F}_p[[X]] \cdot X$, where X is $\mathbb{Z}_p$ written multiplicatively and $\mathbb{F}_p[[X]]$ is the completed group algebra of X discussed in §7.4.

12

Some graded algebras

In Chapter 11 we defined the *modular dimension series* $(D_n(G))$ of an arbitrary group G, and derived some of its properties; we also stated, but did not prove, the theorem of Jennings, that $D_n(G)$ is precisely the kernel of the natural homomorphism of G into the unit group of $k[G]/I^n$, where k is any field of characteristic p and I is the augmentation ideal. We now resume the study of this series, from a somewhat different point of view: our main focus will be on the Lie algebra

$$L(G) = \bigoplus_{n=1}^{\infty} D_n/D_{n+1},$$

where $D_n = D_n(G)$, briefly introduced in §11.2.

Now let

$$\mathrm{gr}(k[G]) = \bigoplus_{n=0}^{\infty} I^n/I^{n+1}$$

be the graded k-algebra associated to the chain of ideals (I^n) in $k[G]$; when $k = \mathbb{F}_p$, $\mathrm{gr}(k[G])$ is the algebra denoted S^* in §7.4. It is easy to see (this will be explained in detail in §12.2) that Jennings's theorem is equivalent to the existence of a *well-defined, injective* linear mapping

$$\theta : L(G) \to \mathrm{gr}(k[G])$$

such that for $\bar{g} = gD_{n+1} \in D_n/D_{n+1}$,

$$\theta(\bar{g}) = (g - 1) + I^{n+1} \in I^n/I^{n+1}.$$

This suggests an intimate connection between the group-theoretic structure of G, as reflected in the Lie algebra $L(G)$, and the ring-theoretic structure of the group algebra $k[G]$, as reflected in the graded associative algebra $\mathrm{gr}(k[G])$. A special case of this connection was established in

§7.4, where we saw that if G is a uniform pro-p group (and $k = \mathbb{F}_p$) then $\mathrm{gr}(k[G])$ is a polynomial ring.

In this chapter, we consider the general case. As well as proving Jennings's theorem, we show that $L(G)$ has the structure of a *restricted Lie algebra* over $\mathbb{F}_p$, and that the mapping θ extends to an isomorphism of the *universal restricted enveloping algebra* of $L(G)$ onto $\mathrm{gr}(\mathbb{F}_p[G])$; this is a direct generalisation of Theorem 7.24. These results, due to Lazard, are proved in §12.2; Section 12.1 gives a self-contained introduction to the relevant concepts.

In the course of the proof, we obtain an explicit formula for the dimensions of the graded components I^n/I^{n+1}. This is used in §12.3 to derive yet another characterisation for the pro-p groups of finite rank:

12.1 Theorem *Let G be a finitely generated pro-p group. Then G has finite rank if and only if the sequence*

$$\left(\dim_{\mathbb{F}_p}\left(I^n/I^{n+1}\right)\right)$$

grows polynomially.

In fact, the proof will show that if G has infinite rank then $\dim_{\mathbb{F}_p}(I^n/I^{n+1})$ must grow much faster than any polynomial in n; a remarkable group-theoretic application of this is described in Interlude E.

12.1 Restricted Lie algebras

Throughout this section, k denotes an arbitrary field of characteristic p. By a *k-algebra* we mean an associative ring with identity, containing k as a subring. Each k-algebra A can be made into a Lie algebra over k, by setting

$$(a,b) = ab - ba;$$

the resulting Lie algebra will be denoted A_L. By 'a Lie subalgebra of A' we always mean a Lie subalgebra of A_L.

12.2 Definition Let A be a k-algebra and L a Lie subalgebra of A. Then L is said to be *restricted* if for each element $a \in L$, also $a^p \in L$.

Thus a restricted Lie subalgebra of A has, in addition to the binary operation $(\cdot,\cdot)$, a unary operation, $[p]$, given by

$$a^{[p]} = a^p.$$

More generally, a Lie algebra L over k, with an additional unary operation $[p]$, is called a *restricted Lie algebra* if there exist a k-algebra A and a Lie algebra monomorphism $\theta : L \to A_L$ such that $\theta(a^{[p]}) = \theta(a)^p$ for all $a \in L$. In this case A (more precisely, the pair (θ, A)) is called a *restricted enveloping algebra* of L. It is *universal* if it has the following universal property: for any restricted Lie algebra homomorphism $\phi : L \to B_L$, where B is a k-algebra, there exists a unique k-algebra homomorphism $\phi^* : A \to B$ such that $\phi = \phi^* \circ \theta$. (A Lie algebra homomorphism between two restricted Lie algebras is called *restricted* if it preserves the operation $[p]$; here B_L is supposed to be endowed with the operation $b^{[p]} = b^p$.)

We shall abbreviate 'universal restricted enveloping algebra' to *universal envelope*. It is easy to see that if it exists, it is essentially unique (Exercise 1). In order to construct it, we start with a non-empty set X, and consider the free associative algebra $k\langle X \rangle$ on the set X; this is the ring of polynomials in the non-commuting variables $x \in X$, discussed in §6.3. Recall that $k\langle X \rangle$ has a k-basis $W(X)$ consisting of all monomials $w = x_1 \ldots x_n$, with $x_i \in X$ and $n \geq 0$; the identity element 1 of $k\langle X \rangle$ is identified with the 'empty monomial', where $n = 0$, and monomials are multiplied by concatenation.

Now let L be a restricted Lie algebra, and choose X to be a basis for L. For all $x, y, z \in X$ there exist $\lambda_{xy}^z, \mu_x^z \in k$ such that

$$(x, y) = \sum_{z \in X} \lambda_{xy}^z z, \quad x^{[p]} = \sum_{z \in X} \mu_x^z z; \tag{1}$$

for each x and y, only a finite number of the λ_{xy}^z and μ_x^z are non-zero. We take J to be the two-sided ideal of $k\langle X \rangle$ generated by all elements of the form

$$xy - yx - \sum_{z \in X} \lambda_{xy}^z z, \quad x^p - \sum_{z \in X} \mu_x^z z$$

with $x, y \in X$, and put

$$U = U_X(L) = k\langle X \rangle / J.$$

Write $\bar{c} = c + J \in U$ for each $c \in k\langle X \rangle$; since X is a basis of L, there is a unique k-linear mapping $\xi : L \to U$ such that $\xi(x) = \bar{x}$ for all $x \in X$. We claim now that (ξ, U) *is a universal envelope of* L.

There are four conditions to be verified. The first is that ξ is a Lie algebra homomorphism: this is clear from the construction, since the Lie bracket operation is bilinear. The second is that ξ is *restricted*: this is harder to verify, since the operation $[p]$ is not in general linear, and we postpone this for the moment. To verify the universal property, suppose we are given some restricted Lie algebra homomorphism $\phi : L \to B_L$, as above. The mapping $x \mapsto \phi(x)$ from X to B extends to a k-algebra homomorphism $\widetilde{\phi}$, say, from $k\langle X \rangle$ to B. Then for $x, y \in X$,

$$\widetilde{\phi}(xy - yx) = (\widetilde{\phi}(x), \widetilde{\phi}(y)) = (\phi(x), \phi(y)) = \phi((x, y))$$

$$= \sum_{z \in X} \lambda_{xy}^z \phi(z) = \sum_{z \in X} \lambda_{xy}^z \widetilde{\phi}(z) = \widetilde{\phi}\big(\sum_{z \in X} \lambda_{xy}^z z\big),$$

so $xy - yx - \sum_{z \in X} \lambda_{xy}^z z \in \ker \widetilde{\phi}$. Similarly, $x^p - \sum_{z \in X} \mu_x^z z \in \ker \widetilde{\phi}$, and so $J \subseteq \ker \widetilde{\phi}$. Hence $\widetilde{\phi}$ induces a k-algebra homomorphism $\phi^* : U \to B$, and it is clear that $\phi = \phi^* \circ \xi$. It is also clear that ϕ^* is the unique k-algebra homomorphism with this property. The injectivity of ξ follows from the universal property: for our definition of 'restricted Lie algebra' means that there exists a $\theta : L \to A_L$ which is *injective*, and then $\theta = \theta^* \circ \xi$ implies that ξ is injective. For future reference, we note that this also implies that the restriction of θ^* to $\xi(L)$ is injective, so we have $\xi(L) \cap \ker \theta^* = 0$.

Before proceeding, we state an important lemma:

12.3 Lemma *Let H be a Lie subalgebra of the k-algebra A, and let X be a basis of H. If $x^p \in H$ for all $x \in X$, then H is restricted.*

Proof Let $a, b \in A$, and let T be an indeterminate. In the polynomial ring $A[T]$ we have

$$(aT + b)^p = a^p T^p + b^p + \sum_{i=1}^{p-1} s_i(a, b) T^i, \qquad (2)$$

where $s_i(a, b)$ is a sum of terms $c_1 c_2 \dots c_p$ with each c_i equal to either a or b. We may formally differentiate (2) with respect to T: this is merely a device for equating coefficients of equal powers of T on each side of the equation. The result is

$$\sum_{i=0}^{p-1} (aT + b)^i a (aT + b)^{p-i-1} = \sum_{i=1}^{p-1} i s_i(a, b) T^{i-1}. \qquad (3)$$

Now we quote the following elementary identity, valid in every algebra over a field of characteristic p:

$$\sum_{i=0}^{p-1} u^i v u^{p-i-1} = (v,_{p-1} u) \qquad (4)$$

(for the proof, see Exercise 2). Putting $aT + b = u$ and $a = v$ in (3), this gives

$$(a,_{p-1} (aT + b)) = \sum_{i=1}^{p-1} is_i(a,b)T^{i-1}.$$

Equating coefficients of T^{i-1} we find

$$s_i(a,b) = i^{-1} \sum (a, c_1, \dots, c_{p-1}), \tag{5}$$

the sum being over all $(p-1)$-tuples $\mathbf{c}$ where $c_j = a$ for $i-1$ values of j and $c_j = b$ for $p - i$ values of j. Returning to (2), put $T = 1$: this now gives

$$(a + b)^p = a^p + b^p + \sum_{i=1}^{p-1} s_i(a, b). \tag{6}$$

Now each element of H is of the form $c = \lambda_1 x_1 + \cdots + \lambda_n x_n$, with $\lambda_i \in k$ and $x_i \in X$ for each i. We argue by induction on n to show that $c^p \in H$. If $n = 1$ we have

$$c^p = \lambda_1^p x_1^p \in H$$

by hypothesis. Suppose $n > 1$, and put $a = \lambda_1 x_1 + \cdots + \lambda_{n-1}x_{n-1}, b = \lambda_n x_n$, so $c = a + b$. The inductive hypothesis says that both a^p and b^p are in H, and (5) shows that $s_i(a,b) \in H$ for $i = 1, \dots, p-1$. It follows by (6) that $(a + b)^p \in H$ also. Thus $c^p \in H$ as required, and the lemma follows.

We can now complete the proof that (ξ, U) is a universal envelope of the restricted Lie algebra L. We have to show that if $c \in L$ then $\xi(c^{[p]}) = \xi(c)^p$. Let $\theta : L \to A_L$ be an injective restricted homomorphism, as above. Then

$$\theta^* \xi(c^{[p]}) = \theta(c^{[p]}) = \theta(c)^p$$
$$= (\theta^* \xi(c))^p = \theta^*(\xi(c)^p),$$

so $\xi(c^{[p]}) - \xi(c)^p \in \ker \theta^*$. Now for each $x \in X$, we have $\xi(x)^p = \xi(\sum_{z \in X} \mu_x^z z) \in \xi(L)$, from the definition of $\xi : L \to U = k\langle X \rangle / J$ (see (1)); it therefore follows from the above lemma that $\xi(c)^p \in \xi(L)$, so in fact we have $\xi(c^{[p]}) - \xi(c)^p \in \xi(L) \cap \ker \theta^* = 0$, as required.

We have established

12.4 Proposition *Let L be a restricted Lie algebra over k, with basis X. Then the k-algebra $U_X(L)$ is a universal envelope of L.*

Since ξ is injective, we may as well consider L as being contained in U, whereupon ξ becomes the inclusion mapping $L \to U$. We shall sometimes write x instead of $\bar{x}$, for $x \in X$. The relations (1) are then identities in U, where $\overline{(x,y)} = xy - yx$ and $x^{[p]} = x^p$. Using these, we can extract from $\overline{W(X)}$ a much smaller spanning set for U. To this end, we start by endowing the set X with a *total ordering*, written $<$. A monomial

$$w = x_1^{f_1} \ldots x_n^{f_n}$$

is said to be *restricted* if $x_1 < x_2 < \ldots < x_n$ and $1 \le f_i \le p - 1$ for each i (here, x^f means $x.x \ldots x$ with f factors).

12.5 Lemma *The images in U of the restricted monomials span U as a vector space over k.*

Proof Let V be the subspace of U spanned by the restricted monomials. It will suffice to prove that if $w \in W(X)$ then $\bar{w} \in V$. This is done by induction on the degree of w. The idea is simple: if w is not already restricted, use (1) to reduce the degree of any x^f with $f \ge p$ occurring in w, and to replace yx by xy if $y > x$ and yx occurs in w; this is done at the cost of adding to w a linear combination of monomials of strictly smaller degrees. We leave it as an exercise for the reader to verify the details.

The following easy consequence is the key to the main results of the next section:

12.6 Lemma *Let A be a k-algebra, L a finite-dimensional Lie algebra over k, and $\phi : L \to A$ a Lie algebra homomorphism. Suppose that $\phi(L)$ is a restricted Lie subalgebra of A and that $\phi(L)$ generates A as a k-algebra. Let $n = \dim_k(L)$. Then*

(i) $\dim_k(A) \le p^n$;

(ii) *if $\dim_k(A) = p^n$ then ϕ is injective, so L is a restricted Lie algebra; and (ϕ, A) is a universal envelope of L; and*

(iii) *if $\dim_k(A) = p^n$ and X is an ordered basis for $\phi(L)$ then the (images in A of the) restricted monomials on X form a basis for A.*

Proof The Lie algebra $\phi(L)$ has a basis $X = \{x_1, \ldots, x_m\}$, where

$$m = n - \dim_k(\ker \phi).$$

We order the set X as written, and let $U = U_X(\phi(L))$. Since the number of restricted monomials on X is exactly p^m, Lemma 12.5 shows that $\dim_k(U) \leq p^m$. Now by Proposition 12.4 there exists a k-algebra homomorphism $\psi : U \to A$ which extends the inclusion mapping of $\phi(L)$ into A; then $\psi(U)$ is a k-subalgebra of A which contains $\phi(L)$, and as $\phi(L)$ generates A it follows that $\psi(U) = A$. Thus

$$\dim_k(A) \leq \dim_k(U) \leq p^m \leq p^n,$$

giving (i).

Now suppose that $\dim_k(A) = p^n$. The above inequalities are then equalities. Hence $\dim_k(\ker \phi) = n - m = 0$, so ϕ is injective. Since $\dim_k(\ker \psi) = \dim_k(U) - \dim_k(A)$, this also shows that ψ is injective, hence an isomorphism of U onto A. As ψ restricts to the identity map on $\phi(L)$, and U is a universal envelope of $\phi(L)$, it follows that (ϕ, A) is a universal envelope of L; thus (ii) holds. Part (iii) is clear, since A is spanned by the (images of the) restricted monomials.

Remark It is more usual to *define* a restricted Lie algebra to be a Lie algebra L with a unary operation $[p]$ such that for all $\lambda \in k$ and $a, b \in L$ the following hold:

$$\begin{aligned}
(\lambda a)^{[p]} &= \lambda^p a^{[p]} \\
(a, b^{[p]}) &= (a, _p b) \\
(a + b)^{[p]} &= a^{[p]} + b^{[p]} + \sum_{i=1}^{p-1} s_i(a, b),
\end{aligned} \tag{7}$$

where $s_i(a, b)$ is given by (5), above (in fact the third condition follows from the other two). That L in this case satisfies our definition follows from the 'restricted Poincaré–Birkhoff–Witt Theorem', which says that *the restricted monomials on X form a k-basis for the universal envelope U of L*, when X is an ordered basis for L (see Jacobson (1962), Chapter V §7, whence we have borrowed the proof of 12.3). In particular, it implies that the monomials of degree 1 are linearly independent, and hence that L is embedded in U. Conversely, every restricted Lie algebra in our sense satisfies (7): see Exercise 3. As we shall see in the next section, the Lie algebras that we are concerned with come ready equipped with an embedding into an associative algebra, and we don't need to quote the PBW Theorem; indeed our approach provides a self-contained proof for it in the special case at hand.

12.2 Theorems of Jennings and Lazard

In this section, G denotes an arbitrary group and k a field of characteristic p. As in Chapter 11, the series of subgroups $D_n = D_n(G)$ is defined to be the *fastest descending series with $D_1 = G$ such that*

$$D_i^p \leq D_{pi}$$
$$[D_i, D_j] \leq D_{i+j}$$

for all i and j.

We write

$$I = I_G = (g-1)k[G]$$

for the augmentation ideal of $k[G]$, and for each $n \geq 1$ put

$$K_n = K_n(G) = (1 + I^n) \cap G;$$

thus K_n is the kernel of the natural homomorphism of G into the unit group of $k[G]/I^n$.

12.7 Lemma $D_n \leq K_n$ *for all $n \geq 1$.*

Proof It is clear that $K_1 = G$ and that $K_{n+1} \leq K_n$ for all n, so it will suffice to check that

$$K_i^p \subseteq K_{pi}, \quad [K_i, K_j] \subseteq K_{i+j}$$

for all i and j. Let $x \in K_i$, $y \in K_j$. Then

$$x^p - 1 = (x-1)^p \in (I^i)^p = I^{pi},$$

$$[x, y] - 1 = x^{-1}y^{-1}(xy - yx)$$
$$= x^{-1}y^{-1}((x-1)(y-1) - (y-1)(x-1)) \in I^{i+j}.$$

Thus $x^p \in K_{pi}$ and $[x, y] \in K_{i+j}$, and the result follows.

Now put $L_n = D_n/D_{n+1}$ for each n. As we observed in §11.2, each L_n is an elementary abelian p-group; we consider it as a vector space over $\mathbb{F}_p$, and put

$$L = L(G) = \bigoplus_{n=1}^{\infty} L_n.$$

This is a Lie algebra over $\mathbb{F}_p$; the Lie bracket operation is defined on homogeneous elements $a = xD_{i+1} \in L_i, b = yD_{j+1} \in L_j$ by

$$(a, b) = [x, y]D_{i+j+1},$$

and extended to L by linearity (see §11.2). We also showed in §11.2 that for each i, there is a well-defined mapping $[p] : L_i \to L_{pi}$ given by $a^{[p]} = x^p D_{pi+1}$. As the notation is meant to suggest, this mapping does indeed extend to a unary operation on L, making L into a restricted Lie algebra; to prove this, however, we need to take a little detour.

Let $x, y \in G$. The identity

$$xy - 1 = (x - 1) + (y - 1) + (x - 1)(y - 1),$$

together with Lemma 12.7, shows that if $x \in D_n$ and $y \in D_{n+1}$ then $xy - 1 \equiv x - 1 \pmod{I^{n+1}}$. We may therefore define a mapping $\theta_n : L_n = D_n/D_{n+1} \to I^n/I^{n+1}$ by setting

$$\theta_n(xD_{n+1}) = (x - 1) + I^{n+1}.$$

The same identity, now taking both x and y in D_n, then shows that θ_n is a homomorphism, hence an $\mathbb{F}_p$-linear map.

The bracket operation on L extends by linearity to the k-vector space

$$L^k = L \otimes_{\mathbb{F}_p} k = \bigoplus_{n=1}^{\infty} L_n \otimes_{\mathbb{F}_p} k,$$

which thereby becomes a Lie algebra over k. For each n, the $\mathbb{F}_p$-linear map θ_n extends uniquely to a k-linear map $\theta_n : L_n \otimes k \to I^n/I^{n+1}$; we put these together to obtain a k-linear map

$$\theta = \bigoplus \theta_n : L^k = \bigoplus L_n \otimes k \to \bigoplus I^n/I^{n+1} = \mathrm{gr}(k[G]).$$

We can now state the main result of this section:

12.8 Theorem (i) *The operation $a \mapsto a^{[p]}$ defined on the subset $\bigcup_{n=1}^{\infty} L_n$ extends to a unary operation on L^k, making L^k into a restricted Lie algebra over k.*

(ii) *The mapping $\theta : L^k \to \mathrm{gr}(k[G])$ is a restricted Lie algebra monomorphism.*

(iii) *$(\theta, \mathrm{gr}(k[G]))$ is a universal envelope of L^k.*

The theorem of Jennings is an easy consequence of the fact that θ is injective:

12.9 Theorem $D_n = K_n$ for all $n \geq 1$.

Proof This is by induction on n. The claim is trivial for $n = 1$; let $n \geq 1$ and suppose that $D_n = K_n$. Then $D_n \geq K_{n+1} \geq D_{n+1}$, by Lemma 12.7, so

$$\ker \theta_n \cap L_n = (D_n \cap K_{n+1})D_{n+1}/D_{n+1} = K_{n+1}/D_{n+1}. \tag{8}$$

If we assume that θ is injective, as stated in part (ii) of Theorem 12.8, we have $\ker \theta_n \leq \ker \theta = 0$; hence $D_{n+1} = K_{n+1}$ and the result follows by induction.

For later reference, we note that this argument is reversible: if $D_{n+1} = K_{n+1}$ then certainly $D_n \geq K_{n+1} \geq D_{n+1}$, so (8) holds and it follows that $\theta_n|_{L_n}$ is injective. As θ_n is the linear extension of this map to $L_n \otimes k$, this shows that θ_n is injective. Thus we have

12.10 Lemma *If $D_{n+1} = K_{n+1}$ then the mapping θ_n is injective.*

We now embark on the proof of Theorem 12.8. Much of the argument is essentially formal; this part is summed up in the following lemma:

12.11 Lemma (i) $\theta(L^k)$ *generates* $\mathrm{gr}(k[G])$ *as a k-algebra.*
(ii) θ *is a Lie algebra homomorphism.*
(iii) $\theta(a^{[p]}) = \theta(a)^p$ *for all $a \in \bigcup_{n=1}^{\infty} L_n$.*
(iv) $\theta(L^k)$ *is a restricted Lie subalgebra of* $\mathrm{gr}(k[G])$.

Proof (i) The k-algebra $\mathrm{gr}(k[G])$ is clearly generated by the subspace I/I^2, which in turn is spanned by the set $\{(g-1) + I^2 \mid g \in G\} = \theta(L_1) \subseteq \theta(L^k)$.
(ii) Let $x \in D_i$ and $y \in D_j$. Then

$$(x-1)(y-1) - (y-1)(x-1) = ([x,y]-1) + (yx-1)([x,y]-1)$$
$$\equiv [x,y] - 1 \pmod{I^{i+j+1}},$$

since $(yx-1) \in I$ and $[x,y] - 1 \in D_{i+j} - 1 \subseteq I^{i+j}$ by Lemma 12.7. It follows that if $a = xD_{i+1} \in L_i$ and $b = yD_{j+1} \in L_j$, then $\theta(a)\theta(b) - \theta(b)\theta(a) = \theta((a,b))$. This gives the result, since elements of this form span L^k.
(iii) Let $a = xD_{i+1} \in L_i$. Then

$$\theta(a^{[p]}) = (x^p - 1) + I^{pi+1}$$
$$= (x-1)^p + I^{pi+1} = \theta(a)^p.$$

(iv) Part (ii) implies that $\theta(L^k)$ is a Lie subalgebra of $\mathrm{gr}(k[G])$; it has a k-basis consisting of elements of the form $\theta(a)$ where $a \in \bigcup_{n=1}^{\infty} L_n$. The result therefore follows from (iii), by Lemma 12.3.

Next, we consider the special case where G is a finite p-group.

12.12 Lemma *Suppose that G has finite order p^n. Then*
(i) $D_{s+1} = 1$ and $I^{t+1} = 0$ for some s and t.
(ii) $\dim_k(L^k) = n$ and $\dim_k(\mathrm{gr}(k[G])) = p^n$.

Proof (i) The first claim is immediate from Lemma 11.13. The second follows from Lemma 7.1 when $k = \mathbb{F}_p$; this implies the general case, since it is clear that the set $(G-1)^i$ spans I^i over k for each natural number i.

(ii) It follows from (i) that $|G| = \prod_{i=1}^{s} |D_i/D_{i+1}| = \prod_{i=1}^{\infty} |D_i/D_{i+1}|$. Hence

$$n = \sum_{i=1}^{\infty} \dim_{\mathbb{F}_p}(D_i/D_{i+1}) = \sum_{i=1}^{\infty} \dim_{\mathbb{F}_p}(L_i)$$
$$= \dim_{\mathbb{F}_p}(L) = \dim_k(L^k).$$

The second claim follows similarly from the fact that $\dim_k(k[G]) = |G| = p^n$.

Proof of Theorem 12.8 for finite p-groups Suppose that $|G| = p^n$. We apply Lemma 12.6, taking $A = \mathrm{gr}(k[G])$, L^k for L and θ for ϕ. Lemma 12.11 shows that the hypotheses are all satisfied, and Lemma 12.12 shows that $\dim_k(A) = p^n$. Part (ii) of Lemma 12.6 therefore tells us that θ is injective; with Lemma 12.11 this shows that L^k is a restricted Lie algebra, with the 'correct' unary operation $[p]$; and Lemma 12.6(ii) also shows that (θ, A) is a universal envelope of L^k. All parts of Theorem 12.8 are therefore established.

The following corollary will be used in the next section:

12.13 Corollary *Suppose that*

$$|D_i/D_{i+1}| = p^{b_i} \quad for \ i = 1, \dots, r,$$

where each b_i is finite. Then

$$\dim_k(I^r/I^{r+1}) = c_r$$

where c_r is the coefficient of T^r in the polynomial

$$F(T) = \prod_{i=1}^{r} \left(\frac{T^{pi} - 1}{T^i - 1} \right)^{b_i}.$$

Proof The hypothesis implies that $G/D_{r+1} = \widetilde{G}$, say, is a finite p-group. Write $\sim: k[G] \to k[\widetilde{G}]$ for the natural epimorphism. Then $\ker(\sim) = (D_{r+1} - 1)k[G] \subseteq I^{r+1}$ by Lemma 12.7, so $\sim$ induces an isomorphism of $k[G]/I^{r+1}$ onto $k[\widetilde{G}]/\widetilde{I}^{r+1}$. We may therefore replace G by $\widetilde{G}$ without altering the values of $b_1, \ldots, b_r$ and c_r, and so assume that in fact G is a finite p-group, and hence that Theorem 12.8 is valid. In particular, we then have

$$\dim_{\mathbb{F}_p}(\theta(L_i)) = \dim_{\mathbb{F}_p}(L_i) = b_i$$

for each i.

For $i = 1, \ldots, r$ let $X_i = \{x_{i1}, \ldots, x_{ib_i}\}$ be an $\mathbb{F}_p$-basis for $\theta(L_i)$. Then $X = \bigcup_{i=1}^{r} X_i$ is a k-basis for $\theta(L^k)$; we order it lexicographically. Lemma 12.6(iii), with Lemma 12.12(ii), shows that the restricted monomials on X form a basis for $\mathrm{gr}(k[G])$. Now recall that $\theta(L_i) \subseteq I^i/I^{i+1}$; it follows that each monomial on X is a *homogeneous* element of $\mathrm{gr}(k[G])$, and hence, in particular, that the restricted monomials which happen to lie in I^r/I^{r+1} form a *basis* for I^r/I^{r+1}. Thus c_r is equal to the number of such restricted monomials. This is just the number of distinct expressions of the form

$$\prod_{i=1}^{r} \prod_{j=1}^{b_i} x_{ij}^{n_{ij}} \qquad (0 \le n_{ij} \le p - 1)$$

such that

$$\sum_{i=1}^{r} \sum_{j=1}^{b_i} i n_{ij} = r.$$

To see that this number is exactly the coefficient of T^r in $F(T)$, multiply out the identity

$$F(T) = \prod_{i=1}^{r}(1 + T^i + T^{2i} + \cdots + T^{(p-1)i})^{b_i}.$$

Remark The reduction to finite p-groups in this proof was made so that we could quote Theorem 12.8 before proving it in full generality; thus the applications to pro-p groups that we make in the next section do not depend on the rather lengthy arguments with which we are about to conclude the present section.

The proof of Theorem 12.8 in general is a matter of reduction to the case of finite p-groups. We begin with the

Proof of Theorem 12.8 (i) *and* (ii) The main task is to prove that $D_n = K_n$ for all n. Suppose this done. Then Lemma 12.10 shows that θ is injective, and the result follows from Lemma 12.11, parts (ii), (iii) and (iv).

It remains to prove that $K_n \subseteq D_n$ for all n. We fix $n > 1$, and show first that it suffices to consider the case where G is *finitely generated*. Let $x \in K_n$. Then $x - 1 \in I^n$, so $x - 1$ is a finite sum

$$x - 1 = \sum_{i=1}^{s} \lambda_i (g_{i1} - 1) \dots (g_{in} - 1),$$

for certain elements $g_{ij} \in G$ and $\lambda_i \in k$. Thus $x \in K_n(H)$ where H is the subgroup $\langle x, g_{i1}, \dots, g_{in} \mid i = 1, \dots, s \rangle$ of G. If our claim is true for all finitely generated groups, we have $K_n(H) = D_n(H)$; but it is easy to see that $D_n(H) \subseteq D_n(G)$, so $x \in D_n(G)$ as required.

We may assume, then, that G is finitely generated. In this case, $G_n/D_n = \widetilde{G}$, say, is a finite p-group, by Exercise 11.2; so $K_n(\widetilde{G}) = D_n(\widetilde{G}) = 1$, by the case done above. Now write $\sim: k[G] \to k[\widetilde{G}]$ for the natural epimorphism. As in the proof of Corollary 12.13, above, we see that $\sim$ induces an isomorphism of $k[G]/I^n$ onto $k[\widetilde{G}]/\widetilde{I}^n$. It follows that for $g \in G$,

$$g \in K_n \iff g - 1 \in I^n$$
$$\iff \widetilde{g} - 1 \in \widetilde{I}^n$$
$$\iff \widetilde{g} \in K_n(\widetilde{G}) = 1$$
$$\iff g \in D_n.$$

Thus $K_n = D_n$, and we are done.

We remark that Jennings's Theorem, Theorem 12.9, is now established in complete generality. The reader who is not particularly interested in universal envelopes may therefore safely skip the rest of this section, which is devoted to the proof of Theorem 12.8(iii).

We start by fixing, for each $n \geq 1$, an $\mathbb{F}_p$-basis X_n for $L_n = D_n/D_{n+1}$, and give it a total ordering $<$. We then put $X = \bigcup_{n=1}^{\infty} X_n$, and order X by putting elements of X_m before elements of X_n when $m < n$. We define the *weight* of a monomial $w = x_1 \dots x_s$ to be

$$\mathrm{wt}(w) = n_1 + \dots + n_s$$

where $x_i \in L_{n_i}$ for each i, and put

$$\widetilde{\theta}(w) = \theta(x_1) \dots \theta(x_s) \in \mathrm{gr}(k[G]).$$

Note that then $\widetilde{\theta}(w) \in I^r/I^{r+1}$ where $r = \mathrm{wt}(w)$.

Finally, we denote by $\mathcal{B}_r$ the set of *all restricted monomials of weight r on X.* We shall prove

12.14 Proposition *For each $r \geq 0$, the restriction of $\widetilde{\theta}$ to $\mathcal{B}_r$ is injective, and $\widetilde{\theta}(\mathcal{B}_r)$ is a linearly independent subset of I^r/I^{r+1}.*

Part (iii) of Theorem 12.8 follows easily from this. In view of Proposition 12.4, it will be enough to show that the induced k-algebra homomorphism $\theta^* : U_X(L^k) \to \mathrm{gr}(k[G])$ is an isomorphism. Now Lemma 12.11 shows that θ^* is surjective; that θ^* is injective is then an immediate consequence of Proposition 12.14 and Lemma 12.5, which says that $U_X(L^k)$ is spanned by the images of the restricted monomials. As remarked at the end of Section 12.1, we also obtain the further consequence that these images are in fact linearly independent in $U_X(L^k)$: this is the 'Poincaré–Birkhoff–Witt Theorem' for the restricted Lie algebra L^k.

Before proving Proposition 12.14 we extract a simple reduction lemma:

12.15 Lemma *Let H_0 be a finitely generated subgroup of G, and let $x_i = h_i D_{n_i+1} \in X_{n_i}$ for $i = 1, \ldots, q$, with $x_1 < x_2 < \ldots < x_q$. Then there exist a finitely generated subgroup H of G, containing H_0, and for each n an ordered $\mathbb{F}_p$-basis Y_n for $L_n(H) = D_n(H)/D_{n+1}(H)$, such that*

(i) $h_i \in D_{n_i}(H)$ *for $i = 1, \ldots, q$;*

(ii) *for each i, the element $y_i = h_i D_{n_i+1}(H)$ is in Y_{n_i};*

(iii) $y_1 < y_2 < \ldots < y_q$, *where $\bigcup Y_n$ is ordered so that elements of Y_m precede elements of Y_n when $m < n$.*

Proof For each n, the group D_n is the union of its subgroups $D_n(H)$ as H ranges over the finitely generated subgroups of G. We can therefore find a finitely generated subgroup H of G, containing H_0, such that (i) is satisfied. Now for $n \neq n_1, \ldots, n_q$, let Y_n be any ordered basis for $L_n(H)$. Suppose $n = n_r = n_{r+1} = \cdots = n_{r+s}$. The inclusion $D_n(H) \to D_n$ induces an $\mathbb{F}_p$-linear mapping $\pi_n : L_n(H) \to L_n$, and then $\pi_n(y_i) = x_i$ for $i = r, r+1, \ldots, r+s$; since $x_r, \ldots x_{r+s}$ are linearly independent, so are $y_r, \ldots y_{r+s}$, and we may extend the set $\{ y_r, \ldots y_{r+s} \}$ to a basis Y_n of $L_n(H)$, ordered so that $y_r < y_{r+1} < \ldots < y_{r+s}$. In this way conditions (ii) and (iii) are satisfied.

Proof of Proposition 12.14 Let $w_1, \ldots, w_m$ be distinct elements of $\mathcal{B}_r$, and suppose that

$$\sum_{i=1}^{m} \lambda_i \widetilde{\theta}(w_i) = 0, \tag{9}$$

where $\lambda_1, \ldots, \lambda_m \in k$. We have to show that $\lambda_1 = \ldots = \lambda_m = 0$. Say $w_i = x_{i1} \ldots x_{is_i}$, where $x_{ij} = h_{ij} D_{n_{ij}+1} \in X_{n_{ij}}$. Then (9) is equivalent to

$$\sum_{i=1}^{m} \lambda_i \prod_{j=1}^{s_i} (h_{ij} - 1) \in I^{r+1}.$$

Since each element of I^{r+1} is a finite linear combination of terms of the form $(g_1 - 1) \ldots (g_{r+1} - 1)$, there exists a finitely generated subgroup H_0 of G such that

$$\sum_{i=1}^{m} \lambda_i \prod_{j=1}^{s_i} (h_{ij} - 1) \in I_{H_0}^{r+1}. \tag{10}$$

Now enlarge H_0 to a finitely generated subgroup H as in Lemma 12.14, taking $x_1, \ldots, x_q$ to be the elements x_{ij} in increasing order; and let Y be the ordered basis of $L(H)$ given in that lemma. Retracing the argument from (10) to (9), we find that

$$\sum_{i=1}^{m} \lambda_i \widetilde{\theta_H}(v_i) = 0,$$

where the $v_i = y_{i1} \ldots y_{is_i}$ are distinct restricted monomials of weight r on Y, and $\widetilde{\theta}_H$ is to the pair (H, Y) what $\widetilde{\theta}$ is to (G, X). Replacing (G, X) by (H, Y), we may therefore suppose that in fact G is finitely generated.

Now put $\widetilde{G} = G/D_{r+1}$; this is a finite p-group, by Exercise 11.2. We have $L_n(\widetilde{G}) = L_n$ for $n = 1, \ldots, r$ and $L_n(\widetilde{G}) = 0$ for all $n > r$, so the set $\widetilde{X} = X_1 \cup \ldots \cup X_r$ is a basis for $L^k(\widetilde{G})$. The special case of Theorem 12.8 proved above shows that $\mathrm{gr}(k[\widetilde{G}])$ is a universal envelope for $L^k(\widetilde{G})$; it follows by Lemma 12.6(iii) that the images of the restricted monomials on $\widetilde{X}$ are distinct and form a basis for $\mathrm{gr}(k[\widetilde{G}])$. Identifying $\widetilde{I}^r/\widetilde{I}^{r+1}$ with I^r/I^{r+1} as before, we see in particular that $\widetilde{\theta}(w_1), \ldots, \widetilde{\theta}(w_r)$ are distinct and linearly independent; hence $\lambda_1 = \ldots = \lambda_m = 0$ as required. This completes the proof.

12.3 Poincaré series: 'l'alternative des gocha'

We return now to pro-p groups, and assume henceforth that G *is a finitely generated pro-p group.* As before, we write

$$D_n = D_n(G), \quad L_n = D_n/D_{n+1}.$$

Recall (§11.3) that each D_n is now an open subgroup of G; hence each of the numbers

$$b_n = \dim_{\mathbb{F}_p}(L_n)$$

is finite. We shall be applying the results of the preceding section, taking $k = \mathbb{F}_p$; thus I will now denote the augmentation ideal of $\mathbb{F}_p[G]$. For each $n \geq 0$ we write

$$c_n = \dim_{\mathbb{F}_p}(I^n/I^{n+1}).$$

Following Lazard's whimsical terminology, we define a formal power series

$$\mathrm{gocha}(G; T) = \sum_{n=0}^{\infty} c_n T^n$$

(the name alludes to 'Golod and Shafarevich'). Writing

$$P_n(T) = \frac{1 - T^{pn}}{1 - T^n},$$

we may restate Corollary 12.13 in the succinct form

12.16 Theorem *Let G be a finitely generated pro-p group. Then*

$$\mathrm{gocha}(G; T) = \prod_{n=1}^{\infty} P_n(T)^{b_n}.$$

In Chapter 11, we showed that the growth rate of the sequence (b_n) determines whether or not G has finite rank. It is now a mere formality to translate this into a criterion using the c_n instead of the b_n; the result, however, has interesting applications.

Suppose first that G has infinite rank. Theorem 11.4 says that then $D_n \neq D_{n+1}$ for all n, so we have $b_n \geq 1$ for all n. Let us define a partial ordering on formal power series by setting $\Phi \geq \Psi$ *if and only if each coefficient of* $\Phi - \Psi$ *is non-negative.* In this sense, we have $P_n(T) \geq 1 + T^n$ for each $n \geq 1$; with Theorem 12.6 this gives

$$\text{gocha}(G; T) = \prod_{n=1}^{\infty} P_n(T)^{b_n}$$

$$\geq \prod_{n=1}^{\infty} P_n(T) \geq \prod_{n=1}^{\infty} (1 + T^n)$$

$$= 1 + \sum_{n=1}^{\infty} p(n) T^n;$$

here, $p(n)$ denotes *the number of partitions of* n. Thus we have

12.17 Proposition *If* G *has infinite rank then*

$$c_n \geq p(n) \geq 2^{[\sqrt{n}]} \quad \textit{for all } n > 1.$$

For the proof of the second inequality, see Exercise 5 (we have given a crude lower bound for $p(n)$, which is easy to prove; in fact

$$p(n) \sim \frac{1}{4\sqrt{3}n} e^{\pi \sqrt{2n/3}};$$

see Hardy (1940)). Since $2^{[\sqrt{n}]}$ of course grows faster than any polynomial in n, this establishes one direction of Theorem 12.1.

The other direction of Theorem 12.1 has already been essentially established in Chapter 7; indeed, if G has finite rank then Exercise 7.5 gives

$$c_n \leq \dim_{\mathbb{F}_p} (\mathbb{F}_p[G]/I^{n+1}) \leq Cn^d$$

for all n, where $d = \dim(G)$ and C is a positive constant. (Note that when G is *uniform*, the exact value of c_n is given in Exercise 7.4: in this case,

$$c_n \sim \frac{n^{d-1}}{(d-1)!}.)$$

However, it is interesting to pursue the finite rank case a little further, using the more explicit information available from Chapter 11. Assume from now on that G is a pro-p group of finite rank, and dimension d. Proposition 11.3 says that $b_r = 0$ for some r. Now recall Proposition 11.15. Putting

$$g_n = \dim_{\mathbb{F}_p} L_n^*,$$

where $L_n^* = \gamma_n(G)D_{n+1}/D_{n+1}$, we have

$$g_{n+1} \le g_n \le b_n \quad \text{for all } n,$$
$$b_n = g_n \qquad \text{if } p \nmid n,$$
$$b_n \le b_m + g_n \quad \text{if } n = pm.$$

It follows that $g_n = 0$ for all $n \ge r$, and hence that

$$b_n = 0 \qquad \text{if } p \nmid n \text{ and } n \ge r,$$
$$b_n \le b_{n/p} \quad \text{if } p \mid n \text{ and } n \ge r.$$

Thus $b_n = 0$ unless $n < r$ or $n = p^i h$ for some $i \ge 0$ and h in the range $r/p \le h < r$. For each h in this interval, the sequence of non-negative integers $b_h \ge b_{ph} \ge b_{p^2h} \ge \dots$ must become stationary at some point. We can therefore choose s so large that for every $n \ge s$,

$$\text{either} \qquad p \nmid n \qquad \text{and } b_n = 0$$
$$\text{or} \qquad n = p^i t, \quad \text{where } s \le t < ps, \text{ and } b_n = b_t.$$

Now for each $t \ge 1$, we have

$$\prod_{i=0}^{\infty} P_{p^i t}(T) = (1 - T^t)^{-1}.$$

In the present case, therefore, Theorem 12.16 becomes

12.18 Proposition *If G is a pro-p group of finite rank, then for sufficiently large s we have*

$$\text{gocha}(G; T) = \prod_{n=1}^{s-1} P_n(T)^{b_n} \cdot \prod_{t=s}^{ps-1} (1 - T^t)^{-b_t}.$$

Note that each $P_n(T)$ is a polynomial, taking the value p at 1. It follows that $\text{gocha}(G; T)$ is a rational function having a pole of order

$$\sum_{t=s}^{ps-1} b_t = m, \quad \text{say,}$$

at 1. What is the value of m? First of all, it is clearly independent of the choice of s, provided s is sufficiently large. Choose s to be large and not divisible by p. Then $b_s = 0$, so $D_s = D_{s+1}$, and hence D_s is powerful, by Theorem 11.5. Since, by Lemma 11.16, D_s is open in G, and may be taken to lie inside any open subgroup of G provided s is large enough, it follows that $\mathrm{d}(D_s) = \dim(G)$. Now Lemma 11.22(ii) shows that $D_{ps} = \overline{D_s^p} = \Phi(D_s)$; consequently

$$p^m = |D_s : D_{ps}| = |D_s : \Phi(D_s)| = p^{\mathrm{d}(D_s)}.$$

Thus the answer to our question is $m = \dim(G)$; and we have a new characterisation for the dimension of a pro-p group of finite rank:

12.19 Corollary *Let G be a pro-p group of finite rank. Then* $\dim(G)$ *is the order of the pole at 1 of the rational function* $\mathrm{gocha}(G; T)$.

Notes

For the theory of restricted Lie algebras, see Jacobson (1962), Chapter 5, §7.

Theorem 12.9 is due to Jennings (1941); for another account see Passman (1977).

Theorem 12.8 and the results of §12.3 are from [L], Appendice, §3. A short proof that $\mathrm{gr}(k[G])$ is the restricted universal envelope of $L(G)$, and an analogue for fields of characteristic zero, is given by Quillen (1968); although it uses the language of Hopf algebras, at its heart lies a simple counting argument that we have adapted here as Lemma 12.6; a similar account is given in Passi (1979).

Exercises

1. Let L be a restricted Lie algebra and (θ, A), (ϕ, B) universal envelopes for L. Show that $\theta^* \circ \phi^* = \mathrm{Id}_A$ and $\phi^* \circ \theta^* = \mathrm{Id}_B$. Deduce that ϕ^* is a k-algebra isomorphism from A onto B such that $\phi^*\theta(x) = \phi(x)$ for all $x \in L$.

2. (i) Prove that

$$(Y - X)^p = Y^p - X^p$$

$$(Y - X)^{p-1} = \sum_{i=0}^{p-1} X^i Y^{p-i-1}$$

for commuting indeterminates X, Y over $\mathbb{F}_p$.

  (ii) Let A be a k-algebra and $v \in A$. Define $\xi, \eta : A \to A$ by $\xi(u) = uv$, $\eta(u) = vu$. Show that there is a homomorphism from $\mathbb{F}_p[X, Y]$ into the ring of k-linear transformations on A sending X to ξ and Y to η. Deduce that for all $v \in A$,

$$(v,_p u) = (v, u^p)$$

$$(v,_{p-1} u) = \sum_{i=0}^{p-1} u^i v u^{p-i-1}.$$

3. Let A be a k-algebra and L a Lie subalgebra of A such that $x^p \in L$ for all $x \in L$. Prove that the identities (7) hold in L, where $x^{[p]} = x^p$.
  [Use the proof of Lemma 12.3, and Exercise 2.]

4. Let H and N be restricted Lie algebras, and $\phi : H \to N$ a Lie algebra homomorphism. Suppose that $\phi(x^{[p]}) = \phi(x)^{[p]}$ for every x in some spanning set of H. Prove that ϕ is a restricted Lie algebra homomorphism.
  [*Hint*: Look at the proof of Lemma 12.3.]

5. Let n be a positive integer. A *partition* of n is a sequence of non-negative integers $b_1 \geq b_2 \geq \ldots$ such that $\sum_{i=1}^{\infty} b_i = n$. Show that if $k \geq a_1 > a_2 > \ldots > a_r \geq 1$ then

$$\left(n - \sum_{i=1}^{k} a_i, a_1, a_2, \ldots, a_r, 0, \ldots\right)$$

is a partition of n provided $k^2 + 3k \leq 2n$. Deduce that the number $p(n)$ of partitions of n satisfies $p(n) \geq 2^{\lfloor \sqrt{n} \rfloor}$ if $n \geq 9$.
  Verify that this holds also for $n = 2, \ldots, 8$.

Interlude D

The Golod–Shafarevich inequality

Let G be a pro-p group, and suppose that $d(G) = d$ is finite. A presentation

$$G = \langle X; R \rangle$$

by generators and relations is called *minimal* if $|X| = d$. In §4.6 we defined $t(G)$ for a finitely generated pro-p group G by

$$t(G) = \inf \left\{ |R| \mid G \text{ has a minimal presentation } \langle X; R \rangle \right\},$$

and gave the lower bound $d(d-1)/2$ for $t(G)$ when G is *powerful*. Roughly speaking, what this means is that to force a d-generator pro-p group to be powerful, one needs to impose at least $d(d-1)/2$ relations; this is intuitively plausible, since we have to make all the generators commute with one another 'modulo pth powers'. In 1964, Golod and Shafarevich gave a lower bound for the number of relations needed to ensure that a pro-p group be *finite*. This bound is also quadratic in d. An examination of their method, and various improvements of it due to Gaschütz, Vinberg and Roquette, shows that a similar lower bound obtains provided only that the pro-p group G is rather 'narrow', in the sense that the factors in the dimension subgroup series do not grow too fast.

The dimension subgroups $D_n = D_n(G)$ were defined in §11.1. For each $n \geq 1$ we put

$$b_n = \dim_{\mathbb{F}_p}(D_n/D_{n+1}).$$

As we saw in §12.3, the sequence (b_n) is closely related to the sequence (c_n) given by

$$c_n = \dim_{\mathbb{F}_p}(I_0^n/I_0^{n+1}),$$

where $I_0 = (G-1)\mathbb{F}_p[G]$ is the augmentation ideal of the group algebra $\mathbb{F}_p[G]$. For any sequence $\mathbf{a} = (a_n)$ of non-negative real numbers we write

$$\rho(\mathbf{a}) = \limsup_{n \to \infty} a_n^{\frac{1}{n}},$$

so $\rho(\mathbf{a})^{-1}$ is the radius of convergence of the power series $\sum a_n X^n$ (in the usual Archimedean sense!). The main result is now

D1 Theorem *Let G be a finitely generated pro-p group, and suppose that $d = d(G) > 1$. If either*

- $\rho(\mathbf{b}) \le 1$ *or*
- $\rho(\mathbf{c}) \le 1$

then

$$t(G) \ge \frac{d^2}{4},\tag{1}$$

and the inequality is strict *unless* $d = 2$ *and* $t(G) = 1$.

The formula (1) is the *Golod–Shafarevich inequality*. If G is finite, the sequence (c_n) is eventually zero, by Lemma 7.1, and we obtain the original theorem of Golod and Shafarevich. If G has finite rank, Exercise 7.5 shows that for all n, $c_n \le Cn^r$, where $r = \dim(G)$ and C is a constant; thus (1) holds if G is a pro-p group of finite rank which is not pro-cyclic. Wilson (1991) showed that this holds under much milder hypotheses on G; and in fact Zel'manov has proved that if the finitely generated pro-p group G does not satisfy (1), then G contains a non-abelian free pro-p subgroup. The proof is hard, and is given in Zel'manov's chapter in [DSS]. Theorem D1 combines results of Koch (1969), Lubotzky (1983) and Lubotzky and Shalev (1994).

In fact the condition $\rho(\mathbf{b}) \le 1$ is *equivalent* to $\rho(\mathbf{c}) \le 1$; we have included it in the statement of the theorem because it refers only to the internal structure of the group G, and can sometimes be more easily verified, as we shall see in Chapter 13.

Let us write $R = \mathbb{F}_p [[G]]$ for the completed group algebra of G, defined in §7.4, and put $I = (G - 1)R$; thus I is the closure of I_0 in R. Using Lemma 7.1 one readily verifies that

$$\dim_{\mathbb{F}_p}(I^n/I^{n+1}) = c_n$$

for each n. Now let $t = t(G)$. Then there is a presentation

$$1 \to N \to F \to G \to 1\tag{2}$$

where F is a free pro-p group on d generators and N is the closure in F of the normal subgroup generated by t elements of F. The next lemma provides the essential link between presentations of G and powers of the ideal I:

D2 Lemma *Provided $d = d(G)$, the sequence (2) gives rise to an exact sequence*

$$R^{(t)} \xrightarrow{\ \alpha\ } R^{(d)} \to I \to 0\tag{3}$$

of R-modules, such that $R^{(t)}\alpha \subseteq I^{(d)}$.

We postpone the proof to the end, and proceed now to deduce Theorem D1. Let us assume that $\rho(\mathbf{c}) \leq 1$, so that the power series

$$P(x) = \sum_{n=0}^{\infty} c_n x^n$$

is convergent for x in the range $(0, 1)$. For $n \geq 0$ we put

$$s_n = \dim_{\mathbb{F}_p}(R/I^{n+1}) = \sum_{j=0}^{n} c_j,$$

and also write $s_{-1} = 0$. Now let $n \geq 1$. From (3) we derive an exact sequence

$$0 \to R^{(t)}/K_n \xrightarrow{\alpha_n} (R/I^n)^{(d)} \to I/I^{n+1} \to 0,$$

where $K_n = (I^n)^{(d)}\alpha^{-1}$. Since $R^{(t)}\alpha \subseteq I^{(d)}$ we see that the second term in this sequence is annihilated by I^{n-1}; its dimension is therefore at most ts_{n-2}. The next two terms have dimensions ds_{n-1} and $s_n - 1$ respectively, since $\dim_{\mathbb{F}_p}(R/I) = 1$. It follows that

$$ts_{n-2} - ds_{n-1} + s_n \geq 1. \tag{4}$$

Now multiply (4) by $x^{n-1} - x^n$, and add up the resulting inequalities. Noting that $s_n - s_{n-1} = c_n$ for each n, we find that

$$P(x)(tx^2 - dx + 1) \geq 1$$

for all $x \in (0, 1)$. As $P(x)$ is positive for such x it follows that

$$tx^2 - dx + 1 > 0$$

for all $x \in (0, 1)$. If $d = 2$ this clearly implies that $t \geq 1$. Suppose that $d \geq 3$. Letting $x \to 1$ from the left we infer that $t \geq d - 1$, which implies that $d/2t < 1$; substituting $x = d/2t$ now gives $-d^2/4t + 1 > 0$.

To complete the proof, it remains to establish

D3 Lemma $\rho(\mathbf{b}) \leq 1$ *if and only if* $\rho(\mathbf{c}) \leq 1$.

Proof Since the mappings θ_n defined in §12.2 are injective, we have $b_n \leq c_n$ for all n, so $\rho(\mathbf{b}) \leq \rho(\mathbf{c})$ in any case.

Suppose now that $\rho(\mathbf{b}) \leq 1$. Theorem 12.16 asserts the power series identity

$$\sum_{n=0}^{\infty} c_n T^n = \prod_{n=1}^{\infty} P_n(T)^{b_n},$$

where $P_n(T) = 1 + T^n + \cdots + T^{(p-1)n}$ for each $n \geq 1$. To show that $\rho(\mathbf{c}) \leq 1$, it will suffice therefore to show that the infinite product on the right is convergent when T is replaced by a real number $x \in (0,1)$. Now

$$P_n(x) = (1 + u_n)^{b_n}$$

where $0 < u_n < x^n/(1 - x^n)$. Let N be so large that $x^n < \frac{1}{2}$. Then $0 < u_n < 1$, and elementary analysis shows that the product $\prod_{n=N}^{\infty} P_n(x)^{b_n}$ is convergent if and only if the series $\sum_{n=N}^{\infty} b_n u_n$ converges. But

$$\sum_{n=N}^{\infty} b_n u_n < \sum_{n=N}^{\infty} (1 - x^n)^{-1} b_n x^n < 2 \sum_{n=N}^{\infty} b_n x^n,$$

and the series on the right does converge if $\rho(\mathbf{b}) \leq 1$. The lemma follows.

As observed by Lubotzky (1983), a variant of the Golod–Shafarevich inequality may be inferred for any finitely generated group whose pro-p completion satisfies the inequality. This depends on

D4 Lemma *Let Γ be a group and $\langle X; R \rangle$ a finite presentation for Γ as abstract group (i.e. $\Gamma \cong F(X)/\langle R^{F(X)} \rangle$ where $F(X)$ denotes the free group on X). Then $\langle X; R \rangle$, considered as a presentation in the category of pro-p groups, is a presentation for the pro-p completion of Γ.*

We leave it to the reader to verify that if $F = \widehat{F(X)}_p$ denotes the free pro-p group on X, then

$$F/\overline{\langle R^F \rangle}$$

has the requisite universal property that characterises the pro-p completion $\widehat{\Gamma}_p$ of Γ.

However, since it is not always the case that $d(\Gamma) = d(\widehat{\Gamma}_p)$, we cannot translate Theorem D1 directly to an inequality for abstract groups. Write

$$d_p(\Gamma) = d(\Gamma/[\Gamma, \Gamma]\Gamma^p).$$

D5 Lemma *If Γ is a finitely generated group then $d(\widehat{\Gamma}_p) = d_p(\Gamma)$.*

To see this, put $G = \widehat{\Gamma}_p$, and note that $G/\Phi(G)$ is the largest elementary abelian p-group quotient of G (by an open normal subgroup). It follows that $G/\Phi(G)$ is isomorphic to the largest elementary abelian p-group quotient of Γ, which is $\Gamma/[\Gamma,\Gamma]\Gamma^p$. This makes it clear that $d(G) = d(G/\Phi(G)) = d_p(\Gamma)$.

D6 Lemma *Let G be a pro-p group and let $\langle X; R \rangle$ be a finite presentation of G. Then there exists a presentation $\langle Y; S \rangle$ of G with $|Y| = d(G)$ and $|S| = |R| - (|X| - |Y|)$.*

Proof Let $F = \widehat{F(X)}_p$ denote the free pro-p group on X; we have an epimorphism $F \to G$ with kernel $\overline{\langle R^F \rangle}$. This epimorphism induces an epimorphism

$$\phi : F/\Phi(F) \to G/\Phi(G).$$

Put $K = \ker \phi$, so K is the subspace of the $\mathbb{F}_p$-vector space $F/\Phi(F)$ generated by the image of R. Let $\{r_1\Phi(F), \ldots, r_k\Phi(F)\}$ be a basis for K with $r_1, \ldots, r_k \in R$. Clearly $k = d(F) - d(G)$ and $R_0 = \{r_1, \ldots, r_k\}$ is a subset of a free basis of F. Thus by Exercise 1.20 (iii) $F^* = F/\overline{\langle R_0^F \rangle}$ is a free pro-p group on $d(F) - k = d(G)$ generators. The image S of $R \backslash R_0$ in F^* gives the desired presentation.

Combining Lemmas D4, D5 and D6 we may deduce that if $\langle X; R \rangle$ is a finite presentation for the group Γ then

$$t(\widehat{\Gamma}_p) \leq |R| - (|X| - d_p(\Gamma)).$$

We therefore have the following corollary to Theorem D1:

D7 Theorem. *Let Γ be a finitely generated group and let $\langle X; R \rangle$ be a finite presentation for Γ. Let p be a prime. If $\widehat{\Gamma}_p$ satisfies the hypotheses of Theorem D1 then*

$$|R| - (|X| - d_p(\Gamma)) \geq d_p(\Gamma)^2/4.$$

This includes all finitely generated groups of finite upper rank (i.e. those whose profinite completion has finite rank). For the subclass of finitely generated nilpotent groups we can choose p so that $d_p(\Gamma) = d(\Gamma)$ (see Exercise 3), and so obtain the following corollary, another direct generalisation of the original result for finite p-groups:

D8 Corollary *If Γ is a finitely generated non-cyclic nilpotent group then*

$$t(\Gamma) \geq d(\Gamma)^2/4.$$

To conclude, here is the

Proof of Lemma D2 We have to turn the exact sequence

$$1 \to N \to F \to G \to 1 \tag{2}$$

into the exact sequence (3) of $R = \mathbb{F}_p\left[[G]\right]$-modules. Let us write $F_n = P_n(F)$ and $G_n = P_n(G)$, and suppose that N is the closure in F of $\langle y_1^F, \ldots, y_t^F \rangle$. Then for each n we have an exact sequence

$$0 \to \sum_{i=1}^{t}(y_iF_n - 1)\mathbb{F}_p\left[F/F_n\right] \hookrightarrow \mathbb{F}_p\left[F/F_n\right] \to \mathbb{F}_p\left[G/G_n\right] \to 0;$$

these fit together to give an exact sequence

$$K \hookrightarrow \mathbb{F}_p\left[[F]\right] \xrightarrow{\beta} \mathbb{F}_p\left[[G]\right] \to 0, \tag{5}$$

where β is a ring epimorphism and $\ker \beta = K$ is the closure in $\mathbb{F}_p\left[[F]\right]$ of the right ideal $\sum_{i=1}^{t}(y_i - 1)\mathbb{F}_p\left[[F]\right]$. Since $\mathbb{F}_p\left[[F]\right]$ is compact (being the inverse limit of the finite rings $\mathbb{F}_p\left[F/F_n\right]$), each finitely generated right ideal is also compact, and hence closed in $\mathbb{F}_p\left[[F]\right]$; consequently $K = \sum_{i=1}^{t}(y_i - 1)\mathbb{F}_p\left[[F]\right]$. Note that $\mathbb{F}_p\left[[G]\right]$ considered as a right $\mathbb{F}_p\left[[F]\right]$-module is annihilated by K.

Now put $I_F = (F - 1)\mathbb{F}_p\left[[F]\right]$. Then $K \subseteq I_F$ and $I_F\beta = (G - 1)\mathbb{F}_p\left[[G]\right] = I$. The epimorphism β therefore induces an epimorphism from I_F/I_F^2 onto I/I^2. Having assumed that $\mathrm{d}(F) = d = \mathrm{d}(G)$, we infer that $I_F/I_F^2 \cong \mathbb{F}_p^d \cong I/I^2$ (see Exercise 1). This implies that $K = \ker \beta \subseteq I_F^2$, and so from (5) we derive an exact sequence

$$0 \to K/K^2 \xrightarrow{\gamma} I_F/I_FK \to I \to 0 \tag{6}$$

with the property that $\mathrm{im}\,\gamma \subseteq I_F^2/I_FK$.

Since K is a t-generator right $\mathbb{F}_p\left[[F]\right]$-module, K/K^2 is a t-generator right $R = \mathbb{F}_p\left[[G]\right]$-module and we have an epimorphism $\pi : R^{(t)} \to K/K^2$. On the other hand, I_F is a free right $\mathbb{F}_p\left[[F]\right]$-module on d generators – for the proof of this see Exercise 2. It follows that $I_F/I_FK \cong \mathbb{F}_p\left[[F]\right]^{(d)}/\mathbb{F}_p\left[[F]\right]^{(d)}K \cong R^{(d)}$. Replacing the middle term in (6) with $R^{(d)}$ we finally obtain the desired exact sequence

$$R^{(t)} \xrightarrow{\alpha} R^{(d)} \to I \to 0,$$

where $\alpha = \gamma \circ \pi$. Moreover, the fact that $\mathrm{im}\,\gamma \subseteq I_F^2/I_FK$ translates into the statement that $R^{(t)}\alpha \subseteq R^{(d)}I = I^{(d)}$, and the proof is complete.

Exercises

1. Let G be a finitely generated pro-p group. Write $I_0 = (G - 1)\mathbb{F}_p[G]$ and $I = (G - 1)\mathbb{F}_p[[G]]$.

(i) Show that the mapping $g \mapsto (g - 1) + I_0^2$ induces an isomorphism $G/\Phi(G) \to I_0/I_0^2$.

(ii) Show that the natural inclusion $\mathbb{F}_p[G] \to \mathbb{F}_p[[G]]$ induces an isomorphism $\mathbb{F}_p[G]/I_0^n \to \mathbb{F}_p[[G]]/I^n$ for each $n \geq 1$. [*Hint:* Use Lemma 7.1.]

(iii) Deduce that $\dim_{\mathbb{F}_p}(I/I^2) = d(G)$.

2. Let F be the (abstract) free group on a finite set $\{x_1, \dots, x_d\}$, and let $I = (F - 1)\mathbb{F}_p[F]$ be the augmentation ideal of $\mathbb{F}_p[F]$.

(i) Prove that

$$I = \bigoplus_{i-1}^{d}(x_i - 1)\mathbb{F}_p[F].$$

[*Hint:* To see that the elements $x_i - 1$ generate I as a right ideal, show that F acts trivially on the right module $\mathbb{F}_p[F]/\sum(x_i - 1)\mathbb{F}_p[F]$. To show that they generate I freely, consider the semi-direct product $H = \mathbb{F}_p[F]^{(d)} \rtimes F$; the mapping $x_i \mapsto u_i \cdot x_i$ $(i = 1, \dots, d)$, where $u_i \in \mathbb{F}_p[F]^{(d)}$ has x_i^{-1} in the ith place and 0 elsewhere, extends to a group homomorphism $\theta : F \to H$. Show that there is an $\mathbb{F}_p[F]$-module homomorphism $\theta^* : I \to \mathbb{F}_p[F]^{(d)}$ such that

$$(y - 1)\theta^* = (y\theta) \cdot y^{-1} \quad \text{for all } y \in F.$$

Deduce that the elements $x_i - 1$ are linearly independent over $\mathbb{F}_p[F]$.]

(ii) Let $\widehat{F}$ be the pro-p completion of F. Write π_n for the natural epimorphism

$$\mathbb{F}_p[[\widehat{F}]] \to \mathbb{F}_p[\widehat{F}/\widehat{F}_n] \cong \mathbb{F}_p[F/F_n],$$

where $\widehat{F}_n = P_n(\widehat{F})$ and $F_n = P_n(F)$. Suppose that $r_1, \dots, r_d \in \mathbb{F}_p[[\widehat{F}]]$ are such that

$$\sum_{i=1}^{d}(x_i - 1)r_i = 0.$$

For each i and n, choose $r_{i,n} \in \mathbb{F}_p[F]$ such that $r_i \equiv r_{i,n} \pmod{\ker \pi_n}$. Deduce from (i) that $r_{i,n} \in (F_n - 1)\mathbb{F}_p[F]$, and hence that $r_i\pi_n = 0$ for each i and all n. Hence show that

$$(\widehat{F} - 1)\mathbb{F}_p[[\widehat{F}]] = \bigoplus_{i-1}^{d} (x_i - 1)\mathbb{F}_p[[\widehat{F}]].$$

3. Let Γ be a finitely generated nilpotent group, and put $\Gamma^{\mathrm{ab}} = \Gamma/[\Gamma, \Gamma]$. Show that $d(\Gamma) = d(\Gamma^{\mathrm{ab}})$. [Consider $\Phi(\Gamma)$.] Deduce that if Γ^{ab} is torsion-free then $d(\Gamma) = d_p(\Gamma)$ for every prime p, while if Γ^{ab} contains an elementary abelian p-subgroup of rank $r_p > 0$ then $d(\Gamma) = d_p(\Gamma)$ for those primes p such that r_p is maximal.

Interlude E

Groups of sub-exponential growth

Let Γ be a group with a finite generating set X; we assume for convenience that $1 \in X$ and that $x \in X \Rightarrow x^{-1} \in X$. We write $W_n = X^{(n)}$ to denote the set of all (values in Γ of) group words of length n on X, and define the *growth function* (relative to X) of Γ by

$$f(n) = f_X(n) = |W_n| \text{ for } n \in \mathbb{N}.$$

This function arises in differential geometry: when Γ is the fundamental group of a Riemannian manifold M, the growth rate of f reflects the curvature properties of M; see Wolf (1968), Milnor (1968). A celebrated theorem of Gromov (1981) asserts that Γ *has polynomial growth* (i.e. there exists s such that $f(n) \leq n^s$ for all n) *if and only if* Γ *is virtually nilpotent*. Gromov's proof uses geometric group theory (largely invented for just this purpose), together with the solution to Hilbert's fifth problem, to show that if Γ has polynomial growth then Γ can be embedded in a (real) Lie group. R.I. Grigorchuk (1989) realised that if the group Γ happens also to be *residually a finite p-group*, then an alternative approach is available: one may embed Γ in its pro-p completion, and use Lazard's solution to the 'p-adic analogue' of Hilbert's fifth problem, namely our Theorem 8.34. The result obtained is weaker than Gromov's in that it applies only to groups which are residually nilpotent; but it is stronger in that it requires a less stringent growth hypothesis.

Henceforth, G will denote a pro-p group, topologically generated by a finite subset X satisfying the conditions above; and $W_n, f(n)$ will refer to words in G and the growth function of G respectively. It is clear that if G is actually the pro-p completion of Γ and Γ is residually a finite p-group, then Γ may be identified with a subgroup of G, and then W_n denotes the same set as before; so the growth function of G is the same as the growth function of Γ.

We use the notation of Section 12.3. Thus

$$I = (G - 1)\mathbb{F}_p[G]$$

denotes the augmentation ideal of the group algebra $\mathbb{F}_p[G]$ and

$$c_n = \dim_{\mathbb{F}_p[G]}(I^n/I^{n+1})$$

for $n > 0$.

E1 Lemma $c_n \leq f(n)$ *for all natural numbers* n.

Proof Write $V_n = \mathbb{F}_p W_n$ for the linear subspace of $\mathbb{F}_p[G]$ spanned by the set W_n. Then $\dim_{\mathbb{F}_p}(V_n) \leq |W_n| = f(n)$, so it will suffice to show that

$$I^n \subseteq V_n + I^{n+1}$$

for each n.

Recall that the mapping $g \mapsto (g-1) + I^2$ induces a homomorphism $\theta : G \to I/I^2$. Since I/I^2 is abelian and has exponent p, it is clear that $\ker \theta \geq \Phi(G)$. Since X generates G modulo $\Phi(G)$, it follows that I/I^2 is additively generated by elements of the form $x - 1$ with $x \in X$; this shows that

$$I \subseteq V_1 + I^2.$$

Now let $n \geq 1$ and suppose that $I^n \subseteq V_n + I^{n+1}$. Put $Z = V_n \cap I^n$. Then

$$I^{n+1} \subseteq ZI + I^{n+1}I \subseteq Z(V_1 + I^2) + I^{n+2} \subseteq V_{n+1} + I^{n+2},$$

since $V_n V_1 = V_{n+1}$. The result follows by induction.

According to Proposition 12.17, if G has infinite rank then $c_n \geq p(n)$ for all $n > 1$, where $p(n)$ denotes the number of partitions of the natural number n. Hence if $f(n) < p(n)$ for some integer $n > 1$ then G has finite rank.

It is now easy to deduce

E2 Theorem *Let* Γ *be a finitely generated residually nilpotent group. Suppose that the growth function* f *of* Γ *relative to some finite generating set satisfies*

$$f(n) < 2^{[\sqrt{n}]} \tag{$*$}$$

for infinitely many positive integers n. *Then*

(i) Γ *has a faithful linear representation over* $\mathbb{C}$;

(ii) Γ *is virtually nilpotent, and has polynomial growth relative to any finite generating set.*

Proof If G_p is the pro-p completion of Γ then the growth function of G_p, relative to the image in G_p of the given generating set of Γ, is dominated by f. Since $p(n) \geq 2^{[\sqrt{n}]}$ for every $n > 1$, by Exercise 12.5, the preceding remarks show that G_p has finite rank. It follows by Proposition B7 that

Γ can be embedded in $\prod_{p \in \pi} G_p$, where π is some finite set of primes. By Theorem 7.19, each G_p is isomorphic to a linear group over $\mathbb{Z}_p$, hence also over $\mathbb{C}$, and (i) follows. (In fact, for (i) it is enough to assume that (∗) holds for just one value of $n > 1$.)

We only sketch the proof of (ii), following Gromov. According to a famous theorem of Tits (1972), *a finitely generated linear group either is virtually soluble or else contains a non-abelian free subgroup.* Observe now that Γ cannot contain such a free subgroup. For if u and v freely generate the subgroup $\langle u, v \rangle$, then for $k \geq 1$ we obtain 2^k distinct elements of the form $z_1 \ldots z_k$ with each z_i equal to either u or v; so if $u \in W_a$, $v \in W_b$ and $c = \max\{a, b\}$ then $|W_{ck}| \geq 2^k$. Let $n \geq c^2$ and put $k = [\sqrt{n}]$: then $ck \leq n$ so

$$f(n) = |W_n| \geq |W_{ck}| \geq 2^k = 2^{[\sqrt{n}]},$$

showing that (∗) holds for only finitely many values of n.

It follows that Γ is virtually soluble. Now Milnor (1968) and Wolf (1968) have shown that for every finitely generated virtually soluble group Γ, one of the following holds: either (a) the growth of Γ is exponential (i.e. bounded below by a function of the form c^n for some $c > 1$), or (b) the growth is polynomial and Γ is virtually nilpotent. Our hypothesis (∗) excludes possibility (a); therefore (b) must hold, and this completes the proof.

13
Analytic groups over pro-p rings

A p-adic analytic group is a group that 'locally' looks like $\mathbb{Z}_p^d$; as we saw in Chapter 8, this implies that such a group has an open subgroup which is a pro-p group. The reason is that $\mathbb{Z}_p$ is – in an obvious sense – a pro-p ring. If we start with an arbitrary pro-p ring R and define an 'R-analytic group' by analogy with the p-adic case, we may reasonably hope to recover in this more general setting at least some of the results known in that case.

However, once we leave the realm of discrete valuation rings, the question of *what should count as an analytic function* becomes less clear-cut, and we find ourselves obliged to impose what may appear to be rather arbitrary restrictions. This issue is discussed briefly in Section 6.

In Section 1 we give our chosen definition of analytic groups over a pro-p ring R. The main result is proved in Section 2: *every R analytic group contains an open R-standard subgroup*. An *R-standard group* is what one gets on endowing the space $(\mathfrak{m}^n)^{(d)}$ with a 'formal group law' ($\mathfrak{m}$ being the maximal ideal of R). The subsequent sections examine the properties of these groups. On the one hand, the formal group law gives rise to a Lie algebra, leading to the beginnings of Lie theory for R analytic groups. On the other hand, an R-standard group is a pro-p group; as in the p-adic case, we can find out a certain amount about its group-theoretic structure by studying the filtration on the group corresponding to the filtration of R by the powers of $\mathfrak{m}$.

13.1 Analytic manifolds and analytic groups

Throughout this chapter $(R, \mathfrak{m})$ will denote *an infinite commutative pro-p ring*; that is, R is a commutative Noetherian complete local ring with maximal ideal $\mathfrak{m}$, and $R/\mathfrak{m}$ is a finite field of characteristic p. We make the standing assumption that the associated graded ring

$$\operatorname{gr}(R) = \bigoplus_{n=0}^{\infty} \mathfrak{m}^n / \mathfrak{m}^{n+1}$$

is an *integral domain* (which implies that R is itself an integral domain, as observed in Proposition 7.27). The significance of this hypothesis was discussed in §6.6; it ensures that *the power series representing an analytic function on a non-empty open set is uniquely determined* (Corollary 6.50), a fact that will be crucial.

As in Chapter 8 we begin with the definition of an R-analytic manifold. We refer back to Chapter 6 for what it means to evaluate a power series with coefficients in R. Recall that $\mathfrak{m}^{(n)}$ denotes the set of n-tuples with entries from $\mathfrak{m}$ whilst $\mathfrak{m}^N$ denotes the Nth power of the ideal $\mathfrak{m}$ (though we still write R^m, K^m for the free modules of rank m over R, K).

If R is a discrete valuation ring (we shall abbreviate this to DVR) then, as explained in §6.6, we can extend the norm defined by the powers of the maximal ideal to a norm on the field of fractions K of R; in this case it makes sense to evaluate power series with coefficients in K. This does not work well if R is not a discrete valuation ring, and in that case we have to restrict power series to have coefficients in the ring R. We fix the following notation.

- K is the field of fractions of R
- $\Lambda = K$ if R is a DVR
- $\Lambda = R$ if R is not a DVR
- $\|x\| = c^k$ if $x \in \mathfrak{m}^k \setminus \mathfrak{m}^{k+1}$, where $c < 1$ is a positive constant
- if R is a DVR, π is a generator for the ideal $\mathfrak{m}$; and for $x \in K$ we have $\|x\| = c^{-n} \|\pi^n x\|$ where $\pi^n x \in R$.
- $\Lambda_0 [[\mathbf{X}]]$, where $\mathbf{X} = (X_1, \dots, X_d)$, denotes the set of all power series

$$F(\mathbf{X}) = \sum_{\alpha \in \mathbb{N}^d} d_\alpha X_1^{\alpha_1} \cdots X_d^{\alpha_d} \in \Lambda [[\mathbf{X}]]$$

such that for some $k \in \mathbb{N}$,

$$d_\alpha \mathfrak{m}^{k\langle\alpha\rangle} \subseteq R \text{ for all } \alpha \in \mathbb{N}^d \setminus \{\mathbf{0}\}.$$

Note that $\Lambda_0 [[\mathbf{X}]]$ consists precisely of those power series in $\Lambda [[\mathbf{X}]]$ that converge on $(\mathfrak{m}^N)^{(d)}$ for some N: this was the content of Lemma 6.45.

13.1 Definition Let U be an open subset of R^n and let $f : U \to R$ be a function. f is Λ-*analytic* on U if for each $\mathbf{y} \in U$ there exist a formal power series $F \in \Lambda_0[[X_1, \dots, X_n]]$ and a positive integer N such that

(i) $\mathbf{y} + (\mathfrak{m}^N)^{(n)} \subseteq U$

(ii) $F(\mathbf{x}) = f(\mathbf{y} + \mathbf{x})$ for all $\mathbf{x} \in (\mathfrak{m}^N)^{(n)}$.

A function $\mathbf{f} = (f_1, \dots, f_m) : U \to R^m$ is Λ-*analytic* on U if f_i is Λ-analytic for $i = 1, \dots, m$.

Again, ultrametric analysis allows analytic continuation, in the sense that a function which can be represented by a power series on a ball around some point is in fact Λ-analytic on the whole ball:

13.2 Lemma *Let* $\mathbf{y} \in R^n$ *and let* $f : \mathbf{y} + (\mathfrak{m}^N)^{(n)} \to \Lambda$ *be a function, where* $N \geq 1$. *Suppose that there exists a power series* $F \in \Lambda[[X_1, \dots, X_n]]$ *such that*

$$f(\mathbf{y} + \mathbf{x}) = F(\mathbf{x}) \text{ for all } \mathbf{x} \in (\mathfrak{m}^N)^{(n)}.$$

Then f *is* Λ-*analytic on the ball* $\mathbf{y} + (\mathfrak{m}^N)^{(n)}$.

Proof We follow the proof of Lemma 8.3. Let $\mathbf{z} \in \mathbf{y} + (\mathfrak{m}^N)^{(n)}$. We must produce a power series $G(\mathbf{X}) \in \Lambda_0[[X_1, \dots, X_n]]$ such that $G(\mathbf{x}) = f(\mathbf{z} + \mathbf{x}) = F(\mathbf{x} + (\mathbf{z} - \mathbf{y}))$ for all $\mathbf{x} \in (\mathfrak{m}^N)^{(n)}$. Suppose that $F(\mathbf{X}) = \sum_{\alpha \in \mathbb{N}^n} d_\alpha \mathbf{X}^\alpha$ and put $\mathbf{a} = \mathbf{z} - \mathbf{y} \in (\mathfrak{m}^N)^{(n)}$.

Observe to begin with that if $\mathbf{x} \in (\mathfrak{m}^N)^{(n)}$ then

$$\lim_{(\alpha, \beta) \in \mathbb{N}^n \times \mathbb{N}^n} d_\alpha \binom{\alpha_1}{\beta_1} \dots \binom{\alpha_n}{\beta_n} a_1^{\alpha_1 - \beta_1} \dots a_n^{\alpha_n - \beta_n} x_1^{\beta_1} \dots x_n^{\beta_n} = 0;$$

for the hypothesis implies that $F(\mathbf{X}) \in \Lambda_0[[\mathbf{X}]]$, and $\binom{\alpha_i}{\beta_i} = 0$ whenever $\beta_i > \alpha_i$. Next, we apply Lemma 6.46 to justify the following rearrangement of power series:

$$F(\mathbf{x} + (\mathbf{z} - \mathbf{y})) = \sum_{\alpha \in \mathbb{N}^n} d_\alpha (x_1 + a_1)^{\alpha_1} \dots (x_n + a_n)^{\alpha_n}$$

$$= \sum_{(\alpha, \beta) \in \mathbb{N}^n \times \mathbb{N}^n} d_\alpha \binom{\alpha_1}{\beta_1} \dots \binom{\alpha_n}{\beta_n} a_1^{\alpha_1 - \beta_1} \dots a_n^{\alpha_n - \beta_n} x_1^{\beta_1} \dots x_n^{\beta_n}$$

$$= \sum_{\beta \in \mathbb{N}^n} c_\beta x_1^{\beta_1} \dots x_n^{\beta_n} = G(\mathbf{x}),$$

say, where

$$c_\beta = \sum_{\alpha \in \mathbb{N}^n} d_\alpha \binom{\alpha_1}{\beta_1} \dots \binom{\alpha_n}{\beta_n} a_1^{\alpha_1 - \beta_1} \dots a_n^{\alpha_n - \beta_n} \in \Lambda.$$

Since the series $G(\mathbf{x})$ is convergent for all $\mathbf{x} \in \left(\mathfrak{m}^N\right)^{(n)}$ it follows that $G(\mathbf{X}) \in \Lambda_0\left[[\mathbf{X}]\right]$.

13.3 Lemma *Every Λ-analytic function is continuous.*

Proof Let $f : U \to R$ be a Λ-analytic function, and let $\mathbf{a} \in U$. To show that f is continuous at $\mathbf{a}$, we may as well assume that $f(\mathbf{a}) = 0$ (replace $f(\mathbf{x})$ by $f(\mathbf{x}) - f(\mathbf{a})$). We then have to show that for each positive integer E, there exists an open neighbourhood $\mathbf{a} + (\mathfrak{m}^D)^{(n)}$ of $\mathbf{a}$ such that $f(\mathbf{a} + (\mathfrak{m}^D)^{(n)}) \subseteq \mathfrak{m}^E$.

According to the hypothesis, there exist a positive integer N, with $\mathbf{a} + (\mathfrak{m}^N)^{(n)} \subseteq U$, and a power series $F(\mathbf{X}) = \sum_{\alpha \in \mathbb{N}^n} d_\alpha \mathbf{X}^\alpha \in \Lambda_0[[X_1, \ldots, X_n]]$, such that $f(\mathbf{a} + \mathbf{x}) = F(\mathbf{x})$ for all $\mathbf{x} \in (\mathfrak{m}^N)^{(n)}$. There exists $k \in \mathbb{N}$ such that $d_\alpha \mathfrak{m}^{k\langle\alpha\rangle} \subseteq R$ for all $\alpha \in \mathbb{N}^n$. Take $D = \max\{N, k+E\}$, and let $\mathbf{x} \in (\mathfrak{m}^D)^{(n)}$. Then $d_\alpha x_1^{\alpha_1} \ldots x_n^{\alpha_n} \in \mathfrak{m}^E$ for each $\alpha \neq 0$, while $d_0 \mathbf{x}^0 = F(\mathbf{a}) = 0$. It follows that $f(\mathbf{a} + \mathbf{x}) = F(\mathbf{x}) \in \mathfrak{m}^E$.

13.4 Lemma *Suppose that $\mathbf{f} : U \to V$ and $\mathbf{g} : V \to W$ are Λ-analytic functions, where $U \subseteq \mathfrak{m}^{(r)}, V \subseteq \mathfrak{m}^{(s)}$ and $W \subseteq \mathfrak{m}^{(t)}$ are non-empty open sets. Then $\mathbf{g} \circ \mathbf{f}$ is Λ-analytic on U.*

Proof The proof of Lemma 8.5 requires only slight modification. Given $\mathbf{y} \in U$, we have to find $h \in \mathbb{N}$ such that, for each $j = 1, \ldots, t$, there exists $H_j(\mathbf{X}) \in \Lambda[[\mathbf{X}]]$ satisfying $g_j(\mathbf{f}(\mathbf{y} + \mathbf{x})) = H_j(\mathbf{x})$ for all $\mathbf{x} \in \left(\mathfrak{m}^h\right)^{(r)}$.

Since $\mathbf{f}$ and $\mathbf{g}$ are Λ-analytic, there exist h_1 and $h_2 \in \mathbb{N}$ such that

(i) for each $i = 1, \ldots, s$, there exists $F_i(\mathbf{X}) = \sum_{\alpha \in \mathbb{N}^r} b_\alpha(i) X_1^{\alpha_1} \ldots X_r^{\alpha_r} \in \Lambda_0[[\mathbf{X}]]$ such that $f_i(\mathbf{y} + \mathbf{x}) = F_i(\mathbf{x})$ for all $\mathbf{x} \in \left(\mathfrak{m}^{h_1}\right)^{(r)}$; and

(ii) if we set $\mathbf{b} = (b_0(1), \ldots, b_0(s)) = \mathbf{f}(\mathbf{y})$, then for each $j = 1, \ldots, t$, there exists $G_j(\mathbf{Y}) = \sum_{\beta \in \mathbb{N}^s} c_\beta(j) \mathbf{Y}^\beta \in \Lambda_0[[\mathbf{Y}]]$ such that $g_j(\mathbf{b} + \mathbf{y}) = G_j(\mathbf{y})$ for all $\mathbf{y} \in \left(\mathfrak{m}^{h_2}\right)^{(s)}$.

By the previous lemma, the function $\mathbf{x} \mapsto F_i(\mathbf{x})$ is continuous for each $i = 1, \ldots, s$. So there exists $h_3 \in \mathbb{N}$ such that

$$\mathbf{f}(\mathbf{y} + \mathbf{x}) \in \mathbf{b} + \left(\mathfrak{m}^{h_2}\right)^{(s)}$$

for all $\mathbf{x} \in \left(\mathfrak{m}^{h_1+h_3}\right)^{(r)}$. Thus

$$g_j(\mathbf{f}(\mathbf{y} + \mathbf{x})) = G_j(z_1, \ldots, z_s)$$

where $z_i = F_i(\mathbf{x}) - b_0(i)$ for $i = 1, \ldots, s$.

We now apply the work done in Chapter 6 on composition of power series. For each i let $E_i(\mathbf{X}) = F_i(\mathbf{X}) - b_0(i)$. Put $H_j(\mathbf{X}) = (G_j \circ \mathbf{E})(\mathbf{X})$ for $j = 1, \dots, t$.

If R is not a DVR then $\Lambda = R$ and all the power series have coefficients in R. It follows by Corollary 6.48 that for all $\mathbf{x} \in (\mathfrak{m}^{h_1+h_3})^{(r)}$ we have

$$g_j(\mathbf{f}(\mathbf{y} + \mathbf{x})) = G_j(E_1(\mathbf{x}), \dots, E_s(\mathbf{x})) = (G_j \circ \mathbf{E})(\mathbf{x}) = H_j(\mathbf{x}). \quad (1)$$

Now suppose that R is a DVR, and let $k \in \mathbb{N}$ be such that for $i = 1, \dots, s$ and all $\alpha \in \mathbb{N}^r \setminus \{\mathbf{0}\}$,

$$\pi^{k\langle\alpha\rangle} \cdot b_\alpha(i) \in R.$$

Let $h_4 = \max(h_1 + h_3, k) + h_2$. Since G_j converges at $(\pi^{h_2}, \dots, \pi^{h_2}) \in (\mathfrak{m}^{h_2})^{(s)}$, we have $\lim_{\beta \in \mathbb{N}^s} \|c_\beta(j)\| \, c^{h_2\langle\beta\rangle} = 0$ (recall that $\|\cdot\|$ is the norm on K and $c = \|\pi\|$). Thus if $\mathbf{x} \in (\mathfrak{m}^{h_4})^{(r)}$ then $\tau_i = \sup\{\|b_\alpha(i)\mathbf{x}^\alpha\| \mid \alpha \neq \mathbf{0}\} \leq c^{h_2}$ for $i = 1, \dots, s$. So the criterion for applying Theorem 6.47 applies, showing that (1) holds for all $\mathbf{x} \in (\mathfrak{m}^{h_4})^{(r)}$.

Thus in either case each $g_j \circ \mathbf{f}$ is represented by the power series H_j on a suitable neighbourhood of $\mathbf{y}$, and the result follows.

13.5 Definition (i) Let X be a topological space and U a non-empty open subset of X. A triple (U, ϕ, n) is an R-*chart* on X if ϕ is a homeomorphism from U onto an open subset of R^n for some $n \in \mathbb{N}$. The *dimension* of the chart is n. The chart (U, ϕ, n) is a *global chart* if $U = X$.

(ii) Two charts (U, ϕ, n) and (V, ψ, m) on a topological space X are *compatible* if the maps $\psi \circ \phi^{-1}|_{\phi(U \cap V)}$ and $\phi \circ \psi^{-1}|_{\psi(U \cap V)}$ are Λ-analytic functions on $\phi(U \cap V)$ and $\psi(U \cap V)$ respectively.

(iii) An R-*atlas* on a topological space X is a set of pairwise compatible R-charts that covers X; i.e. it is a set of the form

$$A = \{(U_i, \phi_i, n_i) \mid i \in I\}$$

with the following properties:

- for each $i \in I$, (U_i, ϕ_i, n_i) is an R-chart on X;
- for all $i, j \in I$, (U_i, ϕ_i, n_i) and (U_j, ϕ_j, n_j) are compatible;
- $X = \bigcup_{i \in I} U_i$.

A is a *global atlas* if for some $i \in I$ the chart (U_i, ϕ_i, n_i) is global.

(iv) Let A and B be atlases on a topological space X. Then A and B are *compatible* if every chart in A is compatible with every chart in B; that is, if $A \cup B$ is an atlas on X.

(v) An *R-analytic structure over R* on a topological space X is an equivalence class of compatible atlases.

That compatibility of atlases is an equivalence relation can be proved as in Chapter 8, using Lemma 13.4. The space X endowed with an R-analytic structure will be referred to as an *analytic manifold* over R, or an *R-manifold*.

13.6 Examples: analytic manifolds over R.

(i) Let $X = K^m$. For each $k \in K^m$ put $U_k = k + R^m$, and define $\phi_k : U_k \to R^m$ by $\phi_k(u) = u - k$. Then $A = \{(U_k, \phi_k, m) \mid k \in K^m\}$ is an atlas on X; the charts are compatible since if $U_k \cap U_{k'}$ is non-empty, then $k - k' \in R^m$ and $\phi_k \circ \phi_{k'}^{-1}|_{\phi_{k'}(U_k \cap U_{k'})}(u) = u + (k' - k)$, clearly an R-analytic function of u. (*Warning*: in asserting that A is an atlas, we implicitly impose a topology on K^m; if R is not a DVR, this is *not* the same as the field topology induced by extending the norm, as in Exercise 6.14: it cannot be, since in the latter topology K is not locally compact.)

(ii) Let $X = \mathrm{GL}_n(K)$; then X has an R-analytic structure defined as follows. Put $U = 1_n + M_n(\mathfrak{m})$, an open subgroup of X. Define $\phi : U \to M_n(\mathfrak{m})$ by $\phi(u) = u - 1_n$. For each $h \in X$, let $V_h = hU$, and define $\phi_h : V_h \to M_n(\mathfrak{m})$ by $\phi_h(x) = \phi(h^{-1}x)$. The charts (V_h, ϕ_h, n^2) are compatible since if $V_h \cap V_{h'}$ is non-empty, then $h^{-1}h' \in U \leq \mathrm{GL}_n(R)$ and $\phi_h \circ \phi_{h'}^{-1}(u) = h^{-1}h'u$; as above, this is clearly an R-analytic function of u. So $A = \{(V_h, \phi_h, n^2) : h \in X\}$ is an R-atlas on X. Note that the topology on X implied by this atlas is not necessarily the one induced from the product topology on $M_n(K) = K^{n^2}$, where K is topologised as in Example (i).

(iii) A discrete space, an open subspace of a manifold, and the product of two manifolds have natural analytic structures exactly as in Examples 8.9. An open subspace of the manifold X, with the induced analytic structure, will be called an *open submanifold* of X.

(iv) Suppose that R has a local subring S, such that $(S, \mathfrak{m} \cap S)$ satisfies our standing hypotheses and such that R is finitely generated and free as a module for S; for example

$$S = R_0 [[T_1, \dots, T_r]], \quad R = R_1 [[T_1, \dots, T_r]]$$

where R_0 is $\mathbb{F}_p$ or $\mathbb{Z}_p$ and R_1 is a finite extension of R_0. Then any R-manifold X has a natural S-analytic structure, obtained as follows. Fix an S-module isomorphism $\sigma : R \to S^e$. To each R-chart $c = (U, \phi, n)$

of X, associate the S-chart $c_S = (U, \sigma \circ \phi, en)$. Three points have to be verified: (a) that $\sigma\phi(U)$ is an open subset of S^{en}, so that c_S really is an S-chart, (b) that if c and c' are compatible R-charts then c_S and c'_S are compatible S-charts, and (c) that a different choice of isomorphism σ produces the same S-analytic structure; the argument is outlined in Exercise 4. The S-manifold $X_{(S)}$ obtained in this manner from X is said to come from X by *restriction of scalars*.

In order to define R-analytic groups, we need the concept of an *analytic function* between two manifolds:

13.7 Definition Let X and Y be two R-manifolds. A function $f : X \to Y$ is *analytic* if there exist R-atlases A and B of X and Y respectively such that, for each pair of charts $(U, \phi, n) \in A$ and $(V, \psi, m) \in B$, the following hold:

(i) $f^{-1}(V)$ is open in X, and

(ii) the composition

$$\psi \circ f \circ \phi^{-1}|_{\phi(U \cap f^{-1}(V))}$$

is a Λ-analytic function from the open set $\phi(U \cap f^{-1}(V)) \subseteq R^n$ into R^m.

As in Chapter 8, the fact that Λ-analytic functions are continuous implies that an analytic function $f : X \to Y$ is continuous. Similarly, one deduces from Lemmas 13.3 and 13.4 that the composition of analytic functions between manifolds is analytic. If S is a subring of R satisfying the hypotheses of Example 13.6(iv), then every R-analytic function $X \to Y$ is S-analytic as a function $X_{(S)} \to Y_{(S)}$; for the proof see Exercise 4.

13.8 Definition Let G be a topological group with an R-analytic structure. Then G is an R-*analytic group* if the functions

$$(x, y) \mapsto xy : G \times G \to G$$
$$x \mapsto x^{-1} : G \to G$$

are analytic. An analytic homomorphism between R-analytic groups is called a *morphism*.

13.9 Examples: R-analytic groups.

(i) Let $G = (K^m, +)$. Then G is an R-analytic group with respect to the analytic structure given in Example 13.6(i).

(ii) Let $G = (K^*, \cdot)$ the multiplicative group of the field. Example

13.6(ii) with $n = 1$ defines an R-analytic structure on this group. If $1 + u \in 1 + \mathfrak{m}$ then $\phi((1 + u)^{-1}) = \sum_{n=1}^{\infty}(-1)^n u^n$, which is an analytic function of $u = \phi(1 + u)$. In general, if $h \in G$ and $u \in \mathfrak{m}$ then $\phi_{h^{-1}}((h(1+u))^{-1}) = \sum_{n=1}^{\infty}(-1)^n u^n$ which is analytic in $\phi_h(h(1+u)) = u$. Multiplication is clearly analytic with respect to this structure. Hence G is an R-analytic group.

(iii) If R is a DVR, then $\mathrm{GL}_n(K)$, with the analytic structure given in 13.6(ii), is an R-analytic group. If not, we can only say that $\mathrm{GL}_n(R)$ is R-analytic.

(iv) Let S be a subring of R satisfying the hypotheses of Example 13.6(iv), and let G be an R-analytic group. Then G is an S-analytic group, when considered as an S-manifold by restriction of scalars; this follows from the remark preceding Definition 13.8. In particular, by Corollary 6.43, it follows that if R has Krull dimension 1 and characteristic p then G is analytic over $\mathbb{F}_p[[T]]$, while if R has Krull dimension 1 and characteristic zero then G is a p-adic analytic group.

Further examples are provided by

13.10 Proposition *Let G be a topological group containing an open subgroup H which is an R-analytic group. Suppose that for each $g \in G$, there exists an open neighbourhood V_g of the identity in H such that*
  (i) $gV_g g^{-1} \subseteq H$ *and*
  (ii) *the function $k_g : V_g \to H$ defined by $x \mapsto gxg^{-1}$ is analytic.*
Then there is a unique R-analytic structure on G inducing the R-analytic structure on H with respect to which G is an R-analytic group.

The proof is exactly as in Proposition 8.15. However, in contrast to the case of p-adic analytic groups, continuous homomorphisms of R-analytic groups are not necessarily analytic (see Exercises 1 and 3); the mere possession of an open R-analytic subgroup is therefore *not* sufficient to ensure that a topological group has the structure of an R-analytic group (see Exercise 3).

In Chapter 8, we defined the *dimension* of a p-adic analytic group by identifying it with an intrinsic group-theoretic invariant, the rank of any uniform open subgroup. This interpretation is no longer available, unless we impose a special condition as we shall do in Section 5; instead we refer to the underlying analytic manifold.

13.11 Lemma *Let X be an R-manifold and (U, ϕ, n), (V, ψ, m) charts of X with $U \cap V \neq \emptyset$. Then $m = n$.*

Proof Choose $u \in U \cap V$. Replacing ϕ by the map $x \mapsto \phi(x) - \phi(u)$, we may suppose that $\phi(u) = 0$; similarly we assume that $\psi(u) = 0$. Then there exists $N \in \mathbb{N}$ such that $\psi^{-1}\left((\mathfrak{m}^N)^{(m)}\right) \subseteq U \cap V$ and $\phi^{-1}\left((\mathfrak{m}^N)^{(n)}\right) \subseteq U \cap V$. Suppose that $m \leq n$. Since the two charts are compatible, there exist power series $F_1(X_1, \ldots, X_n), \ldots,$ $F_m(X_1, \ldots, X_n)$ and $H_1(Y_1, \ldots, Y_m), \ldots, H_n(Y_1, \ldots, Y_m)$ such that for $N' \geq N$ large enough and $\mathbf{x} \in \left(\mathfrak{m}^{N'}\right)^{(n)}$ we have

$$(H_i \circ \mathbf{F})(\mathbf{x}) = H_i(\mathbf{F}(\mathbf{x})) = x_i \quad (i = 1, \ldots, n).$$

By the 'uniqueness of power series', this implies that $(H_i \circ \mathbf{F})(\mathbf{X}) = X_i$ for each i. Now we can write $F_i = a_{i1}X_1 + \cdots + a_{in}X_n + O(2)$ and $H_i = b_{i1}Y_1 + \cdots + b_{im}Y_m + O(2)$, where $O(2)$ stands for a sum of terms of degree at least 2. Then the linear part of $(H_i \circ \mathbf{F})(\mathbf{X})$ is $c_{i1}X_1 + \cdots + c_{in}X_n$ where $(c_{ij}) = (b_{ij})(a_{ij})$, the product of an $n \times m$ matrix with an $m \times n$ matrix. Since $(H_i \circ \mathbf{F})(\mathbf{X}) = X_i$ it follows that (c_{ij}) is the identity matrix, and this is only possible if $m = n$. (This argument will reappear centre-stage in Section 13.3.)

We can therefore make the following

13.12 Definition Let G be an R-analytic group. The *dimension* $\dim(G)$ of G is the common dimension of all charts of G at 1.

13.2 Standard groups

When dealing with power series, it will be convenient to use the following shorthand:

- the expression $O(n)$ stands for any power series in which every term has (total) degree at least n;
- the expression $O'(n)$ stands for any power series in which every term has total degree at least n and has degree at least 1 in each variable.

The formal notation $G \circ \mathbf{F}$ for the composition of power series, introduced in Chapter 6, becomes unwieldy when applied to multiple compositions, and we shall allow ourselves to use a more transparent notation, exemplified by $G(F(\mathbf{X}))$.

We shall need to broaden the definition of standard groups given in Chapter 8:

13.13 Definition Let G be an R-analytic group. Then G is an R-*standard group*, of *level* n and *dimension* d, if

(i) the R-analytic structure on G can be defined by a global atlas $\{(G, \psi, d)\}$ where $\psi = (\psi_1, \dots, \psi_d)$ is a homeomorphism onto $(\mathfrak{m}^n)^{(d)}$ and $\psi(1) = 0$;

(ii) for $j = 1, \dots, d$, there exist formal power series $F_j(\mathbf{X}, \mathbf{Y}) \in R[[\mathbf{X}, \mathbf{Y}]]$, without constant term, such that $\psi(xy) = \mathbf{F}(\psi(x), \psi(y))$ for all $x, y \in G$.

In this case, (G, ψ, d) is a *standard chart*.

We shall use the convention that 'R-standard group' means 'R-standard group of level 1'. While the *dimension* of an R-standard group is just its dimension as an R-analytic group, its *level* is *not* uniquely determined if R is a DVR, as shown in Exercise 5. When R is not a DVR, however, the level is indeed unique: this is proved in Exercise 6.

13.14 Definition Let $F_i(\mathbf{X})$ for $i = 1, \dots, d$ be power series in $R[[X_1, \dots, X_d]]$. Then $\mathbf{F} = (F_1, \dots, F_d)$ is a *formal group law*, of dimension d over R, if

(i) $\mathbf{F}(\mathbf{X}, \mathbf{0}) = \mathbf{X}$ and $\mathbf{F}(\mathbf{0}, \mathbf{Y}) = \mathbf{Y}$, and

(ii) $\mathbf{F}(\mathbf{X}, \mathbf{F}(\mathbf{Y}, \mathbf{Z})) = \mathbf{F}(\mathbf{F}(\mathbf{X}, \mathbf{Y}), \mathbf{Z})$.

13.15 Lemma *Let G be an R-standard group, with standard chart (G, ψ, d). Then a power series $\mathbf{F}$ satisfying condition* (ii) *of Definition 13.13 is a formal group law.*

Proof Given the 'uniqueness of power series', the proof of (i) is precisely as in Lemma 8.27(ii), while (ii) follows from the associative law in G.

The following proposition records some features possessed by any formal group law:

13.16 Proposition *Let $\mathbf{F}$ be a formal group law.*
(i)

$$\mathbf{F}(\mathbf{X}, \mathbf{Y}) = \mathbf{X} + \mathbf{Y} + \mathbf{B}(\mathbf{X}, \mathbf{Y}) + O'(3)$$

where $\mathbf{B}(\mathbf{X}, \mathbf{Y})$ is bilinear in $\mathbf{X}$ and $\mathbf{Y}$.

(ii) *There exists a 'formal inverse' $\mathbf{I}(\mathbf{X}) = -\mathbf{X} + \mathbf{B}(\mathbf{X}, \mathbf{X}) + O(3) \in R[[\mathbf{X}]]$ such that*

$$\mathbf{F}(\mathbf{X}, \mathbf{I}(\mathbf{X})) = \mathbf{0} = \mathbf{F}(\mathbf{I}(\mathbf{X}), \mathbf{X});$$

(iii) *The 'formal commutator'* $\mathbf{C}(\mathbf{X}, \mathbf{Y}) = \mathbf{F}(\mathbf{I} \circ \mathbf{F}(\mathbf{Y}, \mathbf{X}), \mathbf{F}(\mathbf{X}, \mathbf{Y}))$ *satisfies*

$$\mathbf{C}(\mathbf{X}, \mathbf{Y}) = \mathbf{B}(\mathbf{X}, \mathbf{Y}) - \mathbf{B}(\mathbf{Y}, \mathbf{X}) + O'(3). \tag{2}$$

Proof Again, the work has largely been done in Chapter 8. Part (i) is an immediate consequence of property (i) of a formal group law.

(ii) follows by adapting the proof of Theorem 6.37 ('Inverse Function Theorem'). As there, $W^{[j]}$ will stand for the homogeneous part of degree j in a power series W. We want to find power series $I_1, \ldots, I_d$ such that $\mathbf{F}(\mathbf{X}, \mathbf{I}(\mathbf{X})) = 0 = \mathbf{F}(\mathbf{I}(\mathbf{X}), \mathbf{X})$. Put

$$I_i^{[0]} = 0, \qquad I_i^{[1]} = -X_i$$

for $i = 1, \ldots, d$. Then $\mathbf{F}(\mathbf{X}, \mathbf{I}(\mathbf{X})) = 0$ is equivalent to

$$\sum_{j \geq 2} I_i^{[j]}(\mathbf{X}) = -\sum_{k \geq 2} F_i^{[k]}\left(\mathbf{X}, -\mathbf{X} + \sum_{j \geq 2} \mathbf{I}^{[j]}(\mathbf{X})\right). \tag{3}$$

Write $W_i(\mathbf{X})$ for the right-hand side of (3). Then for $t \geq 2$, $W_i^{[t+1]}(\mathbf{X})$ is the same as the homogeneous part of degree $t + 1$ in the polynomial

$$-\sum_{k=2}^{t+1} F_i^{[k]}\left(\mathbf{X}, -\mathbf{X} + \sum_{j=2}^{t} \mathbf{I}^{[j]}(\mathbf{X})\right). \tag{4}$$

Also, $W_i^{[2]}(\mathbf{X}) = -F_i^{[2]}(\mathbf{X}, -\mathbf{X}) = B_i(\mathbf{X}, \mathbf{X})$. So we can construct the $I_i^{[t]}$ recursively by putting $I_i^{[2]} = B_i(\mathbf{X}, \mathbf{X})$ and, for $t \geq 2$, $I_i^{[t+1]} = W_i^{[t+1]}$, since the polynomial (4) depends only on $I_1^{[j]}, \ldots, I_d^{[j]}$ for $j \leq t$.

We can solve $\mathbf{F}(\mathbf{I}'(\mathbf{X}), \mathbf{X}) = 0$ similarly. To show that $\mathbf{I}$ and $\mathbf{I}'$ are the same power series we use the associativity of the formal group law:

$$\mathbf{I}'(\mathbf{X}) = \mathbf{F}(\mathbf{I}'(\mathbf{X}), \mathbf{0}) = \mathbf{F}(\mathbf{I}'(\mathbf{X}), \mathbf{F}(\mathbf{X}, \mathbf{I}(\mathbf{X})))$$
$$= \mathbf{F}(\mathbf{F}(\mathbf{I}'(\mathbf{X}), \mathbf{X}), \mathbf{I}(\mathbf{X})) = \mathbf{F}(\mathbf{0}, \mathbf{I}(\mathbf{X}))$$
$$= \mathbf{I}(\mathbf{X}).$$

(iii) we leave as an exercise in substitution of power series, using the expressions established in (i) and (ii).

Part (ii) shows that the definition of 'standard group' (of level 1) given above is equivalent to the stronger definition in Chapter 8 (where we demanded also that the inverse operation be represented by power series).

Since the coefficients of the power series in a formal group law lie in R, each of these power series converges on the maximal ideal $\mathfrak{m}$ and takes values in $\mathfrak{m}$. Every formal group law of dimension d over R therefore defines a binary operation on the set $\mathfrak{m}^{(d)}$. It is tempting to assert that this makes $\mathfrak{m}^{(d)}$ into a group; but something still needs to checked: namely that identities in formal power series translate into the corresponding identities for our group operation when evaluated on $\mathfrak{m}$.

This is a non-trivial point, as it is possible for both sides of a formal power series identity to converge at some point and yet have different values (see Exercise 6.7). However, Corollary 6.48 prevents this sort of pathology provided that we stay within the ideal $\mathfrak{m}$, and we may infer

13.17 Proposition *Let* $\mathbf{F}$ *be a formal group law of dimension d over* R. *Then* $G = \mathfrak{m}^{(d)}$ *is an R-standard group, with group operation*

$$\mathbf{xy} = \mathbf{F}(\mathbf{x}, \mathbf{y}).$$

and standard chart (G, Id, d).

The only point we have to add to the comments above is that Proposition 13.16(ii) ensures that the inverse operation in this group is also an analytic function, justifying the implicit claim that G is an R-analytic group.

Remark The above discussion in fact shows that a formal group law defines a group structure on $\mathfrak{m}^{(d)}$ for *any* pro-p ring $(R, \mathfrak{m})$, whether or not $\mathrm{gr}(R)$, or indeed R, is an integral domain. Such a group could also be referred to as a 'standard group' (as in Lubotzky and Shalev (1994)). However we prefer to follow Bourbaki and reserve the name for certain types of analytic groups. Lacking a well-defined category of analytic groups over a pro-p ring which is not an integral domain, we do not consider these examples here, although many of the results proved for our standard groups hold equally in the more general context. However, without a concept of 'analytic isomorphism', it is not clear how the concept of dimension should be defined (see Exercise 1).

We leave it as an exercise to verify the following

13.18 Examples: formal group laws.
(i) $F(X, Y) = X + Y$ is a formal group law.
(ii) $F(X, Y) = X + Y + XY$ is a formal group law.

(iii) Let $\mathbf{X} = (X_{ij})$ and $\mathbf{Y} = (Y_{ij})$ be $n \times n$ matrices of indeterminates. Then $\mathbf{F}(\mathbf{X}, \mathbf{Y}) = \mathbf{X} + \mathbf{Y} + \mathbf{X} \cdot \mathbf{Y}$ is a formal group law of dimension n^2, where $(\mathbf{X} \cdot \mathbf{Y})_{ij} = \sum_{\ell=1}^{n} X_{i\ell} Y_{\ell j}$.

13.19 Examples: R-standard groups

(i) $G = (\mathfrak{m}^{(d)}, +)$ is an R-standard group.

(ii) $G = (1+\mathfrak{m}, \cdot)$ is an R-standard group, with standard chart $(G, u \mapsto u - 1, 1)$. The corresponding formal group law is $X + Y + XY$.

(iii) $\mathrm{GL}_n^1(R) = 1 + M_n(\mathfrak{m})$ is an R-standard group, with standard chart $(G, u \mapsto u - 1_n, n^2)$. The corresponding formal group law is given in (iii) above.

(iv) Let $G = \mathrm{SL}_n^1(R) = \mathrm{SL}_n(R) \cap \mathrm{GL}_n^1(R)$; this is the kernel of the natural map $\mathrm{SL}_n(R) \to \mathrm{SL}_n(R/\mathfrak{m})$. Then $\mathrm{SL}_n^1(R)$ has the structure of an $(n^2 - 1)$-dimensional R-standard group, with standard chart

$$(G, (g_{ij}) \mapsto (g_{ij})_{(i,j) \neq (n,n)}, n^2 - 1);$$

see Exercise 9. More generally, any Chevalley group functor gives rise to a standard group over R (Exercise 11).

(v) Let G be a uniform pro-p group. We have seen in Corollary 9.13 that $P_2(G)$ (or $P_3(G)$ if $p = 2$) is a standard group, with standard chart given by the 'co-ordinates of the first kind', which identify G with the Lie algebra $\log(G)$. The corresponding formal group law is $\Phi(p\mathbf{X}, p\mathbf{Y})$ (or $\Phi(4\mathbf{X}, 4\mathbf{Y})$ if $p = 2$), where Φ is the Campbell–Hausdorff series.

(vi) Let H be an R-standard group of level $m > 1$, with standard chart (H, ψ, d), and corresponding formal group law $\mathbf{F}$. According to Proposition 13.17, $\mathbf{F}$ defines a group operation on $\mathfrak{m}^{(d)}$ so that $\mathfrak{m}^{(d)}$ becomes a standard group of level 1. Thus $\psi(H) = (\mathfrak{m}^m)^{(d)}$ is an open subgroup in a standard group of level 1. As ψ is a homeomorphism, we can construct a standard group G of level 1 that contains H as an open subgroup, with the induced analytic structure (identify G with $\mathfrak{m}^{(d)}$ and identify H with $(\mathfrak{m}^m)^{(d)}$ via ψ).

(vii) Assume that R is a DVR, and let G be an R-standard group of level $m > 1$, with standard chart (G, ψ, d). Exercise 5 shows how ψ may be modified so as to turn G into an R-standard group of level 1. (Exercise 6, on the other hand, shows that this can *never* be done if R is not a DVR.)

Further properties of standard groups are examined below. First, let us state the main result.

13.20 Theorem *Let G be an R-analytic group. Then G contains, as an open submanifold, an R-standard subgroup of level h for some h.*

Proof We begin as in the proof of Theorem 8.29. Let A be an atlas defining the manifold structure on G. There exists a chart $(U, \phi, d) \in A$ with $1 \in U$. After making a simple translation of coordinates, we may assume that $\phi(1) = 0$. By hypothesis, the functions $(x, y) \mapsto xy$ and $x \to x^{-1}$ are analytic on $G \times G$ and G respectively. Hence there exist a neighbourhood $(\mathbf{m}^h)^{(d)}$ of the point $\phi(1) = 0$ and power series $F_j \in \Lambda_0[[\mathbf{X}, \mathbf{Y}]]$, $I_j \in \Lambda_0[[\mathbf{X}, \mathbf{Y}]]$ $(j = 1, \dots, d)$ such that (i) $\phi^{-1}(\mathbf{m}^h)^{(d)} \subseteq U$, and (ii) for all $x, y \in \phi^{-1}(\mathbf{m}^h)^{(d)}$ we have

$$\phi_j(xy) = F_j(\phi(x), \phi(y))$$
$$\phi_j(x^{-1}) = I_j(\phi(x)).$$

Now given the 'uniqueness of power series', the proof of Lemma 8.27 is quite formal and so may be applied here, to show that

$$F_j(\mathbf{X}, \mathbf{Y}) = \sum_{\alpha, \beta \in \mathbb{N}^d} c_{j,\alpha,\beta} \mathbf{X}^\alpha \mathbf{Y}^\beta = X_j + Y_j + O(2), \qquad (5)$$

$$I_j(\mathbf{X}) = -X_j + O(2).$$

Suppose first that R is not a DVR. Then $F_j(\mathbf{X}, \mathbf{Y}) \in R[[\mathbf{X}, \mathbf{Y}]]$ and $I_j \in R[[\mathbf{X}]]$ for each j. It follows that $F_j(\lambda, \mu) \in \mathbf{m}^h$ and $I_j(\lambda) \in \mathbf{m}^h$ for all λ and $\mu \in (\mathbf{m}^h)^{(d)}$. This implies that $H = \phi^{-1}(\mathbf{m}^h)^{(d)}$ is closed under both multiplication and inversion, so H is an open subgroup of G. Moreover, $(H, \phi|_H, d)$ is clearly a standard chart, showing that H is a standard group of level h, with corresponding formal group law $\mathbf{F}$.

If R is a DVR, we argue as in the proof of Theorem 8.29, using π in place of p. Since $F_j \in \Lambda_0[[\mathbf{X}, \mathbf{Y}]]$ for each j, there exists a positive integer k such that

$$c_{j,\alpha,\beta} \pi^{k(\langle \alpha \rangle + \langle \beta \rangle)} \in R$$

for all $\alpha, \beta \in \mathbb{N}^d \setminus \{0\}$ and each j; it follows from (5) that provided $N \geq 2k$, the power series

$$\overline{F}_j(\mathbf{X}, \mathbf{Y}) = \pi^{-N} F_j(\pi^N \mathbf{X}, \pi^N \mathbf{Y}) = \sum_{\alpha, \beta \in \mathbb{N}^d} \pi^{-N} c_{j,\alpha,\beta} \pi^{N(\langle \alpha \rangle + \langle \beta \rangle)} \mathbf{X}^\alpha \mathbf{Y}^\beta$$

has coefficients in R. Similarly, provided N is large enough, each of the series

$$\overline{I}_j(\mathbf{X}) = \pi^{-N} I_j(\pi^N \mathbf{X})$$

has coefficients in R. Choosing such a large integer N, we now put $H = \phi^{-1}(\mathfrak{m}^{h+N})^{(d)}$, and define $\psi : H \to (\mathfrak{m}^h)^{(d)}$ by $\psi(x) = \pi^{-N}\phi(x)$. Then for $x, y \in H$, with $\phi(x) = \lambda$ and $\phi(y) = \mu$, we have

$$
\begin{aligned}
\psi_j(x^{-1}) &= \pi^{-N}\phi_j(x^{-1}) = \pi^{-N}I_j(\lambda) \\
&= \overline{I}_j(\pi^{-N}\lambda) = \overline{I}_j(\psi(x)) \in \mathfrak{m}^h, \\
\psi_j(xy) &= \pi^{-N}\phi_j(xy) = \pi^{-N}F_j(\lambda, \mu) \\
&= \overline{F}_j(\pi^{-N}\lambda, \pi^{-N}\mu) = \overline{F}_j(\psi(x), \psi(y)) \in \mathfrak{m}^h.
\end{aligned}
$$

Thus H is a standard group of level h, with standard chart (H, ψ, d) and corresponding formal group law $\overline{\mathbf{F}}$. This completes the proof.

We turn next to the group-theoretic structure of standard groups. We assume in the rest of this section that our standard group G of dimension d and level h has $(\mathfrak{m}^h)^{(d)}$ as its underlying set, and standard chart (G, Id, d) (this is of course just a matter of notation: given a standard chart (G, ϕ, d) we use ϕ to identify G with $(\mathfrak{m}^h)^{(d)}$). We write

$$
G = ((\mathfrak{m}^h)^{(d)}, \mathbf{F})
$$

to mean that $\mathbf{F}$ is the formal group law representing multiplication in G. For any ideal $I \subseteq \mathfrak{m}^h$ we put $G(I) = I$. Each such ideal is closed in R (Atiyah and Macdonald (1969), Corollary 10.19); since $\mathbf{F}$ and the 'formal inverse' $\mathbf{I}$ when evaluated on $I^{(d)}$ converge with sums in $I^{(d)}$, we see that $G(I)$ is a closed subgroup of G. In particular, for each $n \geq h$ we write

$$
G_n = G(\mathfrak{m}^n) = (\mathfrak{m}^n)^{(d)}.
$$

Thus $G_n = ((\mathfrak{m}^n)^{(d)}, \mathbf{F})$ is a standard group of level n; it is an open submanifold and a subgroup of G.

13.21 Lemma *Let $G = ((\mathfrak{m}^h)^{(d)}, \mathbf{F})$ be a standard group, and let I and J be ideals of R contained in $\mathfrak{m}^h$.*
  (i) $G(I) \lhd G$.
  (ii) $[G(I), G(J)] \leq G(IJ)$.
  (iii) $G(I)^p \leq G(I^p + pI)$.

Proof (i) follows from (ii) on putting $J = \mathfrak{m}^h$. Part (ii) is a consequence of Proposition 13.16(iii). This shows that $[x, y] = \mathbf{C}(x, y)$ where the series $\mathbf{C}(\mathbf{X}, \mathbf{Y})$ is in $O'(2)$; hence if $x \in G(I)$ and $y \in G(J)$ then each term of $\mathbf{C}$ when evaluated at (x, y) lies in IJ.

In order to prove (iii), we need to consider the power series that represent the operation $x \mapsto x^p$ in G. For each $m \geq 1$, let $G_i^m(\mathbf{X})$ ($i = 1, \ldots, d$) be the power series constructed in Lemma 9.1: so $G_i^1(\mathbf{X}) = X_i$, and for $m > 1$

$$G_i^m(\mathbf{X}) = F_i(\mathbf{X}, \mathbf{G}^{m-1}(\mathbf{X})).$$

We can write

$$G_i^m(\mathbf{X}) = \sum_{\langle \alpha \rangle \geq 1} c_{i\alpha}(m)\mathbf{X}^\alpha$$

with each $c_{i\alpha}(m) \in R$. Now the proof of Lemma 9.1 remains valid when $\mathbb{Z}_p$ is replaced by any integral domain R of characteristic zero, and shows that

$$c_{i\alpha}(m) = \sum_{j=1}^{\langle \alpha \rangle} \xi_{ij\alpha} \binom{m}{j}, \tag{6}$$

where each $\xi_{ij\alpha} \in R$. If R has characteristic p, the proof still shows that (6) holds for all $\alpha \in \mathbb{N}^d$ such that $\langle \alpha \rangle \leq p-1$ (one of the essential steps, Exercise 9.2, breaks down when applied to polynomials of degree p).

Now let $\mathbf{x} \in G(I) = I^{(d)}$. Then $\mathbf{x}^p = (G_1^p(\mathbf{x}), \ldots, G_d^p(\mathbf{x}))$, and for each i

$$G_i^p(\mathbf{x}) = \sum_{1 \leq \langle \alpha \rangle < p} c_{i\alpha}(p)\mathbf{x}^\alpha + \sum_{\langle \alpha \rangle \geq p} c_{i\alpha}(p)\mathbf{x}^\alpha.$$

Since each $\xi_{ij\alpha}\binom{p}{j} \in pR$, it follows from (6) that the first sum lies in pI; and the second sum clearly lies in I^p. The claim (iii) follows.

It is now easy to derive

13.22 Proposition *Let G be an R-standard group of dimension d and level h, and let n and m be integers not less than h.*

(i) $G_n \lhd G$;

(ii) $[G_n, G_m] \leq G_{m+n}$;

(iii) *If $m \leq n$ then G_n/G_{n+m} is a finite abelian p-group isomorphic to the additive group $(\mathfrak{m}^n/\mathfrak{m}^{n+m})^{(d)}$;*

(iv) $G \cong \varprojlim (G/G_n)$ *and hence G is a pro-p group.*

Moreover, if $\mathrm{char}(R) = p$ then

(v) $(G_n)^p \leq G_{pn}$;

(vi) $G_n \geq D_n(G)$.

Here $D_n(G)$ denotes the nth dimension subgroup of G over $\mathbb{F}_p$, defined in §11.1.

Proof Parts (i), (ii) and (v) are immediate from the preceding lemma, and (vi) follows from (ii) and (v). To prove (iii), observe that for $\mathbf{x}, \mathbf{y} \in (\mathfrak{m}^n)^{(d)}$, we have

$$F(\mathbf{x}, \mathbf{y}) \equiv \mathbf{x} + \mathbf{y} \quad (\mathrm{mod}(\mathfrak{m}^{2n})^{(d)},$$

by Proposition 13.16(i). It follows that the identity map $G_n \to (\mathfrak{m}^n)^{(d)}$ induces a homomorphism from G_n onto the additive group $(\mathfrak{m}^n/\mathfrak{m}^{n+m})^{(d)}$, provided $m \leq n$; the kernel of this homomorphism is G_{n+m}. Finally, (iv) holds because R is complete in the $\mathfrak{m}$-adic topology.

One of the key problems about analytic groups over pro-p rings is *to what extent is the pro-p ring actually determined by the group?* More specifically, can a group have an analytic structure over two distinct commutative pro-p integral domains $(R_1, \mathfrak{m}_1)$ and $(R_2, \mathfrak{m}_2)$?

The answer is 'yes' if R_2 is a finitely generated free module over its subring R_1, as indicated in Example 13.9(iv) (just as a complex Lie group may be regarded as a real Lie group). More challenging is the possibility of a group being analytic over both $\mathbb{F}_p[T_1, T_2]$ and $\mathbb{Z}_p[T]$, both rings of Krull dimension 2 (in Section 13.5, below, we discuss a condition which ensures that the group does at least determine the Krull dimension of the ring). Until we have characterisations of these categories of groups like that for the category of p-adic analytic groups, questions of this sort will be difficult to answer. However, the known characterisations of p-adic analytic groups can be used to show these groups won't admit an analytic structure over a ring significantly different from $\mathbb{Z}_p$:

13.23 Theorem *Let G be an R-analytic group. Suppose also that the topology of G is not discrete. Then G can have the structure of a p-adic analytic group if and only if R is a finitely generated integral extension of $\mathbb{Z}_p$.*

Proof 'If' follows from Example 13.9(iv). To establish the converse, suppose that G is R-analytic, of dimension d. Then $d \geq 1$ since G is not discrete. By Theorem 13.20, G has an open R-standard subgroup H of some level h. Proposition 13.22(iii) shows that for each $n \geq h$,

$$H_n/H_{2n} \cong (\mathfrak{m}^n/\mathfrak{m}^{2n})^{(d)}.$$

Now suppose that G is also p-adic analytic. Then so is H; as H is a pro-p group, it follows by Corollary 8.34 that H has finite rank, r say. As $d \geq 1$, this now implies that

$$\left| \mathfrak{m}^n : p\mathfrak{m}^n + \mathfrak{m}^{2n} \right| \leq p^r$$

for all $n \geq h$. Taking $n = \max\{r+1, h\}$, we see that for some $i \geq 0$ we must have

$$\mathfrak{m}^{n+i} + p\mathfrak{m}^n = \mathfrak{m}^{n+i+1} + p\mathfrak{m}^n. \tag{7}$$

This shows at once that $pR \neq 0$, since R is infinite and the powers of $\mathfrak{m}$ intersect in 0; therefore R has characteristic zero, and by Cohen's Structure Theorem 6.42 R is a finitely generated integral extension of $R_1 = \mathbb{Z}_p[[T_1, \ldots, T_k]]$, for some $k \geq 0$. To complete the proof, we must show that $k = 0$. Since R is integral over R_1, there is a prime ideal P of R such that $P \cap R_1 = pR_1$. Write $\overline{} : R \to R/P$ for the natural map. As $\overline{R}$ is a Noetherian integral domain, Krull's Intersection Theorem tells us that the powers of $\overline{\mathfrak{m}}$ intersect in 0; with (7) this implies that $\overline{\mathfrak{m}} = 0$. Thus $\overline{R}$ is finite; but P was chosen so that $\overline{R}$ contains a copy of $R_1/pR_1 \cong \mathbb{F}_p[[T_1, \ldots, T_k]]$, so k must be zero as required.

13.3 The Lie algebra

In this section, we outline the construction of the Lie algebra associated to an R-analytic group. The method used in Chapter 9 is no longer applicable; instead, we shall simply extract a 'Lie algebra law' from the formal group law. That this procedure actually gives the same Lie algebra in the p-adic case was established in Exercise 9 13.

So let

$$\mathbf{F}(\mathbf{X}, \mathbf{Y}) = \mathbf{X} + \mathbf{Y} + \mathbf{B}(\mathbf{X}, \mathbf{Y}) + O'(3)$$

be a formal group law of dimension d over R, where $\mathbf{B}(\mathbf{X}, \mathbf{Y}) = (B_i(\mathbf{X}, \mathbf{Y}))$ and $B_i(\mathbf{X}, \mathbf{Y})$ is bilinear in $\mathbf{X}$ and $\mathbf{Y}$ for $i = 1, \ldots, d$. To $\mathbf{F}$ we associate the d-tuple of bilinear forms

$$(\mathbf{X}, \mathbf{Y})_{\mathbf{F}} = \mathbf{B}(\mathbf{X}, \mathbf{Y}) - \mathbf{B}(\mathbf{Y}, \mathbf{X}).$$

13.24 Lemma *The polynomials* $(\mathbf{X}, \mathbf{Y}) = (\mathbf{X}, \mathbf{Y})_{\mathbf{F}}$ *satisfy the 'formal Jacobi identity'*

$$(\mathbf{X}, (\mathbf{Y}, \mathbf{Z})) + (\mathbf{Y}, (\mathbf{Z}, \mathbf{X})) + (\mathbf{Z}, (\mathbf{X}, \mathbf{Y})) = 0.$$

Proof We know from Proposition 13.17 that the set $\mathfrak{m}^{(d)}$ becomes a group with multiplication given by evaluating the formal group law $\mathbf{F}$. This group satisfies the *Hall–Witt identity* given in Section 0.3, which we shall use in the following alternative form:

$$\left[a^b, [b, c]\right] \left[b^c, [c, a]\right] \left[c^a, [a, b]\right] = 1. \tag{8}$$

Recall now (Proposition 13.16(ii)) that the commutator formula for a formal group law takes the form

$$\mathbf{C}(\mathbf{X}, \mathbf{Y}) = \mathbf{B}(\mathbf{X}, \mathbf{Y}) - \mathbf{B}(\mathbf{Y}, \mathbf{X}) + O'(3).$$

Let us write $[\mathbf{X}, \mathbf{Y}]$ for the formal power series $\mathbf{C}(\mathbf{X}, \mathbf{Y})$ and $\mathbf{X}^{\mathbf{Y}}$ for the formal power series $\mathbf{F}(\mathbf{I}(\mathbf{Y}), \mathbf{F}(\mathbf{X}, \mathbf{Y}))$. It is easy to see that

$$\mathbf{X}^{\mathbf{Y}} = \mathbf{X} + O(2),$$

and hence to deduce the following formal identities:

$$\left[\mathbf{X}^{\mathbf{Y}}, [\mathbf{Y}, \mathbf{Z}]\right] = (\mathbf{X}, (\mathbf{Y}, \mathbf{Z})) + O(4)$$
$$\left[\mathbf{Y}^{\mathbf{Z}}, [\mathbf{Z}, \mathbf{X}]\right] = (\mathbf{Y}, (\mathbf{Z}, \mathbf{X})) + O(4)$$
$$\left[\mathbf{Z}^{\mathbf{X}}, [\mathbf{X}, \mathbf{Y}]\right] = (\mathbf{Z}, (\mathbf{X}, \mathbf{Y})) + O(4).$$

Since $F(\mathbf{X}, \mathbf{Y}) = \mathbf{X} + \mathbf{Y} + O(2)$, it follows that

$$\mathbf{F}(\left[\mathbf{X}^{\mathbf{Y}}, [\mathbf{Y}, \mathbf{Z}]\right], \mathbf{F}(\left[\mathbf{Y}^{\mathbf{Z}}, [\mathbf{Z}, \mathbf{X}]\right], \left[\mathbf{Z}^{\mathbf{X}}, [\mathbf{X}, \mathbf{Y}]\right])) \tag{9}$$
$$= (\mathbf{X}, (\mathbf{Y}, \mathbf{Z})) + (\mathbf{Y}, (\mathbf{Z}, \mathbf{X})) + (\mathbf{Z}, (\mathbf{X}, \mathbf{Y})) + O(4).$$

Since (8) holds for all $a, b, c \in \mathfrak{m}^{(d)}$, the first line of (9) must be identically zero by the 'uniqueness of power series'. Hence in particular the terms of degree 3 sum to zero, and the lemma follows.

Thus, in an obvious sense, $(\mathbf{X}, \mathbf{Y})_{\mathbf{F}}$ is a 'formal Lie algebra law', and so defines a Lie algebra structure on R^d; the resulting Lie algebra will be denoted $(R^d, (\, , \,)_{\mathbf{F}})$. With a view to examining the functorial properties of this construction, we make the following

13.25 Definition Let $\mathbf{F}$ and $\mathbf{F}'$ be formal group laws, of dimensions d and e respectively. A *formal morphism* $\mathbf{F} \to \mathbf{F}'$ is an e-tuple of power series

$$\Theta = (\Theta_1, \ldots, \Theta_e) \in \Lambda_0 \left[\left[X_1, \ldots, X_d\right]\right]^{(e)},$$

with zero constant terms, such that

$$(\Theta \circ \mathbf{F})(\mathbf{X}, \mathbf{Y}) = \mathbf{F}'(\Theta(\mathbf{X}), \Theta(\mathbf{Y})). \tag{10}$$

Now let $\mathbf{B}$ and $\mathbf{B}'$ be the bilinear parts of $\mathbf{F}$ and $\mathbf{F}'$ respectively, and write

$$\Theta_i(\mathbf{X}) = \sum_{j=1}^{d} \theta_{ij} X_j + \Theta_i^{[2]}(\mathbf{X}) + O(3),$$

where $\Theta_i^{[2]}$ is quadratic. Examining the terms of degree 2 in (10) we find that

$$\sum_{j=1}^{d} \theta_{ij} B_j(\mathbf{X}, \mathbf{Y}) + \Theta_i^{[2]}(\mathbf{X} + \mathbf{Y})$$

$$= B_i'(\sum_{j=1}^{d} \theta_{ij} X_j, \sum_{j=1}^{d} \theta_{ij} Y_j) + \Theta_i^{[2]}(\mathbf{X}) + \Theta_i^{[2]}(\mathbf{Y})$$

for $i = 1, \dots, e$. Interchanging $\mathbf{X}$ and $\mathbf{Y}$ and subtracting the resulting identities now gives

13.26 Proposition *Let $\mathbf{F}$ and $\mathbf{F}'$ be formal group laws and let Θ be a formal morphism $\mathbf{F} \to \mathbf{F}'$. Then*

$$\Theta^{[1]}((\mathbf{X}, \mathbf{Y})_{\mathbf{F}}) = (\Theta^{[1]}(\mathbf{X}), \Theta^{[1]}(\mathbf{Y}))_{\mathbf{F}'},$$

where $\Theta^{[1]}(\mathbf{X})$ denotes the linear part of $\Theta(\mathbf{X})$.

Suppose now that $H = ((\mathfrak{m}^n)^{(d)}, \mathbf{F})$ and $H' = ((\mathfrak{m}^m)^{(e)}, \mathbf{F}')$ are two standard groups and $f : H \to H'$ is an analytic homomorphism. Then there exist a neighbourhood $(\mathfrak{m}^N)^{(d)}$ of 0 in H and an e-tuple $\Theta(\mathbf{X})$ of power series in $\Lambda_0 [[X_1, \dots, X_d]]$ such that $f(\mathbf{x}) = \Theta(\mathbf{x})$ for all $\mathbf{x} \in (\mathfrak{m}^N)^{(d)}$. The fact that f is a group homomorphism, together with the 'uniqueness of power series', implies that Θ is a formal morphism $\mathbf{F} \to \mathbf{F}'$. Write Df for the R-linear mapping $R^d \to R^e$ given by evaluating $\Theta^{[1]}$. Then the last proposition shows that

$$Df : (R^d, (\,,\,)_{\mathbf{F}}) \to (R^e, (\,,\,)_{\mathbf{F}'})$$

is a Lie algebra homomorphism.

Suppose that $g : H' \to ((\mathfrak{m}^t)^{(c)}, \mathbf{F}'')$ is another analytic homomorphism, represented by the formal morphism $\Psi : \mathbf{F}' \to \mathbf{F}''$. The proof of Lemma 13.4 shows that $g \circ f$ is represented by $\Psi \circ \Theta$; and it is very easy to see that $(\Psi \circ \Theta)^{[1]} = \Psi^{[1]} \circ \Theta^{[1]}$; hence

$$D(g \circ f) = Dg \circ Df$$

(we have proved the chain rule!). Thus the assignment

$$((\mathfrak{m}^n)^{(d)}, \mathbf{F}) \mapsto (R^d, (\,,\,)_{\mathbf{F}})$$
$$f \mapsto Df$$

is a functor, from the category of 'standard groups with given standard chart' to the category of Lie algebras over R (given a standard chart on an arbitrary standard group G, we use it to identify G with some $((\mathfrak{m}^n)^{(d)}, \mathbf{F})$).

Given two standard charts on a standard group G, we thus obtain two Lie algebras: but they are isomorphic via Df, where f is the identity map on G considered as an analytic isomorphism between the two standard structures. Up to isomorphism, therefore, our functor associates a unique Lie algebra $L(G) = (R^d, (\,,\,)_{\mathbf{F}})$ to each standard group $G \cong ((\mathfrak{m}^n)^{(d)}, \mathbf{F})$.

We are now ready to define the Lie algebra of an R-analytic group. Let G be an R-analytic group. Then G contains an open R-standard subgroup H of some level (Theorem 13.20). We define the Lie algebra of G to be

$$\mathcal{L}(G) = L(H) \otimes_R K,$$

and have to verify that this is independent of the choice of H.

Suppose, then, that H' is another open R-standard subgroup of G. Then

$$H_m \leq H'_n \leq H$$

for some m and n (in the notation of the previous section). Let $\alpha : H_m \to H'_n$ and $\beta : H'_n \to H$ be the inclusion mappings. Both are analytic homomorphisms, hence induce Lie algebra homomorphisms

$$L(H_m) \overset{D\alpha}{\to} L(H'_n) \overset{D\beta}{\to} L(H).$$

Now it is clear from the definition that $L(H_m) = L(H)$ and that $D(\beta \circ \alpha)$ is the identity map on this Lie algebra; as $D(\beta \circ \alpha) = D\beta \circ D\alpha$, by the functoriality established above, it follows that $D\alpha$ is injective. Hence $D\alpha$ induces an injective Lie algebra homomorphism from $L(H) \otimes_R K$ into $L(H'_n) \otimes_R K = L(H') \otimes_R K$; since these Lie algebras are of the same finite dimension, it follows that they are isomorphic.

13.4 The graded Lie algebra

Throughout this section, G will denote a standard group of the form $(\mathfrak{m}^{(d)}, \mathbf{F})$; for each $n \geq 1$ we write $G_n = (\mathfrak{m}^n)^{(d)}$, as in Section 2. We defined the Lie algebra $L = L(G)$ to be the R-module R^d with the operation

$$(\mathbf{x}, \mathbf{y}) \mapsto (\mathbf{x}, \mathbf{y})_{\mathbf{F}} = \mathbf{B}(\mathbf{x}, \mathbf{y}) - \mathbf{B}(\mathbf{y}, \mathbf{x}).$$

The corresponding *graded Lie algebra* is

$$\mathrm{gr}L(G) = \bigoplus_{n=1}^{\infty} \mathfrak{m}^n L/\mathfrak{m}^{n+1}L = \bigoplus_{n=1}^{\infty} L_n$$

where $L_n = \mathfrak{m}^n L/\mathfrak{m}^{n+1}L$ for each n. As usual, the binary operation is defined on homogeneous components by

$$\left(\mathbf{x} + \mathfrak{m}^{n+1}L, \mathbf{y} + \mathfrak{m}^{m+1}L\right) = (\mathbf{x}, \mathbf{y})_{\mathbf{F}} + \mathfrak{m}^{n+m+1}L \in L_{n+m}$$

for $\mathbf{x} \in \mathfrak{m}^n L$ and $\mathbf{y} \in \mathfrak{m}^m L$, and extended by bilinearity. This makes $\mathrm{gr}L(G)$ into a graded Lie algebra over the field $R/\mathfrak{m}$.

In group-theoretic terms, we have $\mathfrak{m}^n L = G_n$ and $\mathfrak{m}^m L = G_m$, and the Lie bracket $L_n \times L_m \to L_{n+m}$ can be expressed in the form

$$(xG_{n+1}, yG_{m+1}) = [x, y] G_{m+n+1} \quad (x \in G_n, y \in G_m);$$

to see that this is equivalent to the above, observe that if $x = \mathbf{x} \in \mathfrak{m}^n L$ and $y = \mathbf{y} \in \mathfrak{m}^m L$ then

$$[x, y] = \mathbf{C}(\mathbf{x}, \mathbf{y}) = \mathbf{B}(\mathbf{x}, \mathbf{y}) - \mathbf{B}(\mathbf{y}, \mathbf{x}) + O'(3)$$
$$\equiv (\mathbf{x}, \mathbf{y})_{\mathbf{F}} \pmod{G_{m+n+1}}.$$

Now let

$$L_0 = L_0(G) = L/\mathfrak{m}L.$$

Thus $L_0 = (R/\mathfrak{m})^{(d)}$ is a d-dimensional Lie algebra over the finite field $R/\mathfrak{m}$; its Lie bracket is the reduction modulo $\mathfrak{m}$ of the bilinear mapping $\mathbf{B}(\mathbf{X}, \mathbf{Y}) - \mathbf{B}(\mathbf{Y}, \mathbf{X})$. As vector spaces over $R/\mathfrak{m}$ we may identify $\mathrm{gr}L(G)$ with $L_0(G) \otimes_{R/\mathfrak{m}} \mathrm{gr}(\mathfrak{m})$, where

$$\mathrm{gr}(\mathfrak{m}) = \bigoplus_{n \geq 1} \mathfrak{m}^n/\mathfrak{m}^{n+1}$$

is the maximal ideal of the graded ring $\mathrm{gr}(R)$. The fact that $(\,,\,)_{\mathbf{F}}$ is bilinear over R now implies

13.27 Proposition $\mathrm{gr}L(G)$ *is isomorphic as a Lie algebra over* $R/\mathfrak{m}$ *to* $L_0(G) \otimes_{R/\mathfrak{m}} \mathrm{gr}(\mathfrak{m})$.

The Lie algebra $\mathrm{gr}L(G)$ therefore does not retain much information about the group G. For example (see Exercise 10)

$$L_0\left(\mathrm{SL}_n^1(R)\right) \cong \mathfrak{sl}_n(R/\mathfrak{m});$$

since $\mathrm{gr}(\mathbb{Z}_p) \cong \mathrm{gr}(\mathbb{F}_p[[T]])$ it follows that

$$\mathrm{gr}L\left(\mathrm{SL}_n^1(\mathbb{Z}_p)\right) \cong \mathrm{gr}L\left(\mathrm{SL}_n^1(\mathbb{F}_p[[T]])\right).$$

Nevertheless, $\mathrm{gr}L(G)$ has its uses, as we shall see in the following section.

13.5 R-perfect groups

Standard groups over $\mathbb{Z}_p$ have the nice property that their lower p-series coincides with the filtration induced by the powers of the maximal ideal $\mathfrak{m}$. For pro-p rings R of more general type, the link between the group-theoretic structure of an R-standard group and its analytic structure is much weaker; the additive group of $\mathbb{F}_p[[T]]$, for example, bears no trace of the underlying ring (beyond its characteristic!).

On the other hand, one might expect that the structure of a complicated group like $\mathrm{SL}_n^1(R)$ should to some extent reflect the structure of the ring R. Although this group, being a pro-p group, cannot be perfect, it is in some sense close to the perfect group $\mathrm{SL}_n(K)$ (a group is *perfect* if it is equal to its derived group: thus perfect groups are at the opposite extreme to abelian groups). In general, we do not know how to 'swell up' an R-standard group G into a 'K-standard group'; but there is no difficulty at the level of Lie algebras.

Rather than going 'up' to the Lie algebra $\mathcal{L}(G)$ over K, however, it turns out to be sufficient to impose a condition on the finite Lie algebra $L_0(G)$ defined above: this has the advantage of being recognisable in a small finite quotient of G, and is reminiscent of our definition of powerful pro-p groups.

As before, we shall identify a d-dimensional R-standard group G with $\mathfrak{m}^{(d)}$, and for $n \geq 1$ write $G_n = (\mathfrak{m}^n)^{(d)}$.

13.28 Definition Let G be an R-standard group of positive dimension. Then G is R-*perfect* if $[G, G] = G_2$.

It is easy to see that G is R-perfect if and only if $L_0(G)$ is a *perfect Lie algebra*: that is, $(L_0(G), L_0(G)) = L_0(G)$.

Important examples of R-perfect groups are the R-standard groups $\mathrm{SL}_n^1(R)$ with $p \neq 2$ or $n \neq 2$: for the Lie algebra $L_0(\mathrm{SL}_n^1(R)) = \mathfrak{sl}_n(R/\mathfrak{m})$ is perfect unless $p = n = 2$ (see Exercise 10); more generally, most Chevalley groups similarly give rise to R-perfect groups, as explained in Exercise 11.

The following proposition shows that in any R-perfect group, just as in the p-adic case, the filtration defined by the analytic structure is rigidly determined by the group structure.

13.29 Proposition *Let G be an R-perfect group of dimension d. Then*
(i) *G is a finitely generated pro-p group with*

$$\mathrm{d}(G) = \dim_{\mathbb{F}_p} (G/G_2) = d \cdot \dim_{\mathbb{F}_p} \left(\mathfrak{m}/\mathfrak{m}^2\right),$$

and the following hold, for all $n, m \geq 1$:
(ii) *$[G_m, G_n] = G_{m+n}$;*
(iii) *$G_n = \gamma_n(G) = P_n(G)$;*
(iv) *if $pR = 0$ then $G_n = D_n(G)$.*

Proof Proposition 13.22(iii) tells us that $G/G_2 \cong (\mathfrak{m}/\mathfrak{m}^2)^{(d)}$. As $G_2 = [G, G]$ by hypothesis, it follows that in fact $G_2 = \Phi(G)$, and (i) is clear.

(ii) Since $L_0(G)$ is perfect, so is the graded Lie algebra $\mathrm{gr}L(G) = \bigoplus_{n=1}^{\infty} L_n \cong L_0(G) \otimes_{R/\mathfrak{m}} \mathrm{gr}(\mathfrak{m})$. This implies that $(L_n, L_m) = L_{n+m}$ for all n and m. Translating this back to the group we see that $[G_m, G_n] G_{m+n+1} = G_{m+n}$ for all n and m.

Now let $k \geq 1$ and suppose inductively that $[G_m, G_n] G_{m+n+k} = G_{m+n}$ for all n and m. Then

$$[G_m, G_n] G_{m+n+k+1} = [G_m, G_n] [G_m, G_{n+1}] G_{m+(n+1)+k}$$
$$= [G_m, G_n] G_{m+(n+1)}$$
$$= G_{m+n}.$$

It follows by induction that for each m and n we have

$$G_{m+n} = \bigcap_{k \geq 1} [G_m, G_n] G_{m+n+k}$$
$$= [G_m, G_n],$$

since $[G_m, G_n]$ is closed in G by Exercise 1.24.

Part (iii) follows from (i) and Proposition 13.22(iii). When R has characteristic p, Proposition 13.22(vi) shows that $G_n \geq D_n(G)$ for all n. Since $D_n(G) \geq \gamma_n(G)$ in any case, part (iv) now follows from (iii).

It follows from part (iii) that the *Hilbert–Samuel function*

$$n \mapsto f(n) = \sum_{j=0}^{n} \dim_{R/\mathfrak{m}}(\mathfrak{m}^j/\mathfrak{m}^{j+1})$$

of the local ring R is encoded in the group structure of the R-perfect group G; for if $|R/\mathfrak{m}| = q$ then $|G : G_n| = q^{df(n-1)-d}$. Now there exists a polynomial $H(X)$ over $\mathbb{Q}$ such that for large values of n,

$$f(n) = H(n),$$

and the degree of H is equal to the dimension $\mathrm{Dim}(R)$ of the local ring R (see Atiyah and Macdonald (1969), Theorem 11.14, or Matsumura (1986), Theorem 13.4). Thus we may infer

13.30 Theorem *Let G be an R-perfect group, and suppose that*

$$|G : P_n(G)| = p^{g(n)}$$

for each $n \geq 1$. Then there exists a positive rational number c such that

$$g(n) = cn^{\delta} + O(n^{\delta-1})$$

where $\delta = \mathrm{Dim}(R)$.

This shows that a group cannot be both R-perfect and S-perfect unless $\mathrm{Dim}(R) = \mathrm{Dim}(S)$.

It also has an interesting group-theoretic consequence. In Interlude D we stated the *Golod–Shafarevich inequality*: this asserts for a finitely generated pro-p group G that

$$t \geq \mathrm{d}(G)^2/4$$

if there exists a presentation for G on $\mathrm{d}(G)$ generators and t relations. As our final result, we have

13.31 Theorem *Every R-perfect group satisfies the Golod–Shafarevich inequality.*

Proof Let $d_n = \dim_{\mathbb{F}_p}(D_n(G)/D_{n+1}(G))$. Since $D_n(G) \geq \gamma_n(G) = P_n(G)$ by Proposition 13.29(iii), we have $d_n \leq g(n+1)$ for each n. Theorem 13.30 therefore implies that $d_n = O(n^{\delta})$, and the result follows by Theorem D1.

13.6 On the concept of an analytic function

We conclude this chapter with some remarks about our choice of definition for 'R-analytic manifold'. At first sight, the case distinction made in the definition of a Λ-analytic function may seem rather unnatural; it was imposed because on the one hand we want the definition to reduce to the usual notion when R is a DVR such as $\mathbb{Z}_p$, while on the other hand we need to be able to *compose* analytic functions, and this is problematic if the functions are defined by power series over the field of fractions of a ring which is not a DVR.

A consistent theory could be developed by simply decreeing that $\Lambda = R$ for *every* pro-p ring R, but only at the cost of sacrificing the richness of the theory in Chapters 8 and 9. For example, the two obvious charts $\mathbb{Z}_p \to \mathbb{Z}_p$ and $\mathbb{Z}_p \to p\mathbb{Z}_p$ would no longer be $\mathbb{Z}_p$-compatible; this is why we want to allow division by p to count as a p-adic analytic function. (This particular problem does not arise for a ring like $R = \mathbb{Z}_p[[T]]$: in that case multiplication by a non-unit s maps R onto the subset sR which is not open in R; while a chart must be onto an open subset of R.)

When R is a DVR, we therefore need to allow functions which are Λ-analytic but not R-analytic, such as $\exp : p\mathbb{Z}_p \to \mathbb{Z}_p$. However, as we did in the proof of Theorem 13.20, we can use *rescaling* on the underlying set to arrange that an analytic function be represented by a power series with coefficients in R: for example $\exp(pX) \in \mathbb{Z}_p[[X]]$. This suggests an alternative definition of 'analytic', which combines representability by power series over the ring R with rescaling. This definition avoids the irritating case distinction between DVRs and other pro-p rings, and reduces to our original definition when R is a DVR.

For $\mathbf{k}, \mathbf{x} \in K^n$ we write

$$\mathbf{k} \cdot \mathbf{x} = (k_1 x_1, \dots, k_n x_n).$$

Definition Let U be an open subset of R^n. A function $f : U \to R$ is *analytic* on U if for each $\mathbf{y} \in U$ there exist a formal power series $F \in R[[X_1, \dots, X_n]]$, elements $k_1, \dots, k_n \in K^{(n)}$, and a positive integer N such that

- $\mathbf{y} + (\mathfrak{m}^N)^{(n)} \subseteq U$
- $\mathbf{k} \cdot (\mathfrak{m}^N)^{(n)} \subseteq R^n$
- $f(\mathbf{y} + \mathbf{x}) = F(\mathbf{k} \cdot \mathbf{x})$ for all $\mathbf{x} \in (\mathfrak{m}^N)^{(n)}$.

Any function that is Λ-analytic in the sense of Definition 13.1 is analytic in this new sense. If R is not a DVR this is clear, while if R is a DVR it follows from the definition of $\Lambda_0\,[[\mathbf{X}]]$: given $H(\mathbf{X}) = \sum d_\alpha \mathbf{X}^\alpha \in \Lambda_0\,[[\mathbf{X}]]$ such that $f(\mathbf{y} + \mathbf{x}) = H(\mathbf{x})$ for all $\mathbf{x} \in (\mathfrak{m}^N)^{(n)}$, we take $\mathbf{k} = (\pi^{-k}, \ldots, \pi^{-k})$, where $k \in \mathbb{N}$ satisfies

$$d_\alpha m^{k\langle\alpha\rangle} \subseteq R \quad (\alpha \neq 0);$$

then $F(\mathbf{X}) = H(\pi^k \mathbf{X})$ is in $R\,[[\mathbf{X}]]$ (note that $F(0) = H(0) = f(\mathbf{y}) \in R$).

When R is a DVR, the *converse* is also true, since $F(\mathbf{k} \cdot \mathbf{x}) = H(\mathbf{x})$ where $H(\mathbf{X})$ is a power series over K. Whether the converse holds for an arbitrary pro-p ring R is not clear to us; the following lemma implies that it holds in many cases of interest:

Lemma *Suppose that* $\mathrm{Dim}(R) \geq 2$ *and that* R *is a unique factorisation domain. If* $k \in K$ *and* $k\mathfrak{m}^N \subseteq R$ *for some positive integer* N *then* $k \in R$.

Proof According to Matsumura (1986), Theorem 13.4 or Atiyah and Macdonald (1969), Theorem 11.14, the local ring R contains a *system of parameters* $\{t_1, \ldots, t_d\}$, where $d = \mathrm{Dim}(R) \geq 2$; this means that $\sum_{i=1}^d t_i R = \mathfrak{q}$, say, is an $\mathfrak{m}$-primary ideal. Now suppose that $k = ab^{-1}$ where $a, b \in R$, and that $k\mathfrak{m}^N \subseteq R$. If $k \notin R$ there exists an irreducible element z of R such that $z \mid b$ and $z \nmid a$. From $a\mathfrak{m}^N \subseteq bR$ we infer that z divides both at_1^N and at_2^N. This implies that z divides both t_1 and t_2, and hence that

$$\mathfrak{q} \subseteq zR + \sum_{i=3}^d t_i R \subseteq \mathfrak{m}.$$

Thus R contains an $\mathfrak{m}$-primary ideal that can be generated by $d-1$ elements; this is impossible by the theorem quoted above.

Thus for rings of this type, the definition permits only a very limited kind of rescaling; we can now deduce

Proposition *Suppose that* R *is a unique factorisation domain. Then every* Λ-analytic function is analytic in the new sense, and conversely.

Proof A unique factorisation domain of dimension 1 is a DVR, so what remains to be established is that if R is a UFD of dimension at least 2 and $f : U \to R$ satisfies the definition above, then f is R-analytic. Now the lemma shows that $\mathbf{k} \in R^n$, from which it follows that $F(\mathbf{k} \cdot \mathbf{x}) = H(\mathbf{x})$ where H is a power series with coefficients in R. Thus f is R-analytic.

This applies in particular to every ring R of the form $R_0 [[T_1, \ldots , T_m]]$ where R_0 is a finite field or a finite extension of $\mathbb{Z}_p$

Notes

The definition of an analytic group over a complete discrete valuation ring appears in Serre (1965) and Bourbaki (1989b). The definition over more general pro-p rings is new. The investigation of such groups was suggested in Lubotzky and Shalev (1994), who showed that much of the theory of standard groups over complete discrete valuation rings carries over to the more general setting; they also established Theorem 13.23 for the case where $R = \mathbb{F}_p [[t]]$. The construction of the Lie algebra of a formal group, and its graded Lie algebra, appear in Bourbaki and Serre (loc. cit.).

The concept of R-perfect groups was introduced in Lubotzky and Shalev (1994); Theorem 13.31 (the Golod–Shafarevich inequality) is taken from there. In the same paper, the graded Lie algebra is used to prove the following result on subgroup growth:

Theorem *Let G be an R-perfect group, having a_n open subgroups of index n for each n. Then*

$$a_n \le n^{c \log n}$$

for all n, where c is a constant.

Recent work of Richard Pink (1998) shows that pro-p groups which are linear over $\mathbb{F}_p [[t]]$ have surprisingly restricted structure. In particular, as noted in Barnea and Larsen (a), a non-abelian free pro-p group cannot be embedded in $GL_n(\mathbb{F}_p[[T]])$ for any n.

We have emphasised throughout the book that the theory of p-adic analytic groups is driven by the group theory and that sophisticated concepts of analysis can be avoided. For example, the proof that subgroups and quotients of p-adic analytic groups have a p-adic analytic structure did not require the development of a concept of submanifolds or quotient manifolds. To examine these issues for the more general analytic groups of the present chapter, such concepts would need to be developed. In the case of discrete valuation rings, an account of analytic subgroups and quotients may be found in Bourbaki and Serre (loc. cit.). In particular Bourbaki, Chapter III, §1.6 shows that the quotient of an $\mathbb{F}_p[[t]]$-analytic group by an analytic subgroup is analytic.

Exercises

In Exercises 1–3, F denotes a finite field of characteristic p.

1. Let $R = F[[T]]$ and $\mathfrak{m} = TR$. Let G be the R-standard group $(\mathfrak{m}, +)$. Let d be a positive integer. Exhibit a homeomorphic isomorphism $G \to G \times G \times \cdots \times G$ (d factors). Deduce that (a) the dimension of an R-analytic group is not uniquely determined by its topological group structure, and (b) a continuous homomorphism of R-analytic groups need not be analytic.

[*Hint*: consider the mapping $\sum a_i T^i \mapsto \left(\sum a_{di} T^i, \sum a_{di-1} T^i \cdots , \sum a_{di-d+1} T^i \right)$.]

2. Let $R = F[[T]]$ and $G = SL_2^1(R)$. Show that G contains an element of infinite order; deduce that G has an infinite pro-cyclic pro-p subgroup H. Can H have the structure of an R-analytic group?

[*Hint*: recall Theorem 13.23!]

3. Let $R = F[[T_1, \ldots , T_p]]$ and $\mathfrak{m} = (T_1, \ldots , T_p)R$. Let σ be the continuous F-algebra automorphism of R that sends T_i to T_{i+1} ($i < p$), T_p to T_1. Let A be the R-standard group $(\mathfrak{m}, +)$.

(i) Prove that σ is a continuous automorphism of A, but that σ is not analytic.

(ii) Let G be the semi-direct product $G = A \rtimes \langle \sigma \rangle$. Show that G is a pro-p group, having A as an open subgroup with the induced topology. Prove that there exists no R-analytic group structure on G with respect to which A is an open submanifold.

4. Let $S \subseteq R$ be pro-p rings as in Example 13.6(iv), and suppose that $R = v_1 S \oplus \cdots \oplus v_e S$. Let $n \geq 1$.

(i) Show that for each $\alpha \in \mathbb{N}^n$ there exist elements $\gamma_k(\beta) \in S$ such that

$$\prod_{j=1}^{n} \left(\sum_{i=1}^{e} v_i Y_{ij} \right)^{\alpha_i} = \sum_{k=1}^{e} \sum_{\langle \beta \rangle = \langle \alpha \rangle} v_k \gamma_k(\beta) \prod_{i=1}^{e} \prod_{j=1}^{n} Y_{ij}^{\beta_{ij}}$$

for $k = 1, \ldots , e$.

(ii) For $a \in \Lambda$ (which is either R or the field of fractions of R) write

$$a \cdot v_k = \sum_{\ell=1}^{e} a(k, \ell) v_\ell$$

where each $a(k,\ell) \in \Lambda(S)$ (which is either S or the field of fractions of S). To a power series $F(\mathbf{X}) = \sum a_\alpha \mathbf{X}^\alpha \in \Lambda_0[X_1,\ldots,X_n]$ we associate the e power series

$$F_\ell^*(\mathbf{Y}) = \sum_\beta \left(\sum_{\langle\alpha\rangle=\langle\beta\rangle} \sum_{k=1}^e a_\alpha(k,\ell)\gamma_k(\beta) \right) \prod_{i=1}^e \prod_{j=1}^n Y_{ij}^{\beta_{ij}},$$

$\ell = 1,\ldots,e$. Show that $F_\ell^*(\mathbf{Y}) \in \Lambda_0[Y_{11},\ldots,Y_{en}]$, and that if $\mathbf{x} = \sum v_\ell \mathbf{y}_\ell \in R^n$ and $F(\mathbf{x}) = \sum v_\ell z_\ell$ then $z_\ell = F_\ell^*(\mathbf{y}_1,\ldots,\mathbf{y}_e)$.

(iii) Define $\sigma : R \to S^e$ by $\sigma(\sum v_i s_i) = (s_1,\ldots,s_e)$. Let U be an open subset of R^n. Show that $\sigma^{(n)}(U)$ is an open subset of S^{en}. Deduce from the above that if $f : U \to R$ is Λ-analytic then

$$\sigma \circ f \circ (\sigma^{-1})^{(n)} : \sigma^{(n)}(U) \to S^e$$

is $\Lambda(S)$-analytic.

(iv) Let X and Y be R-manifolds and $f : X \to Y$ an analytic function. Using (iii), show that the construction given in Example 13.6(iv) does define S-manifolds $X_{(S)}, Y_{(S)}$, that the manifold structures so obtained are independent of the choice of isomorphism $\sigma : R \to S^e$, and that the function $f : X_{(S)} \to Y_{(S)}$ is analytic.

5. Suppose that R is a DVR. Let G be an R-standard group of level $n > 1$, with standard chart (G,ϕ,d). Let $\psi : G \to \mathfrak{m}^{(d)}$ be the map $g \mapsto \pi^{-n}\phi(g)$. Show that (G,ψ,d) is a standard chart of level 1, compatible with (G,ϕ,d).

[*Hint*: Let $\mathbf{F}(\mathbf{X},\mathbf{Y}) \in R[[\mathbf{X},\mathbf{Y}]]^{(d)}$ the formal group law associated to (G,ϕ,d). Define $\widetilde{F}_i(\mathbf{X},\mathbf{Y}) = \pi^{-(n-1)}F_i(\pi^{n-1}\mathbf{X},\pi^{n-1}\mathbf{Y})$ for $i = 1,\ldots,d$. Verify that $\widetilde{\mathbf{F}}$ is a formal group law giving the multiplication on G with respect to the chart (G,ψ,d).]

6. Assume that R is not a DVR, and let G be an R-standard group of level $n \geq 1$. (i) Let $I \subseteq \mathfrak{m}^n$ be an ideal of R. Prove that the subgroup $G(I)$ is well defined (i.e. that it does not depend on the choice of a standard chart of level n for G).

(ii) Deduce that the level of G is uniquely determined.

[*Hint*: Note that a Λ-analytic function must map $I^{(d)}$ into $I^{(d)}$.]

7. Assume that R is a DVR. Let G be an R-standard group and I a proper ideal of R. For each standard chart (G,ϕ,d) of level 1, put $G_\phi(I) = \phi^{-1}(I^{(d)})$. Is the subgroup $G_\phi(I)$ independent of ϕ?

[The answer is 'no' if we allow the level to vary, by Exercise 5. The answer is 'yes' if $R = \mathbb{Z}_p$, since then $I = p^n\mathbb{Z}_p$ for some n, and $G(p^n\mathbb{Z}_p) = P_{n-1}(G)$ by Theorem 8.32. What happens for $R = \mathbb{F}_p[[T]]$?]

8. Prove that $L(\mathrm{GL}_n^1(R)) = \mathfrak{gl}_n(R)$. [*Hint*: look at the group law in Example 13.18(iii)!]

9. Let $n \geq 2$. Write $(\mathbf{X})$ for the $n \times n$ matrix (X_{ij}), where the X_{ij} are n^2 independent indeterminates, and $\mathbf{X}'$ for the $(n^2 - 1)$-tuple $(X_{ij})_{(i,j)\neq(n,n)}$.
  (i) Show that

$$\det(1_n + (\mathbf{X})) = (1 + X_{nn})(1 + P(\mathbf{X}')) + Q(\mathbf{X}')$$

where P and Q are polynomials and P has constant term zero. Hence show that there exists a power series H in $n^2 - 1$ variables over $\mathbb{Z}$ such that under the substitution $Z_{nn} = H(\mathbf{X}')$, $Z_{ij} = X_{ij}$ for $(i,j) \neq (n,n)$, we have the power series identity $\det(1_n + (\mathbf{Z})) = 1$.
  (ii) Prove that the mapping $\psi : \mathrm{SL}_n^1(R) \to \mathfrak{m}^{n^2-1}$ that sends $1 + (\mathbf{x})$ to $\mathbf{x}'$ is a homeomorphism.
  (iii) Show that $\psi((1 + \mathbf{x}) \cdot (1 + \mathbf{y})) = \mathbf{F}(\psi(1 + \mathbf{x}), \psi(1 + \mathbf{y}))$, where $\mathbf{F} = (F_{ij})_{(i,j)\neq(n,n)}$ and

$$F_{ij}(\mathbf{X}', \mathbf{Y}') = \mathbf{X}' + \mathbf{Y}' + \sum_{\ell=1}^{n} X_{i\ell}Y_{\ell j} \quad (i \neq n, j \neq n)$$

$$F_{in}(\mathbf{X}', \mathbf{Y}') = \mathbf{X}' + \mathbf{Y}' + \sum_{\ell=1}^{n-1} X_{i\ell}Y_{\ell j} + X_{in}H(\mathbf{Y}') \quad (i \neq n)$$

$$F_{nj}(\mathbf{X}', \mathbf{Y}') = \mathbf{X}' + \mathbf{Y}' + \sum_{\ell=1}^{n-1} X_{i\ell}Y_{\ell j} + H(\mathbf{X}')Y_{nj} \quad (j \neq n).$$

Deduce that ψ defines a standard chart on $\mathrm{SL}_n^1(R)$.
  (iv) Show that $P(\mathbf{X}') = \sum_{i=1}^{n-1} X_{ii} + O'(2)$ and that $Q(\mathbf{X}') = O'(2)$. Deduce that $H(\mathbf{X}') = -\sum_{i=1}^{n-1} X_{ii} + O'(2)$.

10. (i) Prove that $L(\mathrm{SL}_n^1(R)) \cong \mathfrak{sl}_n(R)$. Deduce that $L_0(\mathrm{SL}_n^1(R)) \cong \mathfrak{sl}_n(R/\mathfrak{m})$.
  [*Hint*: For the first part, note that $\mathfrak{sl}_n(R)$ is the set of matrices $(\mathbf{x})$ such that $x_{nn} = -\sum_{i=1}^{n-1} x_{ii}$. Now look at the bilinear part of the group law $\mathbf{F}$ given in the preceding exercise, using part (iv).]
  (ii) Let F be a field of characteristic $\neq 2$ and let $n \geq 2$. Prove that the Lie algebra $L = \mathfrak{sl}_n(F)$ is perfect (i.e. that $(L, L) = L$).

[*Hint*: work out $(e_{ij}, e_{k\ell})$ where e_{ij} is the matrix with 1 in the (i,j)-place and 0 elsewhere. The following exercise generalises this.]

Without developing a theory of closed submanifolds, it is difficult to establish a general theorem that algebraic groups yield R-analytic groups. However, the basic examples of simple algebraic groups, the *Chevalley groups*, do naturally give rise to standard groups. For unexplained terminology and unproved assertions in the following exercise, see Steinberg (1967) or Carter (1989).

11. Chevalley groups. Let Φ be a root system of type $X_l \in \{A_l, B_l, C_l, D_l, E_6, E_7, E_8, F_4, G_2\}$ and let $\Pi = \{\alpha_1, \dots, \alpha_l\}$ be a set of simple roots. Let R be pro-p domain with maximal ideal $\mathfrak{m}$. The *Lie algebra* $L_R(X_l)$ of type X_l over R is the free R-module on the basis $\{h_\alpha \mid \alpha \in \Pi\} \cup \{e_\beta \mid \beta \in \Phi\}$, with the following Lie brackets:

(L1) $(h_\alpha, h_\beta) = 0$,
(L2) $(h_\alpha, e_\beta) = A_{\alpha\beta} e_\beta$,
(L3) $(e_\alpha, e_{-\alpha}) = h_\alpha$,
(L4) $(e_\alpha, e_\beta) = 0$ if $\alpha + \beta \notin \Phi$,
(L5) $(e_\alpha, e_\beta) = N_{\alpha,\beta} e_{\alpha+\beta}$ if $\alpha + \beta \in \Phi$;

here $A_{\alpha\beta} = \frac{2(\alpha,\beta)}{(\alpha,\alpha)}$ is the *Cartan integer* and $N_{\alpha,\beta} = \pm(q+1)$ where q is the largest integer for which $\beta - q\alpha \in \Phi$.

The *universal Chevalley group* $G = \mathcal{G}_R(X_l)$ of type X_l over R is the abstract group generated by symbols $x_\alpha(t)$ where $t \in R$ subject to the following relations (called the Steinberg relations):

(G1) $x_\alpha(t_1)x_\alpha(t_2) = x_\alpha(t_1 + t_2)$ for $\alpha \in \Phi, t_1, t_2 \in R$;
(G2) $x_\beta(s)^{-1}x_\alpha(t)^{-1}x_\beta(s)x_\alpha(t) = \prod x_{i\alpha+j\beta}(N_{i,j;\alpha,\beta}(-t)^i s^j)$ for all $\alpha, \beta \in \Phi$ linearly independent and all $s, t \in R$; here the $N_{i,j;\alpha,\beta}$ are integers with $N_{1,1;\alpha,\beta} = N_{\alpha,\beta}$;
(G3) putting $w_\alpha(s) = x_\alpha(s)x_{-\alpha}(-s^{-1})x_\alpha(s)$, the relation

$$w_\alpha(s)x_\alpha(t)w_\alpha(s)^{-1} = x_{-\alpha}(-s^2 t)$$

for each $\alpha \in \Phi, s \in R^* = R \setminus \mathfrak{m}$ and $t \in R$;
(G4) putting $c_\alpha(s) = w_\alpha(s)w_\alpha(1)^{-1}$, the relation $c_\alpha(s)c_\alpha(t) = c_\alpha(st)$ for each $\alpha \in \Phi, s, t \in R^*$.

The following fact is a consequence of these relations (see Carter (1989) Chapter 12):

$$c_\alpha(s)x_\beta(t)c_\alpha(s)^{-1} = x_\beta(s^{A_{\alpha\beta}}t), \text{ for } \alpha \in \Phi, s \in R^*, t \in R.$$

(i) For each positive integer k let

$$G_k = \left\{ \prod_{\alpha \in \Phi^+} x_\alpha(t_\alpha) \prod_{\beta \in \Pi} c_\beta(1+s_\beta) \prod_{\alpha \in \Phi^-} x_\alpha(t_\alpha) \mid t_\alpha, s_\beta \in \mathfrak{m}^k \right\}.$$

(The products here are ordered according to the height function h on the root system defined by $h(\alpha) = \lambda_1 + \cdots + \lambda_l$ where $\alpha = \lambda_1 \alpha_1 + \cdots + \lambda_l \alpha_l$.)
Prove that G_k is a subgroup of $\mathcal{G}_R(X_l)$.

(ii) Given that the above expression for each element of G_k is unique, prove that G_k is an R-standard group of level k and dimension $|\Phi| + |\Pi|$.

(iii) Prove that the Lie algebra of G_1 is the Lie algebra $L_R(X_l)$.

(iv) Using (L1)–(L5) show that if F is a field then $L_F(X_l)$ is a perfect Lie algebra, unless $\mathrm{char}(F) = 2$ and X_l is A_1 or C_l. Hence deduce that G_1 is an R-perfect group, provided (p, X) is not one of $(2, A_1)$ or $(2, C_l)$.

12. Assume that the ring R has characteristic p. Let $G = (\mathfrak{m}^{(d)}, \mathbf{F})$ be a standard group, write $L = L(G)$, and put $G_n = (\mathfrak{m}^n)^{(d)}$ and $L_n = \mathfrak{m}^n L/\mathfrak{m}^{n+1} L$ for each n, as in Section 13.4. Show that

$$(xG_{n+1})^{[p]} = x^p G_{pn+1}$$

gives a well-defined mapping $[p] : L_n \to L_{pn}$. Prove that this extends to an operation $[p] : \mathrm{gr}L(G) \to \mathrm{gr}L(G)$, making $\mathrm{gr}L(G)$ into a restricted Lie algebra over $\mathbb{F}_p$.

[*Hint:* compare §§12.1,12.2.]

13. Let $\sigma_k(G)$ denote the number of open subgroups of index at most p^k in a pro-p group G. Let $G = SL_2^1(\mathbb{F}_p\,[[t]])$ where $p \geq 3$, or $G = SL_3^1(\mathbb{F}_2\,[[t]])$. Using the theorem quoted in the Notes, show that there is a constant c such that $\sigma_k(G) \leq p^{ck^2}$ for all $k \geq 1$.

Appendix A
The Hall–Petrescu formula

In abelian groups, and in powerful p-groups, the product of nth powers is an nth power. The *Collection Formula* of Philip Hall (also called the Hall-Petrescu formula) provides a substitute for this useful fact.

Let G be any group, x, y any two elements of G, and n a positive integer. Then $(xy)^n$ and $x^n y^n$ are equal (modulo G') so we can write $x^n y^n = (xy)^n c$, with $c \in G'$. The collection formula establishes an expression for c as a product of commutators.

Theorem *Let x and y be elements of a group G, and let n be a positive integer. Then*

$$x^n y^n = (xy)^n c_2^{\binom{n}{2}} \ldots c_i^{\binom{n}{i}} \ldots c_{n-1}^n c_n$$

where $c_i \in \gamma_i(G)$ for each i.

Taking G to be the free group on x and y, we may construe each c_i as a group word, equal to a product of commutators on x and y of length at least i; the formula can then be interpreted as an identity, valid in all groups.

Proof We consider a free group of rank $2n$ with free generators $z_{11}, z_{12}, \ldots, z_{1n}, z_{2n}$, and in this group we consider the product $P = z_{11} z_{12} \ldots z_{1n} z_{2n}$. For each m we write $Z_m = \{z_{1m}, z_{2m}\}$. For any subset S of $\{1, ..., n\}$ with $|S| > 1$, we denote by Z_S the set of all commutators on the z_{ij} which involve generators only from the sets Z_m with $m \in S$, but involve at least one generator from each of these sets. Thus a commutator in Z_S has length $|S|$ at least. (For $S = \{m\}$, a one-element subset, we write $Z_S = Z_m$.) We order the subsets of $\{1, \ldots, n\}$ by requiring that if $|S| < |T|$ then S precedes T, and sets of the same size are ordered lexicographically.

Lemma *With these notations, we have $P = \prod Q_S$, where S runs over the non-empty subsets of $\{1, ..., n\}$, ordered as above, and for each S the element Q_S is a product of elements of Z_S.*

Proof The idea is simply to rewrite P by changing the order of the factors. Since usually $uv \neq vu$, we effect the change by writing $uv = vu[u, v]$. This introduces the new factor $[u, v]$ in the product, and as we continue in the process, we add more and more new factors, which are commutators in the old ones. But if we do the rewriting carefully, it is possible to keep track of the added factors, and this yields the formula of the lemma.

We start by 'collecting to the left' the elements of Z_1. The first generator z_{11} already occurs at the extreme left-hand end of P. We use the transformation $uv = vu[u, v]$ to move z_{21} to the left, until it reaches z_{11}. This means that we replace the product $z_{11}z_{12}...z_{1n}z_{21}$ by the equal product $z_{11}z_{21}z_{12}[z_{12}, z_{21}]...z_{1n}[z_{1n}, z_{21}]$. We have thereby introduced some new commutators into P, but they all belong to sets Z_S (of the form $S = \{1, i\}$) succeeding $\{1\}$ in our ordering.

In general, suppose we have already written P in the form $P = Q_1 \ldots Q_T u_1 \ldots u_r v_1 \ldots v_s$, where each Q_A is a product of elements from Z_A, the sets $1, ..., T$, occur in their right order, the elements u_i belong to Z_S, where S is the subset immediately succeeding T, and the elements v_i belong to either Z_S or to some Z_B with B succeeding S. We then continue by looking for the first element of Z_S to the right of u_r, say this is v_j, and collecting it to the left until it reaches u_r. Since there are no elements of Z_S, or of Z_A with A preceding S, between u_r and v_j, the new commutators that we introduce all belong to sets Z_B with B a successor of S. Having finished collecting v_j, we look for the next element of Z_S. Since no new elements of Z_S are introduced in this stage, we will eventually finish collecting Z_S and obtain an expression $P = Q_1 \ldots Q_T Q_S w_1 \ldots w_t$, where each w_j belongs to some Z_B with B succeeding S. The procedure terminates after finitely many steps with the desired expression for P.

In order to apply the lemma, we want to know what the elements Q_S are. Since all the added elements in the products are commutators, no new factors z_{ij} have been introduced, and therefore $Q_{\{m\}} = z_{1m}z_{2m}$ for each m.

To evaluate the other factors Q_S, we take some subset A, and write P_A for the value of P under the substitution which gives the value 1 to the generators which do not belong to a set Z_m with $m \in A$. Any commutator involving one of these generators also takes the value 1 after this substitution. Therefore the equation $P = \Pi Q_S$ becomes $P_A = \Pi_{B \subseteq A} Q_B$. This equation can be considered as a recursion formula for calculating Q_A, starting with the equality $Q_{\{m\}} = P_{\{m\}}$.

To prove the theorem, we make a different substitution: $z_{1j} = x$, $z_{2j} = y$, and write P^*, P_A^*, Q_A^*, for the values of these words under that substitution. Thus $P^* = x^n y^n$, the expression we are interested in, and in general $P_A^* = x^t y^t$ where $t = |A|$. Thus P_A^* depends only on $|A|$, and therefore Q_A^* also depends only on $|A|$. In particular $Q_{\{m\}}^* = xy$, so $Q_{\{1\}}^* \cdots Q_{\{n\}}^* = (xy)^n$, while the product of all the Q_A^* with $|A| = t$ becomes $c_t^{\binom{n}{t}}$, where c_t denotes the common value of all the Q_A^* with $|A| = t$. Since Q_A is a product of commutators of length at least $|A|$, it follows that $c_t \in \gamma_t(G)$ for each $t \geq 2$. The theorem follows since

$$P^* = \prod Q_S^*.$$

We remark that a similar formula, with a similar proof, holds for the product of any finite number of terms.

Appendix B
Topological groups

Here we prove some facts about topological groups, a little less basic than those of §0.6. They are needed only for Exercise 1.2 and Theorem 9.7.

Throughout, G denotes a Hausdorff topological group. For subsets X and Y of G, XY denotes the set $\{xy | x \in X, y \in Y\}$, $X^{(n)} = X^{(n-1)}X$ for $n > 1$, and $X^{-1} = \{x^{-1} | x \in X\}$. The map $\mu : G \times G \to G$ is $(x, y) \mapsto xy$.

B1. *If $K \subseteq U \subseteq_o G$ and K is compact then there is an open neighbourhood W of 1 in G such that $KW \subseteq U$.*

Proof There exist open sets A_α, B_α in G such that

$$\mu^{-1}(U) = \bigcup_\alpha A_\alpha \times B_\alpha.$$

Since $K \times \{1\} \subseteq \mu^{-1}(U)$ and $K \times \{1\}$ is compact, $K \times \{1\} \subseteq \bigcup_{\alpha \in F} A_\alpha \times B_\alpha$ for some finite set F, and without loss of generality $1 \in B_\alpha$ for each $\alpha \in F$. Now take $W = \bigcap_{\alpha \in F} B_\alpha$.

B2. *Every compact open neighbourhood of 1 in G contains a compact open subgroup of G.*

Proof Let $1 \in K \subseteq_o G$, K compact. Put $X = K^{(2)} \backslash K$. By **B1** there is an open neighbourhood W of 1 with $KW \subseteq G \backslash X$. Put $V = K \cap W \cap (K \cap W)^{-1}$. Then

$$KV \subseteq (G \backslash X) \cap K^{(2)} \subseteq K,$$

from which it follows that $KV^{(n)} \subseteq K$ for all $n \geq 1$. Now let $H = \bigcup_{n=1}^\infty V^{(n)}$. Then H is open in G, and $H = H^{-1} = H^{(2)}$ so H is a subgroup of G. Hence H is also closed, and since H is contained in the compact set K, H is also compact.

B3. *Let x and y be elements of distinct connected components in a compact Hausdorff space V. Then V contains an open compact set K with $x \in K$ and $y \notin K$.*

Proof Denote by $\mathcal{C}$ the set of all compact open subsets of V containing x. Note that $\mathcal{C}$ is non-empty since $V \in \mathcal{C}$. Put $D = \bigcap \mathcal{C}$. It will be enough to show that D is connected: since y does not belong to the same connected component of V as x, it will follow that $y \notin D$, and hence that $y \notin K$ for some $K \in \mathcal{C}$.

Suppose, then, that $D = X \cup Y$ with $X \cap Y = \varnothing$, X and Y both open in D, and suppose that $x \in X$; we must show that Y is empty. Now X and Y are compact, so X and Y are contained in disjoint open sets U and W, say, in V (since V is Hausdorff). Then

$$V \backslash (U \cup W) \subseteq \bigcup_{K \in \mathcal{C}} V \backslash K.$$

Since $V \backslash (U \cup W)$ is compact, there exist $K_1, \dots, K_n \in \mathcal{C}$ such that $V \backslash (U \cup W) \subseteq V \backslash \bigcap_{i=1}^{n} K_i$. Hence $U \cup W \supseteq \bigcap_{i=1}^{n} K_i = K$, say. Now $x \in K \cap U$, $K \cap U$ is open, and $K \cap U = K \backslash W$ is compact, so $K \cap U \in \mathcal{C}$. Therefore

$$Y \subseteq D \cap W \subseteq K \cap U \cap W = \varnothing.$$

B4. *Suppose that G is totally disconnected (i.e. each connected component is a singleton). Let V be a neighbourhood of 1 in G, whose closure $\overline{V}$ is compact. Then V contains a compact open subgroup of G.*

Proof By **B2**, it suffices to show that V contains a compact open neighbourhood of 1 in G. Now let U be an open neighbourhood of 1 contained in V. By **B3**, each element x of $\overline{V} \backslash U$ lies in a compact, relatively open subset $K(x)$ of $\overline{V}$ with $1 \notin K(x)$. Since $\overline{V}$ is compact we have $\overline{V} = U \cup K(x_1) \cup \dots \cup K(x_n)$ for some $x_1, \dots, x_n \in \overline{V} \backslash U$. Now put

$$W = U \backslash \bigcup_{i=1}^{n} K(x_i) = \overline{V} \backslash \bigcup_{i=1}^{n} K(x_i).$$

Then W is the required compact open neighbourhood of 1 contained in V.

B5. *If G is compact and totally disconnected then G is profinite.*

Proof Every neighbourhood of 1 now has compact closure, hence by **B4** contains an open subgroup. Thus the open subgroups form a base for the neighbourhoods of 1 in G.

B6. *Let N be a closed normal subgroup of G. If both N and G/N are totally disconnected, then so is G.*

Proof Write G^0 for the connected component of 1 in G. Then $G^0 N/N$ is connected, so $G^0 \subseteq N$. Therefore $G^0 = N^0 = \{1\}$. The result follows since for each $x \in G$, the connected component of x is $xG^0 = \{x\}$.

B7. *Suppose that G is locally compact and totally disconnected. Then every neighbourhood of 1 in G contains a compact open subgroup of G.*

Proof Suppose that V is a neighbourhood of 1. Since G is locally compact, 1 has a compact neighbourhood W, say. Then the closure $\overline{V \cap W}$ of $V \cap W$ is a compact neighbourhood of 1, hence by **B4** contains a compact open subgroup of G.

B8. *Let N be a closed normal subgroup of G. If N and G/N are both locally compact and totally disconnected, then so is G.*

Proof G is totally disconnected, by **B6**. By **B7**, N has a compact, relatively open subgroup K. Then $K = N \cap U$ for some open subset U of G, and without loss of generality $KU = U$. Then UN/N is open in G/N, so, by **B7**, G/N has a compact open subgroup C/N with $C \subseteq UN$. Put $V = C \cap U$. Then V is open in G and $1 \in V$. If we show that V is compact, it will follow that G is locally compact, since then for each $x \in G$, xV is a compact neighbourhood of x.

Suppose then that $V = \bigcup R_\alpha$ with each R_α open. We must show that V is covered by finitely many of the R_α. Let $v \in V$. Then $Kv \subseteq V$ since $K \le C$ and $KU = U$, so as Kv is compact we have

$$Kv \subseteq \bigcup_{\alpha \in \mathcal{N}(v)} R_\alpha = S(v), \quad \text{say,}$$

for some finite set of indices $\mathcal{N}(v)$. By **B1**, there is a compact neighbourhood $W(v)$ of 1 in G with $KvW(v) \subseteq S(v)$.

Now $C = UN \cap C = NV$ and $N \cap V = N \cap U = K$. The natural map of C/N onto V/K is continuous, so V/K is compact. Since V/K

is covered by the open sets $KvW(v)/K$ as v runs over V, there exist $v_1, \ldots, v_m \in V$ such that $V/K = \bigcup_{i=1}^{m} Kv_iW(v_i)/K$. Then

$$V \subseteq \bigcup_{i=1}^{m} S(v_i) = \bigcup_{i=1}^{m} \bigcup_{\alpha \in \mathcal{N}(v_i)} R_\alpha,$$

giving the result.

Bibliography

Books

[L] M. Lazard (1965) *Groupes analytiques p-adiques*. Inst. Hautes Études Scientifiques, Publ. Math. **26**, 389-603.

[DSS] M. P. F. du Sautoy, D. Segal and A. Shalev (eds.) 'New horizons in pro-*p* groups', *in preparation*.

M. F. Atiyah and I. G. Macdonald (1969) 'Introduction to Commutative Algebra.' *Addison-Wesley, Reading, Mass.*

N. Bourbaki (1983) 'Algèbre commutative, Chapitres 8 et 9.' *Masson, Paris.*

N. Bourbaki (1989a) 'Commutative algebra, Chapters 1-7' *Springer-Verlag, Berlin-New York.*

N. Bourbaki (1989b) 'Lie groups and Lie algebras, Chapters 1-3.' *Springer-Verlag, Berlin-New York.*

R. W. Carter (1989) 'Simple groups of Lie type.' *John Wiley and Sons, London-New York.*

J. W. S. Cassels (1986) 'Local fields.' *London Math. Soc. Student Texts 3, Cambridge Univ. Press, Cambridge.*

J. W. S. Cassels and A. Fröhlich (1967) 'Algebraic number theory.' Academic Press, London.

J. Dieudonné (1967) 'Topics in local algebra', *Notre Dame lectures 10, Univ. of Notre Dame Press,*

M. D. Fried and M. Jarden (1986) 'Field arithmetic.' *Springer-Verlag, Berlin-Heidelberg-New York.*

A. Fröhlich and M. J. Taylor (1991) 'Algebraic number theory.' *Cambridge Studies in Advanced Mathematics 27, Cambridge Univ. Press, Cambridge.*

P. Hall (1969) 'Nilpotent groups.' *Queen Mary College Math. Notes, London* Also in 'Collected works of Philip Hall', pp. 415–462, *Oxford Univ. Press, Oxford* (1988).

G. H. Hardy (1940) 'Ramanujan. Twelve lectures on subjects suggested by his life and work.' *Cambridge University Press, Cambridge.*

P. J. Higgins (1974) 'An introduction to topological groups.' *London Math. Soc. Lecture Note Series 15, Cambridge Univ. Press, Cambridge.*

N. Jacobson (1962) 'Lie algebras.' *Interscience Publishers, New York–London–Sydney.*

H. Matsumura (1989) 'Commutative ring theory.' *Cambridge Studies in Advanced Mathematics 8, Cambridge Univ. Press, Cambridge.*

I. B. S. Passi (1979) 'Group rings and their augmentation ideals.' *Lecture notes in maths. 715, Springer-Verlag, Berlin–Heidelberg–New York.*

D. S. Passman (1977) 'The algebraic structure of group rings.' *Wiley-Interscience, New York.*

V. P. Platonov and A. S. Rapinchuk (1994) 'Algebraic groups and number theory.' *Academic Press, London.*

W. H. Schikhof (1984) 'Ultrametric calculus.' *Cambridge Studies in Advanced Mathematics 4, Cambridge Univ. Press, Cambridge.*

D. Segal (1983) 'Polycyclic groups.' *Cambridge Univ. Press, Cambridge.*

J.-P. Serre (1965) 'Lie algebras and Lie groups.' *W. A. Benjamin, New York.*

J.-P. Serre (1997) 'Galois cohomology.' *Springer-Verlag, Berlin–Heidelberg–New York.*

R. P. Steinberg (1967) 'Lectures on Chevalley groups.' *Yale University.*

M. R. Vaughan-Lee (1993) 'The restricted Burnside problem,' 2nd edition. *Oxford Univ. Press, Oxford.*

L. C. Washington (1982) 'Introduction to cyclotomic fields.' *Springer-Verlag, New York–Heidelberg–Berlin.*

B. A. F. Wehrfritz (1973) 'Infinite linear groups.' *Springer-Verlag, Berlin–Heidelberg–New York.*

J. S. Wilson (1998) 'Profinite groups.' *Oxford Univ. Press, Oxford.*

Other references

Y. Barnea and M. Larsen (a) *A non-abelian free pro-p group is not linear over a local field.* J. Algebra (to appear).

Y. Barnea and A. Shalev (1997) *Hausdorff dimension, pro-p groups and Kac–Moody algebras.* Trans. American Math. Soc. **349**, 5073–5091.

N. Blackburn (1958) *On a special class of p-groups.* Acta Math. **100**, 49–92.

S. Donkin (1987) *Space groups and groups of prime-power order, VIII. Pro-p groups of finite coclass and p-adic Lie algebras.* J. Algebra **111**, 316–342.

M. P. F. du Sautoy (1990) *Finitely generated groups, p-adic analytic groups and Poincaré series*. Bull. American Math. Soc. **23**, 121–126 (also Interlude C in this volume).

M. P. F. du Sautoy (1993) *Finitely generated groups, p-adic analytic groups and Poincaré series*. Annals of Math. **137**, 639–670.

R. I. Grigorchuk (1989) *On the Hilbert–Poincaré series of graded algebras associated with groups*. Math. USSR Sbornik **180**, 207–226.

M. Gromov (1981) *Groups of polynomial growth and expanding maps*. Inst. Hautes Etudes Scientifiques Publ. Math. **53**, 53–78.

I. Ilani (1995) *Analytic pro-p groups and their Lie algebras*. J. Algebra **176**, 34–58.

S. A. Jennings (1941) *The structure of the group ring of a p-group over a modular field*. Trans. American Math. Soc. **50**, 169–187.

H. Koch (1969) *Zum Zatz von Golod–Šafarevič*. Math. Nachr. **42**, 321–333.

V. A. Kreknin (1963) *Solvability of Lie algebras with a regular automorphism of finite period*. Soviet Math. Doklady **4**, 683–685.

C. R. Leedham-Green (1994a) *Pro-p groups of finite coclass*. J. London Math. Soc. **50**, 43–48.

C. R. Leedham-Green (1994b) *The structure of finite p-groups*. J. London Math. Soc. **50**, 49–67.

C. R. Leedham-Green and S. McKay (1976) *On p-groups of maximal class, I*. Quarterly J. Math. Oxford **27**, 297–311.

C. R. Leedham-Green and S. McKay (1978) *On p-groups of maximal class, II, III*. Quarterly J. Math. Oxford **29**, 175–186, 281–299.

C. R. Leedham-Green and M. F. Newman (1980) *Space groups and groups of prime-power order, I*. Arch. Math. (Basel) **35**, 193–202.

A. Lubotzky (1983) *Group presentation, p-adic analytic groups and lattices in* $SL_2(\mathbb{C})$. Annals of Math. **118**, 115–130.

A. Lubotzky (1988) *A group-theoretic characterization of linear groups*. J. Algebra **113**, 207–214.

A. Lubotzky and A. Mann (1987a) *Powerful p-groups. I: finite groups*. J. Algebra **105**, 484–505.

A. Lubotzky and A. Mann (1987b) *Powerful p-groups. II: p-adic analytic groups*. J. Algebra **105**, 506–515.

A. Lubotzky and A. Mann (1991) *On groups of polynomial subgroup growth*. Invent. Math. **104**, 521–533.

A. Lubotzky and A. Shalev (1994) *On some Λ-analytic pro-p groups*. Israel J. Math. **85**, 307–337.

A. Mann (1990) *Some applications of powerful p-groups*. In 'Groups – St. Andrews 1989', vol. 2, *London Math. Soc. Lecture Notes Series 160*, Cambridge Univ. Press, Cambridge.

C. Martinez (1994) *On power subgroups of pro-finite groups.* Trans. American Math. Soc. **345**, 865–869

J. Milnor (1968) *Growth of finitely generated solvable groups.* J. Differential Geometry **2**, 447–449.

A. Neumann (1988) *Completed group algebras without zero-divisors.* Arch. Math. Basel **51**, 496–499.

R. Pink (1998) *Compact subgroups of linear algebraic groups.* J. Algebra **206**, 438–504.

D. Quillen (1968) *On the associated graded ring of a group ring.* J. Algebra **10**, 411–418.

D. M. Riley (1993) *Analytic pro-p groups and their graded group rings.* J. Pure Appl. Algebra **90**, 69–76.

D. Segal (1990) *Residually finite groups.* In 'Groups, Canberra 1989.' *Lecture notes in maths. 1456, Springer-Verlag, Berlin–Heidelberg–New York.*

D. Segal (1996) *A footnote on residually finite groups.* Israel J. Math. **94**, 1–5.

D. Segal (1999) *Some remarks on p-adic analytic groups.* Bull. London Math. Soc. **31**, 149–153.

A. Shalev (1992a) *Characterization of p-adic analytic groups in terms of wreath products.* J. Algebra **145**, 204–208.

A. Shalev (1992b) *Growth functions, p-adic analytic groups, and groups of finite coclass.* J. London Math. Soc. **46**, 111–122.

A. Shalev (1993a) *On almost fixed-point-free automorphisms.* J. Algebra **157**, 271–282.

A. Shalev (1993b) *Polynomial identities in graded group rings, restricted Lie algebras and p-adic analytic groups.* Trans. American Math. Soc. **337**, 451–462.

A. Shalev (1994) *The structure of finite p-groups: effective proof of the coclass conjectures.* Invent. Math. **115**, 315–345.

A. Shalev and E. I. Zel'manov (1992) *Pro-p groups of finite coclass.* Math. Proc. Camb. Phil. Soc. **111**, 417–421.

J. Tits (1972) *Free subgroups in linear groups.* J. Algebra **20**, 250–270.

J. S. Wilson (1991) *Finite presentations of pro-p groups and discrete groups.* Invent. Math. **105**, 177–183.

J. A. Wolf (1968) *Growth of finitely generated solvable groups and curvature of Riemannian manifolds.* J. Differential Geometry **2**, 421–446.

E. Zel'manov (1992) *On periodic compact groups.* Israel J. Math. **77**, 83–95.

Index

Printed in the United States
By Bookmasters